Biomechanics IV

International Series on Sport Sciences

Series editors: **Richard C. Nelson and Chauncey A. Morehouse**

This new series, published by University Park Press, will present professional reference works derived from congress and symposium proceedings and advanced-level texts of special interest to researchers, clinicians, students, physical educators, and coaches involved in the growing field of sport sciences. The Series Editors are Professors Richard C. Nelson and Chauncey A. Morehouse of The Pennsylvania State University.

The series will cover all aspects of sport sciences, including physiology, biomechanics, medicine, psychology, sociology, history, philosophy, and education. Individual volumes will feature important original contributions, written by leading international authorities, on specific current topics of primary interest to the world-wide community of sport scientists. One of the highlights of this series will be the publication in English of the highly significant research conducted by scientists from many countries. Because many of these authors normally publish their work in languages other than English the series volumes will be a rich resource for material often difficult if not impossible to obtain elsewhere.

In addition to assisting individuals involved directly with sports-related activities, the *International Series on Sport Sciences* will serve as a valuable source of authoritative, up-to-date information. It will include original scientific studies of interest to those concerned with: (1) the improvement of sports performance, (2) the health and safety of amateur and professional athletes, (3) the role of sport sciences in rehabilitation and public health, (4) the historic, philosophic, and sociologic aspects of sport, (5) the education and training of future sport scientists and professionals, and (6) the free exchange of knowledge about these topics across international boundaries.

Published:

Nelson and Morehouse: **BIOMECHANICS IV** (Fourth International Seminar on Biomechanics)

In Preparation:

Clarys and Lewillie: **SWIMMING II** (Second International Symposium on Swimming)
Komi: **BIOMECHANICS V** (Fifth International Congress on Biomechanics)

International Series
on Sport Sciences, Volume 1

BIOMECHANICS IV

Proceedings of the Fourth International Seminar
on Biomechanics, University Park, Pennsylvania

Editors: **Richard C. Nelson** and **Chauncey A. Morehouse**
The Pennsylvania State University

UNIVERSITY PARK PRESS
Baltimore • London • Tokyo

University Park Press

International Publishers in Science and Medicine
Chamber of Commerce Building
Baltimore, Maryland 21202

Printed in the United States of America by Universal Lithographers, Inc.
Typeset by The Composing Room of Michigan

Library of Congress Cataloging in Publication Data

International Seminar on Biomechanics, 4th, University
    Park, Pa., 1973. Biomechanics IV.
    (International Series on sport sciences, v. 1)
    Contains most of the papers presented at the seminar.
    1. Human mechanics—Congresses. I. Nelson, Richard
C., 1932-   ed. II. Morehouse, Chauncey A., ed.
III. Title. IV. Series [DNLM: 1. Biomechanics—
QP303.I5 1973   612'.76   74-18095
ISBN 0-8391-0525-8

Contents

BIOMECHANICS OF SPORT

Jumping and Vaulting

Kicking and Throwing

Swimming and Diving

ORTHOPEDICS AND REHABILITATION

RESEARCH ON MUSCLE CONTRACTIONS

ADVANCES IN INSTRUMENTATION AND TECHNIQUES

Photographic Methods

Other Techniques

MISCELLANEOUS

SEMINAR ACTIVITIES

Preface

This volume contains most of the papers presented at the Fourth International Seminar on Biomechanics held at The Pennsylvania State University, University Park, Pennsylvania, August 26–31, 1973. These proceedings continue in the tradition established by the previous three seminars. They include 83 original contributions by leading researchers in biomechanics from 18 different countries throughout the world.

The program of this international meeting was quite diverse, as expressed by the principal theme, "Contributions of Biomechanics to the Improvement of Human Performance," and a variety of disciplines therefore was represented. For convenience, the contents of this publication were divided into the following major subject headings: Fundamental Human Movements, Biomechanics of Sport, Electromyography, Orthopedics and Rehabilitation, Research on Muscle Contractions, Advances in Instrumentation and Techniques, and Miscellaneous. Where applicable, these main headings have been subdivided. An added feature of the volume is a section on Seminar activities in which the major events of the Seminar are briefly described. Several photographs showing the site of the Seminar and some of the more than 200 participants are included here.

The editors wish to thank the authors for their excellent cooperation in the preparation, submission, and editing of their manuscripts. Unfortunately, it was necessary to restrict the length of manuscripts in order to keep the volume to a reasonable size. For more detailed information, individuals are encouraged to contact the principal author, whose address appears at the end of each article.

This volume serves to document the overall program of the Seminar and provides an additional reference to literature in biomechanics research. We hope that it will help to stimulate ideas for further research by highlighting the work of fellow researchers in other countries.

RICHARD C. NELSON, Ph.D.

CHAUNCEY A. MOREHOUSE, Ph.D.

Biomechanics I V

Opening session

Welcome

Welcome from The Pennsylvania State University

Dean Robert J. Scannell

College of Health, Physical Education and Recreation

I am honored, both personally and professionally, to welcome the participants of this Seminar. I think the manner in which this young science of Bio-mechanics and this even younger International Group have developed is truly a wonderful thing. The degree to which scientists from a variety of nations and a variety of cultures have worked together to develop this field of study is a true example to us all; an example of international cooperation at its best. We welcome you all to Penn State and hope that we can make your stay with us both personally and professionally satisfying.

The Faculty of the College of Health, Physical Education and Recreation is honored to serve as your hosts and we all hope that you will make your requests known so we can be of help to you. I should also point out that we have very extensive facilities for research, for sport, and for recreation on the Campus. I hope that you can find time in your very busy schedule of the next several days to see these facilities.

With this brief welcome from the College, I would like now to introduce our chief academic officer, the Provost of The Pennsylvania State University, to officially welcome you on behalf of the University Administration: Dr. Russell E. Larson.

Provost Russell E. Larson

On behalf of President Oswald and the Faculty and Administration, I would like to welcome you to the Campus of The Pennsylvania State University.

Dr. Richard Nelson, Chairman of the Seminar, informs me that there are representatives from almost 20 countries and nearly every section of the United States in a variety of disciplines attending this Fourth International

Meeting. With such an international audience, you might expect to see Dr. Henry Kissinger walk in at any moment.

I understand this is the first time that this international seminar has been held in a location other than a European country, and we at Penn State are proud and pleased you have selected our campus, and we are honored to have you as guests of our University.

If you will recall, the last meeting was held in Rome, and Chairman Nelson reminded me to tell you that while you will not find a coliseum or any ancient cathedrals here, you will find peaceful surroundings and a friendly atmosphere. This, we hope, will provide you with the opportunity to become better acquainted, and to enhance your professional interests.

You will also find that beneath this deceptive calm there lies a beehive of activity in the classrooms and laboratories. Dr. Nelson and others will more fully inform you of these activities when you tour the Biomechanics and Human Performance Laboratories and the other research facilities.

When I see this array of professional talent as well as this concentration of interest and concern, I am confident this seminar will prove to be one of the most comprehensive ever to be held on the subject of biomechanics.

I again welcome you to The Pennsylvania State University, commend you on your purpose, and sincerely hope this seminar will provide you with valuable information, assistance, and perspective as you work toward the improvement of human performance.

Welcome from UNESCO

Ernst Jokl, President, Research Committee
UNESCO Int. Council of Sport and Ph. Ed.

It is with much pleasure that I convey to this distinguished gathering the greetings of the UNESCO International Council of Sport and Physical Education under whose auspices the IVth International Seminar on Biomechanics is being held. The Council's President, the Rt. Hon. Philip Noel Baker, has asked me to assure you of his personal interest in this conference. The fact that we are meeting in the United States is chiefly due to the interest taken in the activities of the Council by Dr. Richard Nelson. I would also like to mention my longstanding friendship with the President of Pennsylvania State University, Dr. John Oswald, who for several years headed the University of Kentucky.

Let me say something about the scientific activities of the UNESCO International Council of Sport and Physical Education of which the Committee on Biomechanics is an important branch. The International Council is one of the chief consultative agencies of UNESCO which, as you know, is the Educational and Scientific Division of the United Nations. The Council advises UNESCO in matters pertaining to sport and physical education. It does so within the noble framework of reference of its policy as expressed in the preamble of its charter which says that the conflicts of man originate in the minds of man and that it is thus upon the minds of man that UNESCO must concentrate its efforts to elevate the level and quality of communication of people throughout the world. Insofar as sport occupies a special place in the array of means of such communication and insofar as it transcends the limitations set to communication through language, its relevance in the context of this objective is very great indeed. Biomechanics is one of several scientific and humanistic specialities whose pursuit is encouraged and supported by the UNESCO International Council. By virtue of its association with the Council, the Committee on Biomechanics enjoys a degree of integra-

Author's address: *Dr. Ernst Jokl, Department of Physical Education, University of Kentucky, Lexington, Kentucky 40500 (USA)*

tion with all other subjects whose activities are coordinated by the Council's Research Committee. The Committee on Biomechanics of Sport and Physical Education functions under the able chairmanship of Professor Jurg Wartenweiler of Zurich, Switzerland who was the initiator of the first International Seminar on Biomechanics of Sport in 1967. Ever since he has directed the Committee's affairs, participated in the organization of the 2nd and 3rd International Congresses held in Eindhoven in Holland in 1969 and in Rome, Italy in 1971. Again this year he has assisted Dr. Nelson, Mr. Vredenbregt of Eindhoven, Holland and Professor Marcel Hebbelinck of Brussels, Belgium with the preparation of the seminar which is about to begin.

It is indeed fitting that this Seminar be held here in the United States and I am sure that more than ordinary significance is attached to this meeting on biomechanics of sport and physical education. It has assembled experts in an important branch of science, one in which American research has been outstanding. It has brought together investigators qualified in various specialities able to make a contribution to biomechanics of sport. It has created an opportunity for foreign visitors to acquaint themselves not only with the U.S. and its remarkable technological developments, but also with the hospitality that characterizes the way of life in this country. These foreign visitors will be made aware of new ideas and concepts of biomechanics of physical education, sport and athletics by their American hosts and vice versa. I notice in the audience many members of the American Association for Health, Physical Education and Recreation and of the American Academy of Physical Education. During the next days they will have an opportunity to learn from the scientists who have assembled at University Park how a clearly defined sector of research on physical education and sport must be handled; they will see how appropriate methodologies have been developed for the purpose and how the results obtained are being utilized for the elaboration of new concepts as well as for the application of these concepts in teaching and coaching.

There is no more valuable documentation of research on sport and physical education available than the three magnificent volumes of proceedings which contain the papers read at Zurich in 1967, in Eindhoven in 1969 and in Rome in 1971. The proceedings of the present seminar will no doubt be of the same caliber. The Working Group on Biomechanics of the UNESCO International Council of Sport and Physical Education has set an example which must be followed by other scientific disciplines desirous of serving our profession.

Keynote
address

The manifold implications of biomechanics

J. Wartenweiler
Swiss Federal Institute of Technology, Zürich

As Professor Richard Nelson mentioned in his thoughtful article, "The Contribution of Biomechanics to the Educational Aspects of Sport Science"—published in the journal "Gymnasion"—the international seminars and symposia have contributed to the recent growth of biomechanics.

Richard Nelson himself has had a significant share in promoting this growth and we thank him and his excellent staff for having organized the IVth International Seminar here at Penn State. We thank the authorities of the famous Pennsylvania State University for having accepted our seminar and for hosting us so kindly.

Two things were clear to me when in 1966 I agreed with Prof. Ernst Jokl (President of the Research Committee ICSPE/UNESCO) to undertake the organization of the first seminar in Zurich; a) it would have to be an international seminar and b) all scientific aspects of biomechanics would have to be represented. I think that both aspects were fulfilled: The applications for the first seminar have bestowed a wonderful stamp collection upon me to testify to the importance of biomechanics. I am not certain if my colleagues Schouten, Vredenbregt, Lewillie, Cerquiglini, Venerando, Dal Monte, Nemessuri, and today Richard Nelson, are collecting stamps, but I am sure that they could fill a nice album with stamps from all over the world as I have.

I should like to concentrate my remarks on the second point in question. The fascination of biomechanics lies in the fact that it is an interdisciplinary science. May I try to stipulate the sciences which today are obviously interested in biomechanics.

MOVEMENT STUDIES

The first impression of motion is an optical one. Here we deal with physics. Course, velocity, acceleration and the pertinent forces can be described in physical units. Biomechanics has reliable working methods at its disposal today, thanks to intensive work done with film analysis and electronic motion recording and measuring.

Mathematics is another science with which to work on movement problems. In connection herewith computer technology is used successfully. Exact mathematical movement models were discussed this year in Paris at the "Deuxième Colloque sur la Biomécanique des Mouvements", organized by the SELF.

Thus, many of the problems of motor research techniques are practically solved and many investigations are due concerning movement analysis in sport, everyday life and work, as well as in orthopedics.

FUNDAMENTAL RESEARCH

If we want to come to the core of a movement, the study of forces is of fundamental importance. In measuring these forces—active and passive, static and dynamic—not only physical but also physiological knowledge is required. As the musculature are working under physiological circumstances, the physiological parameters should be taken into consideration with every muscle and movement model.

Next to the mechanical and physiological aspects we shall have to consider that every movement has a psychological context. For a long time, psychologists have been interested in movements. They have been investigating learning processes most intensively and problems of motor skill development, basic motor abilities, movement adaptation, movement concepts, etc.

These terms are used by everyone present. However, everyone has his own definition. This shows how important it is to clarify the terminology of biomechanics and to include psychology in our work.

FIELDS OF APPLIED BIOMECHANICS

If we ask the question of where biomechanics is applied, we will find that the following fields are of special importance.

Sport and Physical Education

It is our aim to improve the analysis of all sport movements so as to understand the techniques of the various disciplines scientifically, and thus enable a better instruction. In sport movement talents should be discovered in time by means of motion tests.

Physical education is in a broad sense interested in a training of movement depending thereby on the comparative movement research. When comparing movements with each other, analogies can be identified, equal or similar structures based on equal or similar elements. These mutual elements enable the sifting and systematic classification of movements. This taxonomy, which also includes the animal movement system, forms the essential basis of physical education. The Philantropes began working on this subject 200 years ago, but it is left to biomechanics to provide the scientific base.

Physical education is not only restricted to sport movement training, but seeks knowledge of the entire human experience and its personality expressed by its various movements. Are there motor types as there are constitutional types from Sheldon and Kretschmer, and psychological types like introverts and extroverts?

Biomechanics is not yet able to provide such a taxonomy as it is still too young to find a technique with which it would be possible to determine the artistic content of a movement. But we should keep in mind that every human emotion can be expressed in movements and that these can be recorded and analyzed biomechanically.

Medicine

If movements vary from the normal standard, we have to reckon with disturbed or pathological forms. The disorder can have its cause in the movement system itself, or in the CNS. In this case orthopedics and rehabilitation are confronted with this problem.

Pathological forms of movement occur especially in cases of cerebral disorders. To eliminate disorders it is important to diagnose them as early as possible. Oseretzky was one of the first to develop a "metric scale for testing the motoric skill." It would be worthwhile to continue these investigations with biomechanical parameters.

Industrial Movements, Ergonomics, Engineering

Earlier motor research work of physiologists was usually limited to industrial movements. Although these processes are increasingly being changed due to automation, ergonomics is still very much interested in biomechanics. Researchers in this area are mainly concerned with the optimal working techniques, the equipment and the industrial environment.

An engineer does not only want to construct machines as a substitute for human beings or as a means of transport, but he also wants to adapt his construction to the requirements of people, be it an electric razor, a flat iron or a car seat. To achieve this he must be familiar with the mechanics of human motions. He also profits from investigating biological systems and their interaction with new mechanical inventions. A seminar on these problems was held in 1967 at the Rock Island Arsenal in the USA.

This is just a short enumeration of the main fields of biomechanics, which shows clearly the various aspects which are of importance. I am very pleased to see the entire variety represented in the program of this IVth International Seminar on Biomechanics.

We all feel that we can learn from each other and that we belong together. This feeling existed from the very beginning of our meetings and led, two years ago in Rome, to the decision to found an international society. This society should act like an orchestra. We are the musicians and we have joined together to play the IVth Symphony of Biomechanics. Each one of us has brought his special instrument; our full score is the program composed by Richard Nelson and his staff. We will now play together; each one of us acting with his own personal individuality and yet all together with the responsibility of playing in tune.

But not only playing is important, listening is also essential. In listening we are able to grasp the meaning of the music, to which we are all contributing in order to achieve a harmonious whole.

Caused by the necessity of specialization we are at the point of losing the general view of our work and life and because of that, we are forgetting to consider other people's individuality, and to be receptive to their ideas.

Our universities have proceeded to offer subjects in all fields of sciences to counteract intellectual stagnation. These international seminars are also multidisciplinary in nature. It would be one of the first duties of our International Society of Biomechanics to guarantee the continuation of these seminars.

Just as the Working Group on Biomechanics acts as a sponsor of many symposia (like Biomechanics of Swimming, Problems of Motion-Biology in Track and Field) the ISB will have many opportunities to coordinate and support local and regional meetings. The ad hoc committee will present its proposal to found an International Society on Biomechanics on Wednesday evening. We shall all look forward to this event and to future developments in biomechanics.

Author's address: *Prof. Dr. Jurg Wartenweiler, Kurse für Turnen und Sport, Swiss Federal Institute of Technology, Plattenstrasse 26, 8032 Zürich, Switzerland*

Fundamental human movements

Lifting
and
standing

Dynamic characteristics of man during crouch- and stoop-lifting

D. W. Grieve
Royal Free Hospital School of Medicine, London

Movements of the back, motions of the body and the load, development of intratruncal pressures and electromyography, and strength of the erector spinae musculature in lifting are reported by Davis, Troup, and Burnard (1965), Davis and Troup (1964), Morris, Benner, and Lucas (1962), Morris, Lucas, and Bresler (1961), Floyd and Silver (1955), and Troup and Chapman (1969), and are discussed in the literature which they cite. The practical reasons for studying lifting are to identify features which are potentially injurious, especially to the back, as discussed by Troup (1965) and Davis (1967), to make recommendations as to safe loads, and to recognize dangerous working conditions.

This study describes lifting in terms of an equivalent three-component model, identifies characteristic features of the performance of the upper and lower parts of the body without recourse to the anatomy, and relates the behavior of the model to the movements of the back. Ways in which predictions of maximal loads might be made on the basis of observations of safe laboratory loadings are discussed.

METHODS

The experiments were performed on young adult males. Most of the results were compiled from two subjects who lifted loads fitted with a horizontal rod handle located 29 cm above the base onto a shelf either 61 or 78 cm above the floor. The subjects stood on a Whitney force platform (Whitney, 1958), and the load (4, 14, or 29 kg) rested initially on a force transducer. An

Thanks are due to the Medical Division of the National Coal Board and to the Central Research Fund of London University for funds and equipment used in these studies.

accelerometer was attached to the load. Galvanometer measurements were taken of the forces at the feet and on the load support and of the acceleration of the load. Measurements were taken at 20-msec intervals, and the forces at the feet and hands and the velocities of the load and the center of gravity of the body were computed throughout the lift. The subjects were filmed by a Telford Type N camera at 50 frames per sec, and the film and the galvanometer measurements were synchronized by means of an electric clapboard which was operated at the start and finish of each lift. The subjects wore plastic markers to indicate the positions of the hip, knee, ankle, shoulder, elbow, and wrist joints in the sagittal plane. Marker pins were attached to the back to indicate the angles of the surface of the back at sacral and first thoracic levels.

A further series of experiments was conducted with 12 subjects; the vertical force at the feet was measured by means of a strain-gauge transducer supporting a plate which was fitted with horizontal restraints at two levels to prevent it from tilting and rotating about the vertical axis. The loads were attached by cord to a spring-loaded tachometer so that their velocity could be monitored.

The methods have been described previously by Gorell (1970) and by Hawkins (1972). The raw data from these studies together with subsequent analysis of film material have provided the basis for this article.

RESULTS

Measured Forces and the Velocities of the Body and Load

A series of lifts to the same height by one subject are featured in Figures 1 and 2. When the crouch-lifts (Figure 1) are compared with the stoop-lifts (Figure 2), similarities are to be found principally in the forces developed at the hands and in the resulting velocities of the load. In all lifts, which were performed as quickly as possible, an impulsive force was applied to the load, which reached a peak within 100 msec of lift-off in most cases. As the load was increased, the ratio of the peak force to the weight of load fell from 3.83 (range, 2.63–5.38) at 4 kg to 2.58 (range, 2.38–2.85) at 14 kg and 1.75 (range, 1.60–2.10) at 29 kg. When the subject lifted 4- and 14-kg loads, the force at the feet rose to a peak of between 50 and 100 percent above body weight and fell markedly below body weight later in the lift. When the subject lifted 29-kg loads, the force at the feet fell from peak to around body weight.

The most obvious differences between the styles of lift lie in the temporal patterns of force development at the feet and in the relative velocities of the body and load. In a crouch-lift, forces are developed at the feet more than 100 msec before lift-off and reach peak at lift-off or very soon thereafter. As

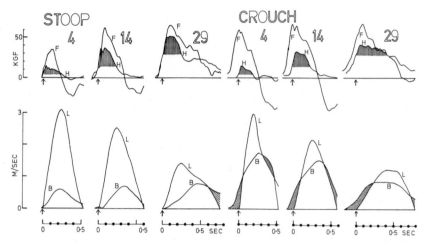

Figures 1 and 2. Lifts of 4-, 14-, and 29-kg loads by one subject through a distance of 61 cm are featured in crouch style (Figure 1) and stoop style (Figure 2). The vertical downward force (*H*) at the hands and the vertical upward force (*F*) at the feet, relative to body weight, are shown at top. The vertical velocity of the load (*L*) and of the center of gravity of the body (*B*) are plotted at bottom. Vertical shading indicates periods during which forces at the hands exceed the weight of the load. Diagonal shading indicates periods during which the center of gravity is travelling upward at a rate faster than that of the load. *Arrows* indicate lift-off.

a result, the body has an upward velocity at lift-off and initially travels upward faster than the load. This state persists until the force at the hands reaches its peak, and the load then travels faster than the body. In a stoop-lift, the load travels upward faster than the body throughout most of the lift, with both starting from rest at lift-off. In the late stages of the lift in both styles, the body may travel upward faster than the load, but the difference is slight. It must be remembered that although the measurements relate to the period preceding and including time at which the load reaches maximal height, the subject then had to place the load on a shelf. The continuing upward movement of the body as the load reached the top of its trajectory may have played some useful role in the placement of the load.

Movements of the Back

The difference of angle in the sagittal plane of marker pins attached to the surface of the back at sacral and first thoracic levels was taken as an index of the posture of the back. Changes of this angular difference during lifting are presented in Figure 3. The back was initially in a more flexed posture (25° more), and much more movement subsequently occurred in a stoop-lift than was found in a crouch-lift. The extension of the back did not take place uniformly throughout the lift but commenced apparently at the moment when the velocity of the center of gravity of the body reached its maximum.

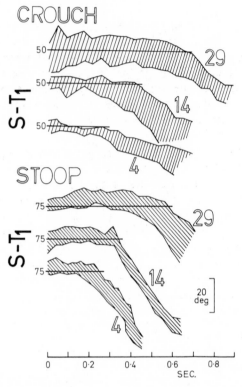

Figure 3. The changes of angle between marker pins placed over the sacrum and the first thoracic vertebra during lifts of 4-, 14-, and 29-kg loads in crouch and stoop style. Each shaded area represents the data of four lifts from two subjects lifting to heights of 61 and 78 cm. *Deg.*, degree of flex.

The back was fixed to a remarkable extent (within a few degrees across the entire lumbar and thoracic regions) while the center of gravity accelerated. In some cases, there was a tendency for the back to flex slightly in the early stages of the lift. The shaded portions in Figure 3 show how consistent the behavior was between the two subjects and that the height of the lift had little effect on the temporal sequence of movements of the back. The angular velocities of the back were derived from the original data and are presented in the "Results."

Analysis of the Lifting in Terms of a Model

Although it is not necessary to postulate a model in order to present the following data, it may help the reader to consider the body as a concentrated mass located at the center of gravity. The mass is connected to the load and to the ground by upper and lower force elements, respectively. The model thus generates a force in the upper element that is equal to the measured force at the hands, and it moves (one end relative to the other) with a velocity equal to the difference between the computed velocities of the load and the center of gravity of the body. The lower element generates a force equal to the measured force at the feet, and its velocity is equal to the computed velocity of the center of gravity. We therefore may plot force-velocity characteristics of both elements during a lift and compare the behavior of the elements in different lifts in a manner similar to that in which a muscle group might be studied.

Force-velocity plots for one subject are shown in Figure 4. The most striking feature is the different use of the elements in the two styles of lift. Borrowing terms from muscle physiology, it can be seen that the upper element is used eccentrically at first (center of gravity travelling upward faster than the load) and then concentrically during crouch-lifts, while it is used concentrically throughout most of a stoop-lift. The principal difference between the two styles of lift in the lower element is that much higher velocities are acquired in a crouch-lift at a given force than are achieved in stoop-lifts. The activation of the element in a stoop-lift takes place under almost isometric conditions, whereas both force and velocity develop to-gether in a crouch-lift.

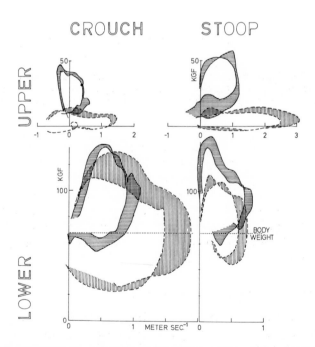

Figure 4. Force-velocity characteristics of the upper and lower force elements (see text for details) in crouch- and stoop-lifting. Each shaded area represents two lifts of different heights. The horizontal and vertical shading represents 29- and 4-kg lifts, respectively. Data are from one subject performing eight lifts.

Figure 5 illustrates all the characteristics of both subjects in all lifts of both crouch and stoop styles, together with the mean conditions under which peak forces were developed when the subjects lifted the various loads.

The rates of work (product of force and velocity) of the two elements were computed throughout each lift and they emphasize the very great differences in the use of the elements in the two styles of lifting, as may be seen in Figure 6. The lower element produced much greater power outputs in crouch-lifts than in stoop-lifts. The upper element absorbed energy shortly before and after lift-off in the crouch-lift, eventually producing a little more than it had absorbed by the later stage of the lift. In stoop-lifting, the upper and lower elements achieved similar rates of power output, except with 4-kg loads.

Lift Impedance

When a voltage is applied to an electrical circuit, a current flows which depends on the impedance of the circuit. By analogy, the term "lift imped-

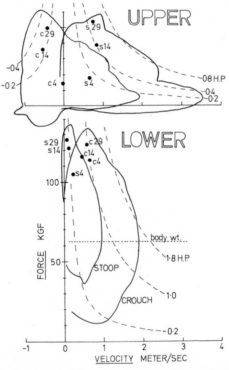

Figure 5. Force-velocity characteristics of the upper and lower force elements. Data for upper and lower sections are for 12 lifts each (three loads, two heights, two subjects). The mean conditions for peak force development for each load are indicated by *closed circles. Dashed lines* indicate rates of work (*H.P.*). Since the subjects had different body weights, the zero-base line of force for the lower element is a compromise to place the body-weight line at a level equal to that of the mean body weight.

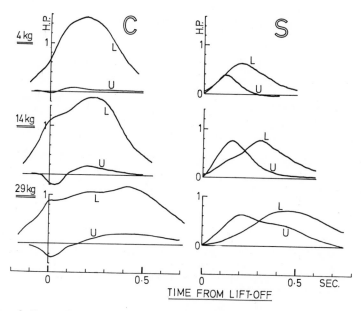

Figure 6. Rates of work of the upper (*U*) and lower (*L*) force elements in crouch- (*C*) and stoop- (*S*) lifts. Each curve represents the mean of four lifts (two heights, two subjects).

ance" expresses the relationship between the force at the feet (relative to body weight) and the velocity of the load.

$$\text{Lift impedance} = [(\text{force at feet} - \text{body weight})/\text{velocity of load}]$$

The results of two sets of experiments are presented in Figure 7, which shows the changes of lift impedance when two subjects lifted 4-, 14-, and 29-kg loads, as well as all the data concerning the two subjects while they lifted to two heights and in both crouch and stoop styles. The temporal patterns of the lift impedance therefore are insensitive to the height and style of lift but are sensitive to the magnitude of the load. The portions of the figure which illustrate the 6-, 15-, and 25-kg lifts apply to crouch-lifts only, but with 12 subjects at each load. This second series was obtained with a vertical-force plate and a load-velocity transducer in a study that was specially designed to measure lift impedance and amply confirm the temporal patterns that are to be found.

When the subject lifts loads of up to 15 kg, the lift impedance is initially infinite (the values have been clipped at $\pm$ 100 kgf m^{-1} sec in Figure 7) and falls rapidly to a small positive value (about 20 kgf m^{-1}sec) after about 100 msec. It then falls more slowly, becoming negative, and finally it becomes a large negative quantity. The change of sign occurs when the force at the feet falls below body weight. The lift impedance becomes negative infinity as the

LIFT IMPEDANCE

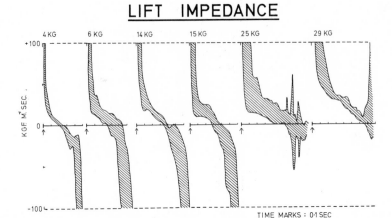

Figure 7. Lift impedance when subject lifts loads ranging from 4 to 29 kg. The areas representing 4-, 14-, and 29-kg loads each contain data from eight lifts (two subjects, two heights, crouch and stoop styles). The areas representing 6-, 15-, and 25-kg loads refer to 12 subjects in crouch-lifts.

load reaches the top of its trajectory if the force at the feet is still below body weight when this occurs. When loads of 25 kg are lifted, the lift impedance falls rapidly at first but does not become strongly negative. In the 29-kg lifts, the fall of lift impedance occurs more slowly and then usually remains positive.

Back Extension as a Function of the Forces Acting at the Feet

It was stated earlier that the back is stabilized during a lift until the body (center of gravity) has acquired its maximal velocity, and then it extends. Examination of the force patterns in the upper force element suggests a relationship between the instantaneous force acting at the feet and the rate of flexion or extension of the back. Figure 8 shows plots of the force at the feet versus the rate of back movement in crouch- and stoop-lifts. The angular velocities were obtained by calculating the differences between the samples of the thoracolumbar angles at 40-msec intervals and subjecting the velocities obtained to three-point smoothing.

It is apparent that, in the stoop-lifts, the back was unable to withstand the stresses associated with forces at the feet that were more than 1.7 times body weight, and that it was necessary to flex the back when forces greater than this existed. In this condition, the hip-shoulder line was almost horizontal. For a given force at the feet in a crouch-lift, the back extends more slowly than it does in a stoop-lift, but it was not possible to make any useful comparisons between the crouch and stoop data because the postures differed.

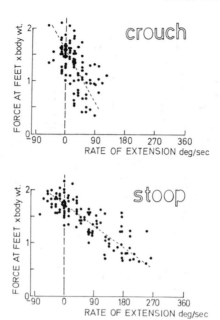

Figure 8. Relationships between the forces at the feet relative to body weight and the rate of angular movement of the back (S to T1) in stoop- and crouch-lifts. Each graph contains data from four lifts (two subjects, two heights). *Deg.*, degrees.

DISCUSSION

The highest stresses during the lifts used in this experiment were found in the first 0.4 sec, and research into potentially injurious stresses should concentrate on this period. Although industrial lifting would be conducted in a more leisurely fashion, it was hoped that a limiting feature would be recognized more easily by requiring maximal effort in the laboratory. It is not clear at present why a subject exerting maximal effort should limit the power or force developed by the upper or lower parts of the body to less than maximum unless some smaller component in the chain is working at full capacity and creating a limitation.

The successful performance of stoop-lifting requires extension of the back, and it would seem reasonable for the subject to extend his back earlier than was observed in a 4- or 14-kg lift in order to achieve a more rapid lift. It is possible that the delayed extension occurs because the strength of the back is initially dependent on ligamentous restaints and intratruncal pressures rather than on erector spinae activity, which is known to be slight at the commencement of these lifts. No data appear to be available concerning either the strength of the erector spinae group as a function of posture or the sensitivity of intratruncal pressure to postural change. Cadaver studies of sarcomere spacings within the fibers of the erector spinae group are planned in order to estimate the competence of the muscles in the stabilized posture.

It may be that the almost incompressible contents of the abdominal cavity would depressurize and throw the tensile stresses entirely on ligamentous and muscular constraints if extension occurred prematurely. In that case, a mismanagement of the muscles of the pelvic floor, anterior abdominal wall, diaphragm, and glottis could be an important factor in injury of the dorsal muscles and ligaments. Repetition of the present experiments during deliberate exhalation therefore should be instructive because if this leads to flexion of the back when large forces are being generated at the feet, it would suggest a fresh approach in industrial training.

Assuming that no dramatic changes in the sequence of events are introduced for loads in excess of 29 kg, it is possible to make a rough estimate of the limiting load that can be lifted in a stoop-lift. The peak forces at the hands in the stoop-lifts always exceeded the weight of the load. Lifting would be impossible if the peak force merely reached the weight of the load, and extrapolation suggests that this would happen with the present subjects with a 40- to 50-kg load. Also, the yield point of the back (flexion instead of extension) occurs in the heaviest lifts under conditions of near equality between the forces at the hands and feet. Thus, the limiting load at the hands for which extension of the back is just possible is also between 40 and 50 kg. These estimates seem rather low in comparison with loads that are accepted by industry, but it is perhaps unlikely that a man would adopt a stoop style when a crouch could be adopted for even a 40-kg load. In any case, many more subjects are needed before extrapolations can be trusted. The experiments described offer the possibility of predicting maximal loads without exposing the subjects to them.

The lift impedance was developed as a relatively simple measure by which the performances in a variety of industrial lifting tasks might be classified. It proved very insensitive to technique when maximal effort was involved, but there is no reason why the observed temporal patterns should be generated under industrial conditions. A quasi-static lift would produce a large positive impedance throughout, whereas an explosive leap into the air while the subject clutches the load would produce a sharp change from positively to negatively infinite impedance. A wide variety may be expected between these extremes when free styles are adopted.

ACKNOWLEDGMENTS

My thanks are due to Messrs. Gorell and Hawkins for access to their experimental data and film material, to Dr. J. D. G. Troup for postural analysis from film, and to Dr. P. R. Cavanagh for assistance with computing.

REFERENCES

Davis, P. R. 1967. Physiotherapy 44—47.
Davis, P. R., and J. D. G. Troup. 1964. Ergonomics 7: 465—474.
Floyd, W. F., and P. H. S. Silver. 1955. J. Physiol. 129: 184—203.
Gorell, L. 1970. Unpublished M.Sc. Ergonomics thesis, University of London, London, England.
Hawkins, J. N. 1972. Unpublished M.Sc. Ergonomics thesis, University of London, London, England.
Morris, J. M., G. Benner, and D. B. Lucas. 1962. J. Anat. 96: 509—520.
Morris, J. M., D. B. Lucas, and B. Bresler. 1961. J. Bone Joint Surg. 43A: 327—351.
Troup, J. D. G. 1965. Lancet 857—861.
Troup, J. D. G., and A. E. Chapman. 1969. J. Biomech. 2: 49—62.
Whitney, R. J. 1958. Ergonomics 1: 101.

Author's address: *Dr. Donald W. Grieve, Department of Anatomy, Royal Free Hospital, Medical School, 8 Hunter Street, London, WCIN, IBP (England)*

Models for lifting activities

M. M. Ayoub
Texas Tech University, Lubbock

R. D. Dryden
University of Texas at Arlington

J. W. McDaniel
Bramson-Givens & Associates, Inc.

In the most basic sense, a model is a representation of reality and must be regarded as a tool to be used in the analysis and/or design of a real system. If models are as complex and difficult to control as reality, there would be no advantage in their use. Fortunately, models usually can be constructed that are simpler than reality but that will predict and explain reality with a reasonable degree of accuracy. These characteristics of simplicity and accuracy provide the greatest advantage with models and the reason for their widespread use in analyzing complex systems. Unfortunately, the simplification of reality, as well as other limitations of models, generates the prime causes of erroneous interpretation of the results of model manipulation. Considerable caution must be exercised in the evaluation and application of the results obtained from models; however, when the pitfalls are avoided, advantages such as efficiency and feasibility of analysis result.

The primary function of a model is to facilitate determination of how changes in one or more aspects of a real entity may affect other aspects of the entity. This feature is especially useful when applied to lifting activities. A need exists for generalized models capable of predicting safe lifting capacities of man and safe stresses imposed on him during the lifting activity. It is logical that these models be divided into two basic groups; and, at this time, the appropriate divisions may be: (a) performance models and (b) biomechanical models. It should be noted that these classifications are not exhaustive.

The performance models attempt to develop correlations between physical parameters and lifting capacities of industrial employees. On the basis of correlations, models have been developed that can predict lifting capacities for several lifting height ranges with reasonable degrees of accuracy and confidence.

The biomechanical models apply Newtonian mechanics to the human frame for the purpose of generating data on the effects of reactive forces and torques on the various joints, links, and other critical parts of the human anatomy. Once these biomechanical models are developed, they can be used for predictive purposes. For example, biological component maximal loads or data on human strength may be used as criteria for developing safety factors. The biomechanical models can cover a large range of lifting postures and configurations; therefore, they are useful in providing data on the stresses at several points on the musculo-skeletal system.

METHODS

Models of the type previously discussed are currently in the developmental and/or verification stage at Texas Tech University. The performance models have utilized a modified psychophysical technique to arrive at the maximal acceptable weight of lift as a function of anthropometry as well as of strength, endurance, and other physical characteristics. Data used in the development of the performance models were obtained by procedures similar to that practiced and verified by Snook and Irvine (1967) and by Snook, Irvine, and Bass (1969).

The models are in the form of linear regression obtained by the technique of stepwise regression. The particular method utilized was the BMDO2R, one of the biomedical computer routines published by the University of California at Los Angeles. A good description of the technique is given by Draper and Smith (1966).

Models for the acceptable weight of lift have been developed for two lifting ranges: floor-to-knuckle height (McDaniel, 1972) and knuckle-to-shoulder height (Dryden, 1973). The models proposed for predicting the maximal acceptable lift are given in Table 1.

RESULTS AND DISCUSSION

Figure 1 shows an example of the observed and predicted acceptable weight of lift for five test subjects using the male model in the range from floor-to-knuckle height. The error of prediction with this model was 8.78 percent. Predictive errors with the remaining models were of similar magnitude. All models predicted with accuracy well within acceptable limits when tested with separate test subjects.

The data generated by this endeavor compare quite well with results published by Snook et al. (1969). Table 2 shows a comparison of data

Table 1. Models proposed for predicting maximal acceptable lift

Floor-to-knuckle height	Knuckle-to-shoulder height

Male: Maximal lift =
  $-172.356 + 0.022 \times (\text{ht})^2$
  $-2.73 \times (\text{st. end.*})^2$
  $+0.021 \times (\text{RPI} \times \text{arm strength})$
  $+0.053 \times (\text{RPI} \times \text{back strength})$
  $-2.513 \times (\text{fitness/dyn. end.})$
	Male: Maximal lift = 0.0
	  $+0.828 \times (\text{chest cir.})$
	  $+0.559 \times (\text{dyn. end.})$

Female: Maximal lift =
  -24.027
  $+0.194 \times (\text{RPI})^2$
  $+0.006 \times (\text{arm strength} \times \text{leg strength})$
	Female: Maximal lift = 0.0
	  $+3.809 \times (\text{RPI})$
	  $-0.002 \times (\text{ht.} \times \text{fitness})$
	  $-0.312 \times (\text{RPI} \times \text{st. end.})$
	  $+0.001 \times (\% \text{ fat} \times \text{fitness})$

*St. end., static endurance; dyn. end., dynamic endurance; cir., circumference.

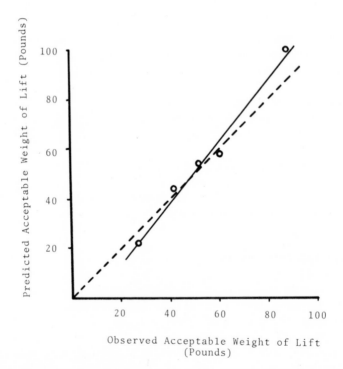

Figure 1. Observed and predicted acceptable weight of lift for five male test subjects in the range from floor-to-knuckle height.

Table 2. Acceptable weight of lift for selected lifting heights

	Percentage of population				
Floor-to-knuckle:					
Snook and Irvine, 1967 (male only)	52	59	50	25	80
Snook et al., 1969 (male only)	37	45	54	62	70
Combined model	20	33	43	53	66
Male (only) model	31	42	53	65	75
Female (only) model	19	27	33	39	46
Knuckle-to-shoulder:					
Snook and Irvine, 1967 (male only)	51	56	62	68	73
Snook et al., 1969 (male only)	34	43	53	62	71
Combined model	21	30	41	52	61
Male (only) model	40	47	55	62	69
Female (only) model	23	25	27	30	31

reported by Snook and Irvine and by Snook et al. with those generated by our studies.

Dynamic biomechanical lifting models are needed urgently. Although the breakthrough by Fisher (1969) in developing a dynamic model was certainly a significant contribution, it does have the disadvantage of difficulty of application. The lack of data concerning the kinematics of lifting could have contributed to this phenomenon. The usual means of collecting dynamic data have been by using either goniometry or several photographic techniques (Ayoub, 1970). Of course, this is quite difficult when one attempts to obtain dynamic data on lifting in the industrial environment. If such dynamic data could be generated easily, they would enhance the usefulness of the application as well as the development of additional dynamic biomechanical models.

The approach currently used to develop the needed data arose from a modification of a displacement-time relationship for arm movement reported by Slote and Stone (1963). After modification, this relationship was used to describe angular displacement-time relationships at the various joints. When the total displacement at the joint and the time of movement was known, it was then possible to generate (a) displacement-time relationships, (b) velocity profiles, and (c) acceleration profiles. Velocity and acceleration profiles are determined by obtaining the first and second derivatives of the angular displacement relationship.

Successful attempts have been made to use these relationships to generate data on the kinematics of lifting. Similar data can be generated and are being compiled at Texas Tech University for the remaining articulations involved in lifting tasks. Figure 2 shows a typical angular displacement-time relationship

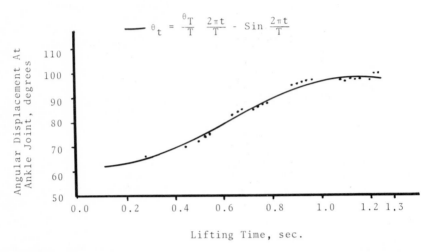

• Experimental

$$\theta_t = \frac{\theta_T}{T}\left[\frac{2\pi t}{T} - \mathrm{Sin}\,\frac{2\pi t}{T}\right]$$

Figure 2. Male subject.

for the ankle joint during a lifting task. Similar results were obtained also for the knee and hip joints. It should be noted that the angular displacement data of Figure 2 were obtained through goniometry; therefore, it appears that such a technique can provide suitable kinematic data on lifting.

In order to determine the effect of accelerations on the forces and torques at various critical anatomical points during the lifting actions, it is necessary to find the translation accelerations at these points. Using free-body diagrams, such as that shown in Figure 3, it is possible to find these accelerations. It also has been determined that the phase relationships (sequence and time of rotation) of the various joints significantly affect the acceleration profiles and therefore the forces caused by the movement (force = mass X acceleration). Such phase differences concerning the start and end movements at the various joints are functions of the method of lift, the magnitude of the burden, and the weight-to-bulk ratio. Current investigations of these variables and their effects on the kinematics of lifting now are being conducted.

CONCLUSIONS

On the basis of the acceleration profiles obtained from this study, it was determined that the kinematics of lifting plays an important role in determining the stress components on the musculo-skeletal system.

Predictive models for performance as well as biomechanical models are

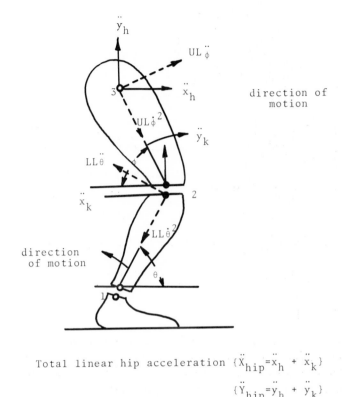

Total linear hip acceleration $\{\ddot{X}_{hip} = \ddot{x}_h + \ddot{x}_k\}$

$$\{\ddot{Y}_{hip} = \ddot{y}_h + \ddot{y}_k\}$$

Figure 3. Free-body diagram; lower extremity.

very useful in gaining insights into the factors and methods of lifting activities that relate to the lifting capacity of man. Both types of models complement one another in providing information concerning the lifting capacity of the individual. Such predictive ability is required urgently in industry for the purposes of more effective placement of employees and better design of work areas. Both types of models have advantages and disadvantages; before any model is used, its limitations and assumptions should be understood.

The proper selection and application of lifting models, whether bio-mechanical or performance, may provide data to improve existing work areas for lifting purposes and are useful in the design of new jobs requiring the manual handling of weights.

REFERENCES

Ayoub, M. M. 1970. Human movement recording for biomechanical analysis. Internatl. J. Product. Res. 10.

Draper, N. R., and H. Smith. 1966. Applied Regression Analysis. John Wiley and Sons, New York.

Dryden, R. 1973. A predictive model for the maximum permissible weight of lift from knuckle to shoulder height. Ph.D. dissertation, Texas Tech Univ., Lubbock, Tex.

Fisher, B. O. 1967. Analysis of spinal stresses during lifting. MSE thesis, Univ. of Michigan, Ann Arbor, Mich.

McDaniel, J. W. 1972. Prediction of acceptable lift capability. Ph.D. dissertation, Texas Tech Univ., Lubbock, Tex.

Slote, L., and G. Stone. 1963. Biomechanical power generated by forearm flexion. Hum. Factors 5: 443–452.

Snook, S. H., and C. H. Irvine. 1967. Maximum acceptable weight of lift. Amer. Ind. Hyg. Assoc. J. 28: 322.

Snook, S. H., C. H. Irvine, and S. S. Bass. 1969. Maximum weight and work loads acceptable to male industrial workers while performing lifting, lowering, pushing, pulling, carrying and walking tasks. Presented at the American Industrial Hygiene Association Conference, Denver, Colo., May 13.

Author's address: *Dr. M. M. Ayoub, Department of Industrial Engineering, College of Engineering, P. O. Box 4130, Texas Technical University, Lubbock, Texas 79409 (USA)*

Biomechanics of lifting

A. Roozbazar

North Carolina State University, Raleigh

The objective of this study was to determine the effects of compressive stress, shear stress, and bending moment on the intervertebral disc between the fourth and fifth lumbar vertebrae during three methods of lifting. This site in the spine was chosen because the axis of movement in flexion and extension passes through the nucleus pulposus of the intervertebral discs (Fick, 1904) and it is the site of the most serious lower back injuries (Chaffin, 1969).

Among the various biomechanical criteria employed to evaluate the burden of load in lifting and carrying, the following are the most popular: (a) increase in intraabdominal pressure (Asmussen and Poulsen, 1968; Davis, 1959a,b; Eie, 1966; Eie and When, 1962), (b) isometrical back muscle strength (Asmussen and Poulsen, 1968; Poulsen and Jørgensen, 1971), (c) stress on L4/L5 disc or L5/S1 (Chaffin, 1969; Martin and Chaffin, 1972), (d) electromyography-tension in the deep back muscles (Basmajian, 1972; Donisch and Basmajian, 1972; Floyd and Silver, 1955), and (e) the amount of intradisc pressure (Nachemson, 1966, 1971).

It is interesting to note that a linear relationship exists between each of these variables and the amount of load lifted. It also has been found that there are possible strong relationships between the variables themselves. For example, an excellent correlation ($r = 0.99$; $p < 0.001$) has been found to exist between the activities of the erector spinae and changes in intraabdominal pressure (Kumar and Davis, 1973).

The three methods of lifting discussed in this chapter are: bent back/extended knees (stooped position), also known as "derrick"-type lifting; back inclined/knees bent; and back vertical/knees bent. The first method is a back-lift and the last two are leg-lift methods. The mechanical analysis in this study was restricted to statics. The three conditions considered were actually instantaneous positions when the loads were held against gravity.

This investigation was supported by U. S. Public Health Service Grant Number 5-TO-1-OH-00089 from the National Institute for Occupational Safety and Health.

METHOD

Average dimensions for a male belonging to the 25–35 age group were obtained from Contini (1972) and Hertzberg (1972). The results yielded a height of 178 cm, a weight of 78 kg, and a length of spine equal to about 52 cm. The mass of the trunk above the intervertebral disc L4/5 was estimated from Dempster (1961) to be around 60 percent of the total body weight, or about 44 kg. On the basis of Dempster's data, this weight was assumed to act on a 3:2 ratio on the spine. This gave a length of 31 cm from the L4/5 disc and 21 cm from the shoulder.

The three methods studied were: (a) vertical back and bent knees, (b) inclined trunk and bent knees, and (c) fully flexed trunk and extended legs. Figure 1 shows these positions. The angle of inclination in Figure 1*b* was chosen as 45°, or midway through a fully extended and fully flexed trunk.

The study was limited to two-handed, symmetrical lifting in the sagittal plane, with very little or no rotation of the trunk. The subject was assumed to breathe normally. He was considered a normal man with no disc degeneration or lower back pain. The amount of weight lifted was 55 kg, which is considered by the International Labor Office (Thacker, 1972) to be the maximal weight a worker should be expected to lift.

The analysis was based on Figure 1*b* ($\alpha = 45°$), which is more general, and then was applied to the other cases ($\alpha = 0°$ and $\alpha = 90°$). Figure 2 shows a modified version of the work of Eie (1966). The effects of the contractions of abdominal wall muscles (rectus abdominis, external oblique, and internal oblique) were neglected here. Asmussen and Poulsen (1968), Bartelink (1957), and Eie (1966) reached the conclusion that the rectus abdominis plays a minor, negligible role in this situation. The effects of the external and

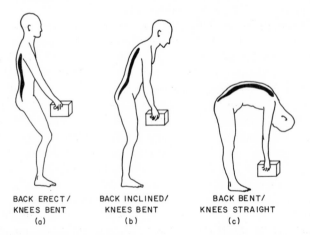

BACK ERECT/   BACK INCLINED/   BACK BENT/
KNEES BENT   KNEES BENT   KNEES STRAIGHT
(a)   (b)   (c)

Figure 1. The three methods of lifting.

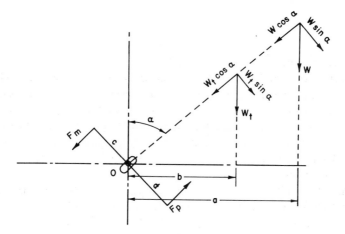

Figure 2. The schematics of forces and moments acting on the trunk during lifting. See text for explanation. (Modified after Eie, 1966)

internal obliques are not known. The psoas muscle has been found to act as a stabilizer of the lumbar spine (Nachemson, 1966), but the effects have yet to be evaluated quantitatively. Considering the data of Asmussen and Poulsen (1968) and Eie and When (1962), the effects of intrathoracic pressure also can be neglected.

The parameters for Figure 2 have been obtained from different sources. The arm of muscle tension (F_m) was assumed to be 6 cm (Troup and Chapman, 1969). The force of increased abdominal pressure (F_p) was assumed to act in the center of the abdomen parallel to the erector spinae force but opposite to it. Its arm was taken to be 11 cm (Morris, Lucas, and Bresler, 1961). The area of L4/L5 lumbar discs was assumed to be about 17.5 cm² (Orne and Liu, 1971). The cross-section of abdomen was taken to be 483 cm² (Eie and When, 1962). The arm of weight lifted was taken to be 37 cm from the disc in the inclined back/bent knee position, and it varied in the other two positions. The intraabdominal pressure for stooped position was assumed to be 100 mm Hg, and for inclined trunk, to be 40 mm Hg. For erect posture, this pressure was considered negligible. The arm of the weight of trunk, arms, and head was $31 \times \sin \alpha$. The arms of the muscle force and intraabdominal force have been assumed to be constant for all three positions.

CALCULATIONS AND RESULTS

By looking at Figure 2 and considering a state of equilibrium, the following two equations can be written as the resultant of forces and moments at the L4/5 disc:

$$-F_c + F_m - F_p + W_t\cos \alpha + W \cos \alpha = 0 \qquad (1)$$

$$F_m \times c + F_p \times d - W_t \times b - W \times a = 0 \qquad (2)$$

where F_c = resultant of the compressive forces on the L4/5 disc, kg; F_m = the extension force of the deep back muscles, kg; F_p = the force of intraabdominal pressure = $p \times 483$, kg; W = weight of the load = 55 kg; a = arm of the weight lifted, cm; W_t = weight of upper body over the fourth-fifth lumbar disc, kg; b = the weight arm of upper part of the body, cm; c = arm of the muscle force, cm; d = arm of the pressure force, cm; and α = angle of inclination of the trunk with vertical, degree.

Also, the following equation has been adopted from Nachemson (1966) for the tangential stress in the annulus fibrosus of the disc:

$$\alpha_{td} = 4 P = 2.7 P_n \qquad (3)$$

where α_{td} = tangential stress on the dorsal part of the disc, kg/cm^2; P = applied load per unit area of disc, kg/cm^2; and P_n = the intradisc pressure actually measured, kg/cm^2.

A sample calculation of the case of the inclined trunk is shown below. F_m is obtained from (2) and substituted into (1) to obtain F_c, the compressive load on the disc:

Fm = [(−40mm Hg × Kg × 483 cm² × 11 cm)/(760 mm Hg × cm²) +
44 kg × 22 cm + 55 kg × 37 cm]/6 cm = 454 kg.

Then from (1)

$$F_c = 454 \text{ kg} + 39 \text{ kg} + 31 \text{ kg} - 25 \text{ kg} = 499 \text{ kg}.$$

The clockwise moment (considered positive) caused by weights about Point 0 = + 44 kg × 22 cm + 55 kg × 37 cm = +3,003 kg − cm. Also, compressive stress on the disc = P = 499 kg/17.5 cm² = 28.5 kg/cm². The tangential stress on the dorsal part of the disc is:

$$\alpha_{td} = 4P = 4 \times 28.5 \text{ kg/cm}^2 = 114 \text{ kg/cm}^2$$

and the pressure in the nucleus pulposus is:

$$P_n = 114 \text{ kg/cm}^2 /2.7 = 42 \text{ kg/cm}^2$$

which is about 1.5 times the applied pressure. The calculations are given in Table 1.

Table 1. Calculations of the effects of moment, compressive stress, and tangential stress on the L4/5 disc

Method	Movement (kg/cm)	Compressive stress kg/cm^2	Tangential stress (on the annulus fibrosus) (kg/cm^2)
Back bent/straight knees	+4224	30	120
Inclined back/bent knees	+3003	28.5	114
Back vertical/bent knees	Negligible	6	24

DISCUSSION AND CONCLUSIONS

The results show the superiority of the back erect/knees bent method over the other two methods. Actually, the other two methods are inside the "damage" zone (28–43 kg/cm^2). It has been shown by the aid of electromyography and by measurement of intradisc and intraabdominal pressure that the leg-lift is much superior to the back-lift. Nachemson (1971) has measured the intradisc pressure in a man lifting 20-kg weights with the back bent/knees straight method and with the back straight/knees bent method. The load on the third lumbar disc in the former case was 380 kg, while that for the latter method was 250 kg.

Clarke (1945) has shown that, with the back-lift, a 50th-percentile male is able to exert a pulling force of about 520 lbs., whereas with the leg-lift he is able to pull 1,480 lbs. Davis (1959a, b) has pointed out the danger of anterior abdominal wall hernias and prolapses of pelvic viscera during stooped lifting because of increased abdominal pressure. With the aid of electromyography, Basmajian (1972), Donisch and Basmajian (1972), and Floyd and Silver (1955) have shown the economy of human erect posture.

An interesting point is that the compressive force on the disc in a stooped position is 524 kg when the role of abdominal pressure has been accounted for. Because the m. erector spinae in this position is inactive (MacConaill and Basmajian, 1969), this load should be supported by the ligaments and other organs of the back. Therefore, the danger of torn ligaments is obvious.

The only similar data for comparison are those of Schürmann and Luginbühl (1972). According to these authors, the loads on the fourth-fifth lumbar disc at the angle of inclination of the trunk of 0°, 45°, and 90° with a weight of 50 kg are 100, 500 (approximately), and 700 kg, respectively. However, it is obvious that they have not considered the role of intraabdominal pressure. If this pressure were disregarded in the present study, the results would be the same.

Thus, it appears that the maximal weight as set by the International Labor Office for male industrial workers can be lifted most efficiently with a vertical back and bent knees. However, if the trunk is kept quite vertical and heavy loads are lifted, then the stability of the body is a problem. In order to solve this, the author suggests the use of the "kinetic method" (Himbury, 1967), but the angle of inclination of the trunk should be as small as possible.

REFERENCES

Asmussen, E., and E. Poulsen. 1968. On the role of intraabdominal pressure in relieving the back muscles while holding weights in a forward inclined position. Dan. Natl. Assoc. Infan. Paral., No. 28.

Bartelink, D. L. 1957. The role of abdominal pressure in relieving the pressure on the lumbar intervertebral discs. J. Bone Joint Surg. 39B(4): 718–725.

Basmajian, J. V. 1972. Electromyography comes of age. Science 176: 603–609.

Chaffin, D. B. 1969. A computerized biomechanical model–development and use in studying gross body actions. ASME Paper No. 69-BHF-5.

Clarke, H. 1945. Analysis of physical fitness index test scores of air crew students at the close of a physical conditioning program. Res. Quart. 16: 192–195.

Contini, R. 1972. Body segment parameters. Part II. Artif. Limbs 16: 1–19.

Davis, P. R. 1959a. Role of the trunk during the lifting of weights. Brit. Med. J. 10: 87–89.

Davis, P. R. 1959b. The causation of herniae by weight lifting. Lancet 22: 155–157.

Dempster, W. T. 1961. Free-body diagrams as an approach to the mechanics of human posture and motion. In: Evans, F. G. (ed.), Biomechanical Studies of the Musculo-Skeletal System, pp. 81–85. Charles C Thomas, Springfield, Ill.

Donisch, E. W., and J. V. Basmajian. 1972. Electromyography of deep back muscles in man. Amer. J. Anat. 133: 25–36.

Eie, N. 1966. Load capacity of the low back. J. Oslo City Hosp. 16: 74–98.

Eie, N., and P. When. 1962. Measurements of the intra-abdominal pressure to weight bearing of the lumbosacral spine. J. Oslo City Hosp. 12: 205–217.

Fick, R. 1904. In: Handbuch der Anatomie und Mechanic der Glenke, Vol. I, p.75. G. Fischer. Cited by Donisch and Basmajian (1972).

Floyd, W. F., and P. H. S. Silver. 1955. The functions of the erectores spinae muscles in certain movements and postures in man. J. Physiol. 129: 184–203.

Hertzberg, H. T. E. 1972. Engineering anthropology. In: Van Cott, H. P., and R. G. Kinkade (eds.), Human Engineering Guide to Equipment Design, pp. 467–584. Joint Army-Navy-Air Force Steering Committee.

Himbury, S. 1967. Kinetic methods of manual handling in industry. Occupational Safety and Health Series No. 10. International Labor Office, Geneva.

Kumar, S., and P. R. Davis. 1973. Lumbar vertebral innervation and intra-abdominal pressure. J. Anat. 114: 47–53.

MacConaill, M. A., and J. V. Basmajian. 1969. Muscles and Movements, A Basis for Human Kinesiology. Williams & Wilkins, Baltimore.

Martin, J. B., and D. B. Chaffin. 1972. Biomechanical computerized simulation of human strength in sagittal-plane activities. AIIE Trans., March, 19–28.

Morris, J. M., D. B. Lucas, and B. Bresler. 1961. The role of the trunk in the stability of the spine. J. Bone Joint Surg. 43A: 327–351.

Nachemson, A. L. 1966. The load of lumbar disks in different positions of the body. Clin. Orthop. No. 45.

Nachemson, A. L. 1971. Low back pain, its etiology and treatment. Clin. Med. Jan., 18–24.

Orne, D., and Y. K. Liu. 1971. A mathematical model of spinal response to impact. J. Biomech. 4: 49–71.

Poulsen, E., and K. Jørgensen. 1971. Back muscle strength, lifting, and stooped working postures. Appl. Ergonom. 2: 133–137.

Schürmann, J., and H. R. Luginbühl. 1972. Lifting, Carrying. *In:* Encyclopedia of Occupational Safety and Health, pp. 777–780. International Labor Office, Geneva.

Thacker, P. V., 1972. Maximum weights. Encyclopedia of Occupational Safety and Health, pp. 829–830. International Labor Office, Geneva.

Troup, J. D. G., and A. E. Chapman. 1969. The strength of the flexor and extensor muscles of the trunk. J. Biomech. 2: 49–62.

Author's address: *Dr. Aziz Roozbazar, Department of Industrial Engineering, 328 Riddick Engineering Laboratories, North Carolina State University, Raleigh, North Carolina, 27607 (USA)*

Biomechanical analysis of the knee joint during deep knee bends with heavy load

B. G. Ariel

University of Massachusetts, Amherst

The knee joint, the largest and most complex synovial joint in the human body, is an anatomical region subject to injuries from activities in various fields including athletics, industry, and recreation. Because this joint is between the longest bones in the body, the femur and the tibia, the forces and moments of force around this joint produce torques of such magnitude that injuries ensue. In athletics, various injuries may occur by overloading the knee joint (Nicholas, 1970; Peterson, 1970). In several studies (Kennedy and Fowler, 1971; Marshall and Olsson, 1971; Newman, 1969; Slocum and Larson, 1968), it was found that the instability of the knee joint was the result of the application of excessive external rotation and abduction forces to a flexed, weight-bearing knee.

The knee joint, described as a hinge joint, is much more complex. It consists of three articulations, the surfaces of which are not mutually adapted to each other, so that movement is not simply gliding (Gray, 1954; Lockhart, Hamilton, and Fyfe, 1959). The quadriceps femoris muscle group is responsible for extension of the knee joint. The four muscles of this group pull through a common tendon and insert via the ligamentum patella, which continues from the patella to the tuberosity of the tibia. The movements of the knee joint are primarily flexion and extension and, in certain positions of the joint, internal and external rotation (Dick, 1969).

The purpose of the present study was to investigate the forces and moments of force acting about the knee joint during a deep knee bend exercise with a heavy load.

METHODS

Twelve experienced weightlifters, ranging in age from 21 to 25 years, served as subjects. Their mean height was 181.5 cm and their mean weight was 90.5

kg. The training period encompassed the academic year from September to May, and data were collected at the beginning and at the end of this period. High-speed cinematography was used to record the subjects performing the deep knee bend. Special tracing equipment permitted direct processing of the data by a high-speed computer. The segments used in the present study form a link system consisting of the shank, thigh, and trunk with their respective weights. The load consisted of the barbell suspended on the shoulders, and this was applied as an external force to the link system. The load was specific for each individual, with a range of from 375 to 650 lbs. and an overall mean load of 445 lbs.

Figure 1 illustrates one instantaneous body position. In each segment, the center of the joint was located and traced by a digitizer and then was sent electronically to the computer. Knowledge of the film speed and the displacement of the joint centers enabled the calculation of velocities of the body segments, and from the velocities it was possible to calculate the segment accelerations. Segment masses, locations of segment centers of gravity, and radii of gyration were obtained from anatomical data (Clauser, McConville, and Young, 1969; Dempster, 1955) and by a displacement method and were utilized in the calculation of forces and moments of force. A special computer program was used for the calculation of moments of force and of dynamic, bone-on-bone, shearing, and compression forces.

A biomechanical analysis of the knee joint may be approached through the use of a model. Examinations were made of X-rays of 50 males ranging in height from 175 to 182 cm. The following parameters were measured and are illustrated in Figure 1:

1. Distance between the knee joint center and the perpendicular line from the tuberosity of the tibia (e).

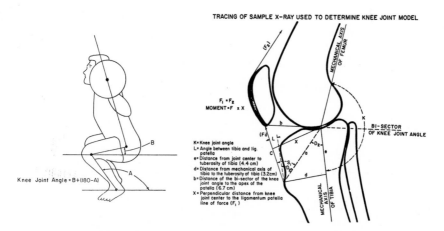

Figure 1. An instantaneous position in the deep knee bend and tracing of sample X-ray used in determining the knee joint model.

2. Distance from the tuberosity of the tibia to the mechanical axis of the tibia (*d*).
3. The bi-sector of the knee joint angle and its distance from the knee joint center to the apex of the patella (*b*).

These parameters enable calculation of the perpendicular distance from the knee joint center to the line connecting the tuberosity of the tibia with the apex of the patella (X = force arm).

An important feature of the computer used in the calculation of forces is its consideration of the forces caused by motion, as well as the forces caused by muscular contraction. Summation of the inertial and muscular forces permitted the calculation of the bone-on-bone, shearing, and compression forces. The bone-on-bone force is the total resultant force derived from summation of the forces caused by the motion and those caused by the muscular contractions. The bone-on-bone force can be partitioned into the shearing and compression forces by considering the mechanical axis of the bones.

RESULTS

The scope of this chapter permits the inclusion of only three subjects as examples of the force analysis. Their performances were indicative of the situations encountered in the biomechanical analysis and were selected merely for illustrative purposes.

Figure 2 presents the horizontal forces caused by motion for two subjects, representing periods before and after a training session. A positive force

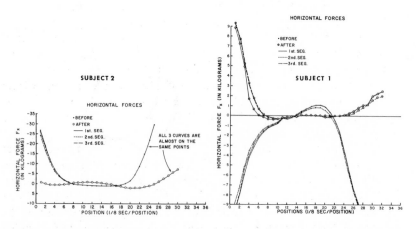

Figure 2. Horizontal force curves for Subjects 1 and 2 representing periods before and after training. *Seg.*, Segment.

SUBJECT 2

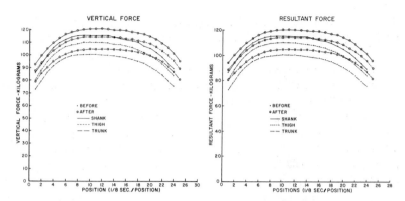

Figure 3. Vertical and resultant force curves representing periods before and after training for Subject 2.

indicates leaning backward and a negative force indicates forward sway. One may observe that, after the training period, Subject 2 was able to demonstrate a horizontal force approaching zero. This reduction in horizontal forces indicates an improved efficiency, since the subject maximized the vertical forces and minimized the horizontal ones. On the other hand, the force curves for Subject 1 illustrate the inefficiency in his lifting technique before and after the training period, as can be seen by the magnitude and direction of the horizontal forces.

Figure 3 illustrates the vertical and resultant forces for Subject 2. Note that, after training, the magnitude of these forces increased, possibly as a result of greater muscular strength and better coordination.

Moments of force (Figure 4) indicate the dominant muscular action. The negative value for Segment 1 indicates a dominant muscular action by the foot flexors; positive values for Segment 2 indicate a dominant muscular action by the knee extensors; and negative values for Segment 3 indicate a dominant muscular action by the hip extensors. The blip at Position 16 for the subject shown at left occurred at the lowest position when the subject bounced, thus causing an abrupt increase in the moment. This increase is the major contributor to a great shearing force and, therefore, possible knee injury.

Table 1 presents the computer output for three different subjects. It was found that the shearing force for all subjects was of greatest magnitude at the beginning of the exercise, when the subject initially bent his knees. However, at this stage, the knee is anatomically more protected. When the knee is bent beyond 90°, it is more vulnerable, and the shearing force may affect the ligaments. Subject 2 was the strongest and the most experienced lifter, with a personal record lift of 650 lbs. The analysis of his movements yielded a

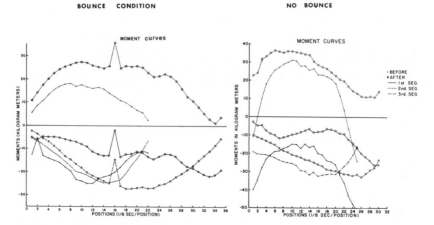

BOUNCE CONDITION

NO BOUNCE

Figure 4. Moment of force curves representing bounce and no-bounce conditions for two different subjects. *Seg.,* segment.

vertical force of 538.7 kg at the lowest point (Position 12), with a shearing force of 12.6 kg. Subject 1 yielded a vertical force of 841.5 kg at his lowest point, with a shearing force of 60.8 kg. The abrupt increase in all his forces at that point was caused by the bounce at the lowest point in the squat and is illustrated in the moment curves.

It was found that some subjects moved their knees forward while performing the squat exercise, whereas other subjects maintained the knees in relatively the same position. The forward movement of the knees while the subject performed the squat was associated with the greatest shearing force (Subject 3). This knee-shifting introduces mechanical factors which influence the magnitude of the shearing forces and may be one of the causes of knee injuries.

SUMMARY

The present study revealed that muscular force is comprised of vertical and horizontal components caused by muscles and motion. At times, a subject may appear to be weak, not because of muscular insufficiency, but because of a reduced vertical component or the inhibiting influence of the shearing forces. For example, the strongest subjects always demonstrated less shearing force than did the weaker subjects. In addition, the shearing forces associated with a bounce condition in the squat exercise may be one of the main causes of knee injuries.

It was found that training elicits improvement exhibited not only by increases in muscular force, but also by biomechanical patterns which result in an increased vertical force and reduced horizontal and shearing forces.

Table 1. Knee joint angle, forces, and moments of force for selected subjects in the deep knee bend with heavy load

Position	Knee angle (degrees)	Moment (kg/m)	Horizontal force (kg)	Vertical force (kg)	Bone-on-bone force (kg)	Shear force (kg)	Compression force (kg)
Subject 1							
1	150.1	10.7	−145.7	−166.7	221.4	104.5	195.2
2	140.7	14.1	−197.5	−222.7	297.6	126.1	269.6
3	131.8	17.5	−252.1	−282.6	278.7	143.3	350.5
4	123.2	20.2	−298.8	−336.0	449.7	150.4	423.8
5	115.3	22.5	−340.5	−387.3	515.7	151.8	492.8
6	107.7	24.4	−377.7	−435.9	576.8	147.4	577.6
7	100.8	25.3	−399.9	−469.2	616.5	136.0	601.4
8	94.5	26.2	−421.7	−504.9	657.9	124.1	646.1
9	88.8	27.0	−440.9	−540.9	697.8	111.4	688.9
10	83.9	27.3	−451.4	−566.4	724.3	97.7	717.7
11	79.5	27.0	−451.8	−578.8	734.2	82.8	729.6
12	75.9	25.9	−437.1	−571.1	719.2	68.1	716.0
13	73.1	25.2	−427.5	−568.9	711.7	57.4	709.3
14	70.9	24.9	−424.0	−573.1	712.9	49.6	711.1
15	69.4	25.3	−432.4	−589.5	731.0	45.4	729.6
16	68.7	35.9	−612.5	−841.5	1040.8	60.8	1039.0
17	68.8	24.8	−422.7	−582.1	719.4	42.5	718.1
18	69.5	25.6	−435.4	−597.3	739.1	46.3	737.7
19	71.0	25.2	−425.9	−581.5	720.7	50.5	719.0
20	73.1	25.6	−430.2	−580.9	722.8	58.2	720.5
21	75.7	25.0	−417.9	−555.2	694.9	65.1	691.9
22	79.2	22.8	−378.0	−492.9	621.2	69.1	617.3
23	83.1	21.0	−345.6	−440.4	559.8	73.3	554.9

(continued on next page).

Table 1 (continued).

Position	Knee angle (degrees)	Moment (kg/m)	Horizontal force (kg)	Vertical force (kg)	Bone-on-bone force (kg)	Shear force (kg)	Compression force (kg)
24	87.6	21.3	-347.5	-431.7	554.2	85.2	547.6
25	92.6	22.2	-359.0	-433.8	563.1	100.8	554.0
26	98.0	21.2	-339.8	-399.0	524.1	108.2	512.8
27	103.9	19.2	-304.5	-348.1	462.5	109.4	449.4
28	110.0	15.6	-245.6	-272.5	366.9	98.2	353.5
29	116.4	12.3	-191.9	-208.2	283.1	85.0	270.1
30	123.0	10.5	-162.7	-172.6	237.3	79.2	223.6
31	129.8	8.0	-123.8	-129.0	178.8	65.9	166.2
32	136.6	5.2	-81.4	-84.0	117.0	47.2	107.0
33	143.3	2.7	-44.9	-46.2	64.5	28.3	57.9
34	150.0	1.0	-21.1	-21.8	30.3	14.5	26.7
35	156.5	.3	-12.3	-12.6	17.6	9.2	15.0
Subject 2							
1	166.5	-15.2	186.0	205.5	277.2	152.8	231.2
2	148.6	-7.3	92.4	99.0	135.5	62.4	120.2
3	132.2	1.3	-25.0	-25.8	36.0	13.8	33.2
4	117.4	9.0	-144.9	-149.2	208.0	63.5	198.1
5	104.4	14.5	-242.9	-250.3	348.8	83.4	338.7
6	93.0	18.8	-329.8	-342.3	475.4	86.1	467.5
7	83.4	20.0	-365.7	-385.2	531.1	70.3	526.4
8	75.4	22.3	-422.3	-453.2	619.5	57.0	616.8
9	69.2	23.0	-447.4	-490.2	663.7	40.4	662.4
10	64.6	22.7	-449.4	-503.3	674.7	25.5	674.2
11	61.8	22.6	-450.5	-515.3	684.4	16.3	684.2
12	60.6	23.2	-460.4	-538.7	708.6	12.6	708.5
13	61.0	21.8	-428.1	-507.7	664.1	13.1	664.0
14	63.0	20.9	-401.3	-483.3	628.2	18.7	627.9

15	66.4	19.7	−368.1	−446.8	578.9	27.1	578.3
16	71.3	17.5	−317.0	−384.7	498.4	35.5	497.2
17	77.6	14.4	−251.2	−304.6	394.8	40.6	392.7
18	85.1	12.9	−216.8	−260.3	338.8	47.7	335.4
19	93.8	11.1	−179.6	−213.0	278.6	51.5	273.8
20	103.5	8.1	−127.2	−147.6	194.8	45.6	189.4
21	114.2	6.1	−93.6	−106.2	141.6	40.9	135.5
22	125.8	2.6	−41.2	−45.2	61.2	21.5	57.3
23	138.0	−5.3	69.6	75.2	102.5	41.7	93.6
24	150.9	−12.7	167.6	174.8	242.2	114.8	213.2
25	164.0	−17.4	225.6	227.4	320.4	173.4	269.4
Subject 3							
1	130.7	20.4	−290.9	−321.7	433.7	161.2	402.6
2	121.6	22.5	−330.1	−371.7	497.2	161.9	470.1
3	113.8	24.1	−362.1	−416.1	551.5	157.8	528.5
4	107.8	25.4	−390.1	−453.2	598.0	153.0	578.1
5	101.9	26.4	−411.6	−489.4	639.5	144.5	622.9
6	97.7	26.3	−414.9	−501.2	650.7	133.3	636.9
7	94.5	26.4	−420.4	−514.0	664.1	125.3	652.1
8	92.2	25.5	−408.7	−504.6	649.4	115.0	639.1
9	90.8	25.8	−414.8	−516.0	662.0	112.6	652.4
10	90.2	26.5	−426.7	−532.2	682.1	114.0	672.5
11	90.5	26.6	−427.5	−533.5	683.7	115.3	673.9
12	91.5	26.9	−431.0	−536.0	687.8	119.4	677.3
13	93.2	26.0	−415.6	−511.4	659.0	120.1	647.9
14	95.6	25.3	−403.1	−488.8	633.6	123.1	621.5
15	98.7	25.6	−406.0	−483.5	631.4	132.6	617.3
16	102.4	25.4	−401.6	−466.4	615.5	140.8	599.2
17	106.7	24.0	−377.9	−426.9	570.2	142.9	552.0
18	111.6	25.0	−391.8	−429.2	581.1	160.1	558.6
19	117.1	23.5	−367.1	−387.3	533.7	162.0	508.5
20	123.3	20.5	−319.4	−323.0	454.2	152.2	428.0
21	130.0	19.2	−298.2	−288.3	414.7	153.1	385.4

REFERENCES

Clauser, C. E., J. T. McConville, and J. W. Young. 1969. Weight, Volume, and Center of Mass of Segments of the Human Body. AMRL Technical Report. Aerospace Medical Research Laboratory, Wright-Patterson Air Force Base, Ohio.

Dempster, W. T. 1955. Space requirements of the seated operator. WADC Tech. Rep. 55: 159.

Dick, F. W. 1969. Rotation and the knee joint. Brit. J. Sports Med. 4: 203–208.

Gray, H. 1954. Anatomy of the Human Body, pp. 380–388. 26th Ed. Lea & Febiger, Philadelphia.

Kennedy, J. C., and J. P. Fowler. 1971. Medial and anterior instability of the knee. J. Bone Joint Surg. 53: 1257–1270.

Lockhart, R. D., G. F. Hamilton, and F. W. Fyfe. 1959. Anatomy of the Human Body, pp. 121–124. Lippincott, Philadelphia.

Marshall, J. L., and S. E. Olsson. 1971. Instability of the knee. J. Bone Joint Surg. 53: 1561–1570.

Newman, P. H., 1969. Athletic injuries of the extensor mechanism of the knee. Brit. J. Sport Med. 4: 209–211.

Nicholas, J. A., 1970. Injuries to knee ligaments. JAMA 212: 2236–2239.

Peterson, T. R. 1970. The cross-body block, the major cause of knee injuries. JAMA 211: 449–452.

Slocum, D. B., and R. L. Larson. 1968. Rotatory instability of the knee. J. Bone Joint Surg. 50: 211–225.

Author's address: *Dr. Gideon B. Ariel, School of Physical Education, University of Massachusetts, Amherst, Massachusetts 01002 USA*

A study of stability of standing posture

K. Kobayashi and M. Miura
University of Nagoya, Nagoya

Y. Yoneda
Aichi University of Education, Aichi

K. Edo
Chukyo University, Nagoya

Mechanical, physiological, and psychological factors affect the stability of a standing posture of the human body. Broer (1966), Bunn (1955), Hellebrandt (1938), Morehouse and Cooper (1950), and other researchers described the well-balanced posture from many points of view. The purpose of this study was to measure the degree of stability of various standing postures when an external force was applied to the body.

METHODS

The subjects were 16 healthy adults (12 men and four women) whose body heights and weights are shown in Table 1. The subjects were asked to resist an external force with a maximal effort in nine different defensive postures that were divided into two groups: one in which the hip or knee joints or both were flexed and the other in which the feet were spread apart in different positions. Positions in the first group were: (a) hip and knee joints immobilized, A (an erect posture), (b) trunk bent forward and knee joints extended, B, and (c) trunk bent forward and knee joints flexed, C (a crouching posture). Positions in the second group were: (a) closed stride, 1, (b) side stride, with feet 30 cm apart, 2, and (c) forward-backward stride, with feet 30 cm apart, 3. Nine postures were devised from the combinations of the three different leg postures (A, B, C) and foot positions (1, 2, 3), as illustrated in Figure 1.

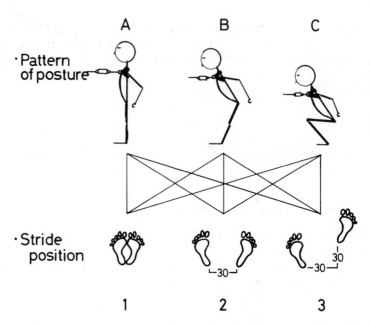

Figure 1. Nine different postures consisting of the combinations of A, B, C and 1, 2, 3.

The subject stood on a mat. The friction between the surface of the mat and the soles prevented the feet from sliding. One end of a rope was secured to the chest of the standing subject, while the other end was attached to a winch that was rotated electrically at the desired speeds. The rope was pulled at constant speeds of 10.6, 32.3, 53.7, 76.8, and 105.4 cm/sec, respectively, and was passed over a pulley in order to pull the subject forward horizontally at armpit height. The rope was pulled continuously at a uniform rate until the subject could not maintain his initial position. The tensile force of the rope was measured with a semiconductive strainmeter. A rigid body (i.e., a

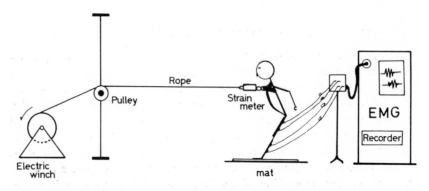

Figure 2. Instrument of experiment. *EMG,* electromyogram.

dummy, 170 cm in height and 63 kg in weight) was pulled by the same method in order to make comparisons with the human body. The electromyograms of the lower extremities were recorded throughout the experiments (Figure 2).

RESULTS

The maximal tensile force varied, depending on the different patterns of the resistant postures and the different speeds of pulling.

The maximal tensile force and the pulling speed

For comparison with the human body, a dummy was pulled opposite the body's center of gravity, the chest, and the head. Quadratic equations between the maximal resistant force and the pulling speed were determined by the method of least squares in the rigid body.

The equation was expressed as:

$$Y = a_0 + a_1 V + a_2 V^2$$

where Y is the maximal resistant force in kg and V is the pulling speed in cm/sec ($0 < V \leqslant 105.4$ cm/sec). The coefficients a_0, a_1, and a_2 were largest when the dummy was drawn at the point opposite the body's center of gravity (Table 1). The same relationship was found in an erect posture with a closed-stride position (Style A_1) for each subject (Table 1). Although a relationship was found in the other postures in which the pulling speed was directly proportional to the resistant force, the quadratic equations could not be obtained. The increments of the resistant force for Styles B and C in relation to the increase in the pulling speeds were relatively small in comparison with the increases for Style A.

The maximal resistant force and the resistant posture

The mean values of the maximal tensile force of 12 subjects are shown in nine different postures in Figure 3. The resistant force increased as the stride position was widened. For example, at a lower speed of 10.6 cm/sec, the maximal resistant force in A_2 and A_3 was 15 and 29 percent more than that in A_1, respectively. The same tendency was found at the higher speed of 105.4 cm/sec. However, the percentage decreased from 15 to 5 percent and from 29 to 12 percent, respectively. Therefore, the relative difference in the tensile force became less with the increase of the pulling speed among A_1, A_2, and A_3. The same effect of the stride position on the maximal resistant force was seen in the posture with the trunk bent forward and in the

Table 1. The characteristics of the subjects and the coefficients of quadratic equation

Subject	Age (yr)	Body height (cm)	Body weight (kg)	Coefficients of quadratic equation		
				a_0	a_1	a_2
Dummy		170.0	63.0			
Center of gravity				6.0465	0.1646	0.0048
Breast				2.5520	0.1062	0.0028
Head				1.5942	0.0369	0.0023
Male						
Y. Y.	26	190.0	98.0	12.1243	0.1422	0.0032
H. T.	26	183.0	88.0	14.2711	0.2333	0.0003
R. Y.	21	176.0	91.5	13.9813	0.0804	0.0020
K. I.	24	176.0	70.0	5.5716	0.2414	0.0003
T. T.	21	175.0	66.0	8.9391	0.0594	0.0011
T. A.	23	172.0	62.0	6.2453	0.2690	0.0001
H. S.	23	169.5	56.5	6.4289	0.1205	0.0026
J. K.	24	168.0	56.0	11.5951	0.0885	0.0015
M. S.	26	167.0	61.0	4.3777	0.2169	0.0011
K. O.	21	165.0	68.5	10.9269	0.1771	0.0016
H. M.	24	162.5	53.5	5.9488	0.1877	0.0005
M. M.	26	156.0	54.0	8.5549	0.0247	0.0016
Female						
J. O.	24	170.0	73.0	4.2223	0.2542	0.0006
S. G.	27	158.0	53.0	8.7356	0.0686	0.0005
K. K.	20	158.0	46.2	8.8632	0.1011	0.0004
K. M.	20	156.0	55.0	4.4740	0.0605	0.0015

crouching posture. At a lower speed, the maximal resistant force in Postures B_1 and C_1 was 64 and 123 percent more than that in Posture A_1, respectively. At a higher speed, the percentage decreased from 64 to 38 and from 123 to 49.

In addition, electromyograms recorded in nine different postures show that the muscle activities of the lower extremities were larger and that more muscle groups became active when the knee and hip joints were flexed, as in Styles B and C (Figure 4).

DISCUSSION

The maximal tensile force of the rope represents the maximal resistant force exerted by the subject against the external force. In other words, the maximal tensile force indicated the degree of stability in the human standing posture. The most stable posture was Style C_3 and the least stable posture was Style A_1 at all pulling speeds.

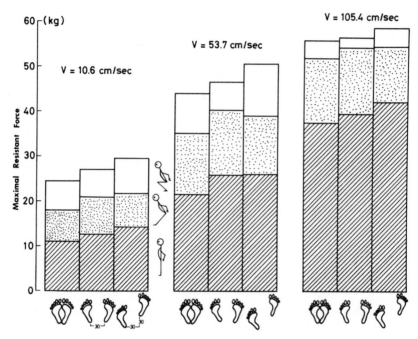

Figure 3. The maximal resistant force in nine different resistant postures and in different drawing speed.

Mechanically speaking, the degree of stability of the rigid body depends on the following factors: (a) the larger area of the base on which the body rests, (b) the shorter distance of the center of gravity of the body above the base, and (c) the greater weight of the body. As for the area of the base, man can widen the base in his posture by holding the side-stride or the forward-backward stride position. The results obtained in this study show that the resistant force in either position was 16 or 29 percent more than that of the closed-stride position, at a drawing speed of 10.6 cm/sec.

For the most stable posture (make the resistant force greatest), the position in which the knee and hip joints were flexed was more effective than that in which the feet were spread. The resistant force in the posture with the hips flexed or in the crouching posture was 64 or 123 percent more stable than that of the erect posture. When man flexes his hip and knee joints, the body's center of gravity moves closer to the floor. Mechanically, this means that the distance of the center of gravity above the base is lessened.

Furthermore, there would seem to be a positive relationship between the body weight and the resistant force (Figure 5). In individual cases, however, this relationship was not in evidence. For this reason, the muscle strength or the techniques used to maintain the equilibrium of the standing posture must be considered. In the case of the stability of the standing posture of a human

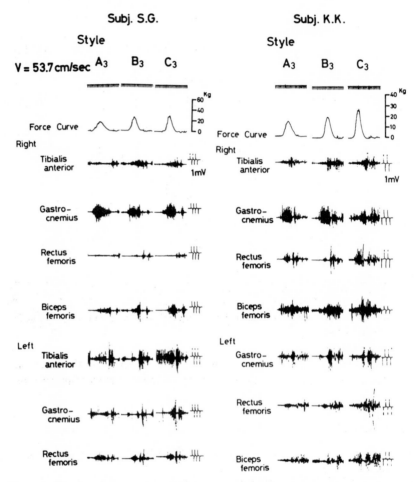

Figure 4. The electromyogram in different resistant postures. *Subj.*, subject.

being, the muscle activity must be taken into account. For example, there is the question of whether man holds a posture so that he can exert muscle strength more efficiently.

Similar experimental equations were found between a human body in an erect posture and a rigid body. This fact shows that man has difficulty in exerting any extra force in an erect posture and, as a consequence, he reacts as if he were a rigid body. Therefore, the individual differences were small in an erect posture, but when the subjects assumed the postures with knee and hip joints flexed, the individual differences appeared larger. This trend became greater as the drawing speed was increased. Consequently, the technique of holding the equilibrium is an important factor in the stability of the standing posture.

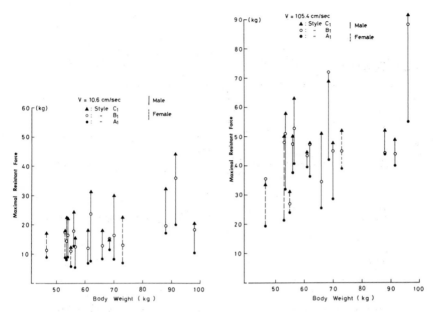

Figure 5. The relationship between the maximal resistant force and body weight in different drawing speed.

In conclusion, in addition to the mechanical principles, the muscular activity has a large influence on the stability of the human standing posture.

REFERENCES

Broer, M. R. 1966. Efficiency of Human Movement. W. B. Saunders, Philadelphia.

Bunn, J. 1955. Scientific Principles of Coaching. Prentice-Hall, New York.

Hellebrandt, F. A. 1938. Standing, a geotropic reflex, the mechanism of the asynchronous rotation of motor units. Amer. J. Physiol. 121: 471–474.

Morehouse, L. E., and J. M. Cooper. 1950. Kinesiology. C. V. Mosby, St. Louis.

Author's address: *Dr. Kando Kobayashi, Department of Physical Education, University of Nagoya, Furo-cho, Chikusa-ku, Nagoya (Japan)*

Postural reactions of man on a slowly moving base

E. J. Willems and M. S. Vranken
University of Leuven, Leuven

Different studies have shown that standing is a complex phenomenon and that the way in which a body reacts to a problem of balance depends largely on the kind of disturbance of the equilibrium. Bass (1938), for instance, concluded from a factor analysis that nine different factors play a role in the conservation of the upright position.

The specific aim of this study was to provide data on the regulation of equilibrium when the body was in a nearly static position on a slowly moving base.[1]

METHODS

Female students between 23 and 25 years of age (N = 51) who were enrolled in a physical education curriculum were placed on a special device to record the displacement of the vertical projection of the center of gravity in a horizontal plane. This apparatus has been described previously by Willems and Swalus (1967).

A hydraulic system installed on the platform beneath the right foot tilted the subjects toward one side or the other at two different speeds and for two displacements (Figure 1).

Simultaneous recordings of some muscle activities were made on an Elema Schönander eight-channel electromyograph, and pictures were taken to analyze the reactions of different segments of the body.

For all the experiments, the subjects were asked to stand at ease on the platform without bending the knees and to look at some drawings on the opposite wall. They were not informed of the purposes of the experiments.

[1] Details on the experimental set-up are available on request.

Figure 1. Apparatus to determine the center of gravity. A, moving part; B, stationary part.

The results of a pilot study with 10 subjects indicated that the rate of compensation of the body to a disturbance of the equilibrium was influenced by: (a) the distance (A) that the right foot was moved in a vertical direction (0.5 and 1 cm); (b) the speed (S) of this displacement (0.5 and 1 cm/min); and (c) the distance (E) between the feet (equal to the distance between the centers of the capita femoris and equal to twice this distance). The angle between the feet was always $22°$, which corresponds approximately to the standing-at-ease position.

The rate of compensation was determined by comparing the real displacement of the body's center of gravity in the frontal plane with the theoretical displacement that would have occurred if the subject were tilted as a whole without any reaction. This theoretical displacement (X) may be calculated from the vertical displacement (A), the distance between the feet (E), and the height (H) of the body's center of gravity above the ankle joint (determined in a supine position on a platform).

A closer examination of the factors influenced by altering A, S, and E indicates that the distance between the feet and the vertical distance determines the degree of tilting of the body and the change of pressure on the feet (ΔP).

Because of the mutual interactions among these factors, no definite conclusions could be drawn. In the following experiments, an attempt has been made to isolate the influences of the different factors.

The experimental situations were chosen so that, in the first series of tests, the variations in pressure on the feet, the rate of pressure variation, and the duration of the experiments remained the same for different distances between the feet.

To be able to compare the reactions of different subjects, it seemed reasonable to determine the variations of pressure on the feet as a certain percentage of the body weight. In this series, a ΔP was chosen arbitrarily as equal to 20 percent of the body weight.

For practical purposes, the velocity of the vertical displacement of the piston of the hydraulic system was not adapted to each individual case but was set consecutively at 0.5, 1, 2, and 4 cm/min. The desired changes in pressure and rate of pressure change then were calculated.

This series of experiments was compared with a second one in which the position of the feet and the rate of pressure variation remained the same but in which the pressure variation on the feet was doubled by doubling the vertical displacement. In this case, the duration of the test also was doubled automatically.

These tests were run for only three different distances between the feet, because the vertical displacement would become too large to be applicable at the largest distance (E_4).

In both series of experiments, the second situation always was chosen in such a way that the distance between the feet (E_2) equalled the distance between the centers of the capita femoris.

The seven different situations are listed in Table 1, with the different parameters calculated for a subject with: (a) a body weight of 50.9 kg; (b) the height of the center of gravity 84 cm above the ankle joints, and (c) a distance between the centers of the capita femoris of 15.67 cm. The sequence of the test is indicated in Figure 2.

Table 1. Example of the different parameters E,S,A,ΔP and t calculated for the seven different experimental situations (first series, subject no. 25)

Situation	E (cm)	S (cm/min)	A (cm)	ΔP	t (sec)
First series					
1 I, E −	11.08	0.5	0.29	10.18 kg	
2 I, E =	15.67	1	0.58	or	0.59
3 I, E +	22.16	2	1.18	20% of	
4 I, E X 2	31.34	4	2.38	body weight	
Second series					
5 II, E −	11.08	0.5	0.58	20.36 kg	
6 II, E =	15.67	1	1.18	or 40% of	1.58
7 II, E +	22.16	2	2.38	body weight	

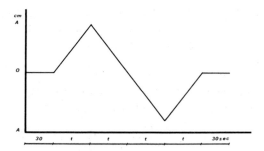

Figure 2. Schema of the movements of the right foot as a function of time (first series of experiments).

Half of the group started the test in this way, and the other half started in the opposite direction. This means that for the first group, after the subjects had been standing for 30 sec on a level base, the right foot first was elevated over a distance, A, and then was lowered to the horizontal and beyond until it was A cm below the horizontal. At the conclusion of the test, the feet were brought back to the initial position. The sequence was reversed for the second half of the group, i.e., the right foot was lowered after the first 30 sec and then was elevated above the horizontal and finally lowered back to the horizontal again.

RESULTS AND DISCUSSION

Figure 3 represents an example of a record of the displacement of the body's center of gravity in the sagittal and frontal planes. On these records, 33 points

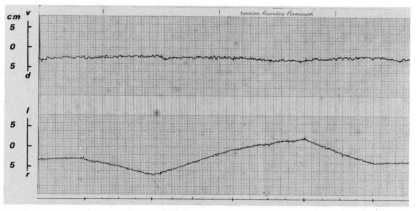

Figure 3. Example of a record representing the excursions of the center of gravity in Situation 3 (1, E +, Subject 4). Upper, in the sagittal plane: v = ventral, d = dorsal. Lower, in the frontal plane: l = left, r = right.

were determined in equal time intervals. The amplitudes of individual sways were not taken into account; rather, a mean value was used.

Figure 4 represents the same facts for the movements in the frontal plane. The distance of the center of gravity from the midposition now is placed on the abscissa, with the corresponding height of the right foot above and below the horizontal on the ordinate. A straight line, T, through the origin represents the theoretical displacement as determined above.

All of these curves for all of the subjects in the different experimental conditions had a similar form. In the first part of the movement, the center of gravity always was shifted in the direction of the lower foot. After the direction of the movement was changed, the center of gravity was moved to the other side, and so on.

Comparing the slope of the curves of the real displacement of the center of gravity with the slope of the theoretical one, in each part of the movement there was a reaction of the subject to restore the initial position of the center of gravity between the feet. The compensation was never immediate and seldom complete.

The amount of this compensation differed largely under the various experimental conditions. In the same situations, there was also a difference between the subjects, but the compensation was smaller when the distance between the feet was larger. The amount of compensation was also more pronounced in the situations in which the pressure variations on the feet were doubled.

Considering these facts and knowing that, within the two series of experiments, the pressure variation on the feet, the rate of variation of this pressure, and the duration of the test remained the same, the following hypotheses were formulated.

When a subject was standing in the normal upright position and he was

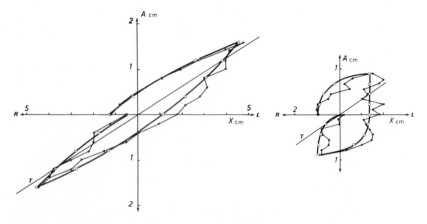

Figure 4. Displacement of the center of gravity in function of the height of the right foot. ● = experimental values; ○ = calculated values; T = theoretical displacement. Left: example from Situation 3 (1, E +, subject 4), $K' = 0.278$; S.F. = 0.66. Right: example from Situation 6 (2, E=, subject 19), $K' = 0.084$; S.F. = 1.45.

brought out of his equilibrium by moving one foot up or down with a constant and slow speed, he reacted to this disturbance by moving his center of gravity in the opposite direction. This compensation was a function of time and depended on the change in pressure on one foot in unit time. The change in pressure was expressed as a percentage of body weight.

These relationships may be expressed in a mathematical formula:

$$P_r = \frac{K \cdot E}{K'} - \left(\frac{K \cdot E}{K'} - P_{r1} \right) e^{(t_1 - t)(K'/E)}$$

$$= \frac{K \cdot E}{K'} - \left(\frac{K \cdot E}{K'} - P_{r_1} \right) \cdot e^{(t_1 - t)(K'/E)}$$

where P_r equals the real load on the foot at time, t; P_{r_1} is the load at time t_1; K' is a factor of proportion that expresses the velocity with which the subject reacts to the overload (in percentage of body weight, G) on one foot; and K is a constant expressing the change of pressure in unit time, $K = G.A.H./E^2 t$. Knowing this load, P_r, at each moment permitted determination of the distance of the center of gravity from the initial position at every time, t, during the experiment.

By repeating this method, the value of the individual "sensitivity" factor, K', could be determined from the experimental data. This was done for the results of the seven different tests performed by 27 subjects. As a criterion for the best-fitting curve, the standard error (S.E.) was calculated as:

$$S.E. = \sqrt{\Sigma d^2 / N - 1}$$

where d indicated the deviation of the experimental values from the calculated ones and N is the number of points. An example of this calculated curve added to the experimental one is shown in Figure 4. The mean values for the 27 subjects are listed in Table 2. Comparison of the mean values of K' indicates that this sensitivity factor remains quite constant over the different situations.

Table 2. Mean, standard deviation, and mean standard fault of the "sensitivity factor" K' obtained in seven experimental situations (27 female subjects) in the first series

Experimental situation	N = 27	Mean	Standard deviation	Mean standard fault
1	I, E −	0.8957	0.5244	3.2897
2	I, E =	0.9228	0.6424	2.2152
3	I, E +	0.9105	1.0020	1.7966
4	I, E × 2	0.4934	0.4081	1.8037
5	II, E −	0.9340	0.8234	3.6663
6	II, E =	0.8277	0.4839	2.5948
7	II, E +	0.6610	0.5509	2.6956

This conclusion certainly is reinforced when one considers the complexity of the factors that exert a possible influence on the results. The subjects were required to maintain a stationary position for a considerable time in seven different situations. The tests were spread over a period of four days. Since the K' indicates a functional parameter, one may expect it to be influenced by several internal and external factors, such as concentration, training, alertness, etc.

ELECTROMYOGRAM AND POSTURAL REACTIONS

In the first part of this study, we described the general reaction of the body to a disturbance of the equilibrium. In this part, we illustrate how the subjects reacted to restore the balance.

To avoid excessive fatigue and influence of irrelevant factors, a shorter test was chosen. In the first situation the right foot was moved in an upward (or downward) direction and then back to the initial position. In a second test, the foot was moved only up (or down). These tests were run once with a distance between the feet that was equal to the distances between the centers of the capita femoris and once with twice this distance. All other parameters remained the same as those described in the first part of this study. There were 25 subjects who participated in these four tests.

The reaction of different segments of the body could be determined from photographs. Reference points were indicated on the body (Figure 5) as follows: (a) head: on front and chin, and (b) shoulders: at just below the

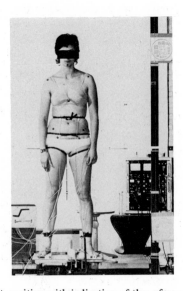

Figure 5. Subject in test position with indication of the reference points and electrodes.

acromioclavicular joint. For a control on the movement of one shoulder, a third point was indicated on the xyphoid process and (c) pelvis: on the front of the iliac spine and on the centers of the capita femoris. The location of the ankle joint centers was known in relation to the platform.

Figure 6 demonstrates the possible reactions of the pelvis caused by the induced movements. The upper five drawings indicate that, when the distance between the feet equals the distance between the centers of the capita femoris, the pelvis is always parallel with the horizontal until one foot is lifted. With twice this distance between the feet (lower five drawings), this is no longer true.

As a general rule, it may be concluded from the experiments that the displacement of the pelvis takes the same direction as the center of gravity; with the legs a small distance apart, however, the distance actually moved is much less than the theoretical one. With the feet spread far apart, this displacement is relatively greater; in four of the 25 cases, it is even greater than the theoretical displacement. This corresponds very well with the data given above. As may be seen in Figure 7, where the general reactions are summarized for the four situations, the movement of the shoulders is, in most of the cases, in the same direction as the movements of the pelvis, but with a certain delay. In this way, the shoulders stay in a more horizontal position.

An electromyogram control was employed, with four pairs of electrodes which were placed on the left and right tensor fasciae latae, gluteus medius, obliquus externus abdominis, and erector spinae (L3).

As a general rule, it may be stated that an increase of the activity of the

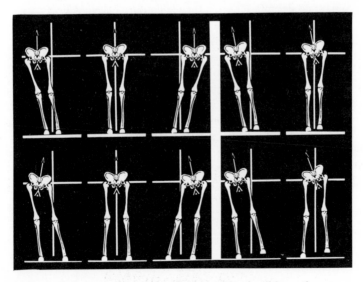

Figure 6. Theoretical movements of the pelvis caused by the tilting action or compensation in different positions of the feet. See text for details.

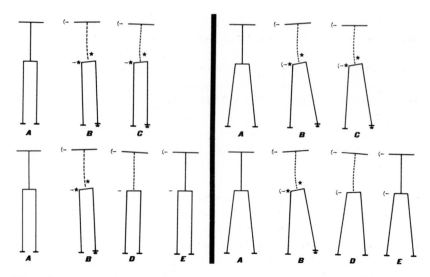

Figure 7. Summary of the mean reactions in four different experimental situations (Series 2). A = starting position; → = direction of the compensatory movements; ★ = increasing activity of the muscle group; ↑ = movement of the foot.

tensor fasciae latae and/or gluteus medius was noted on the left side if the right foot was moved upward. The gluteus from the left side may or may not relax, but the tensor fascia lata maintains its activity or is increased. The action of the erector spinae, but on the opposite side, is analogous to the action of tensor fasciae latae (Figure 7). The obliquus externus abdominis seems to have no definite action in the regulation of the position.

It should be noted that, in a series of trials, no change in the activity of the adductors of the hip joint was detected.

As a general conclusion, when a subject is standing in the normal upright position and he is brought out of his equilibrium by moving one foot up or down with a constant and low speed, he reacts to this disturbance by moving his center of gravity in the opposite direction. This compensation is a function of time and depends on the change in pressure on one foot in unit time. The change in pressure is expressed as a percentage of body weight.

To realize these corrections, the subjects react with movements on two levels: (a) by bringing back the pelvis, and (b) by curving the spine in the opposite direction and tilting the shoulders. Depending on the slope of the pelvis, one or the other reaction is of greater importance. The position of the head was not related to the position of the center of gravity or that of the hip or shoulder.

ACKNOWLEDGMENT

The authors wish to express their gratitude to Mr. A. Spaepen and Ms. L. Van de Water for their suggestions and assistance in this study.

REFERENCES

Bass, I. R. 1938. An analysis of components of tests of semicircular canal function and of static and dynamic balance. Res. Quart. 10: 33–50.
Willems, E. and P. Swalus. 1967. Apparatus for determining the center of gravity of the human body. *In*: J. Wartenweiler, E. Jokl, and M. Hebbelinck (eds.), Biomechanics I, pp. 72–77, S. Karger, Basel, and University Park Press, Baltimore.

Author's address: *Dr. Eustache Willems, Institute of Physical Education, University of Leuven, Tervuurse Vest, 101, B3030, Heverlee (Belgium)*

Walking and running

Facilitation of the IA peroneal reflex by volleys in the IA gastrocnemius afferents

R. Emmers

Columbia University, New York City

During the cycle of normal walking, the ankle is dorsiflexed approximately 0.30 sec before the heel-strike occurs (Murray, 1967). This dorsiflexion (Figure 1,*A*) begins to lessen only slightly during the last 0.10 sec prior to the heel-strike. As the heel touches the ground, however, the position of the ankle changes abruptly, and in approximately 0.05 sec it reaches the endpoint of its plantar flexion (Murray, 1967), also referred to as extension (Figure 1,*B*). A gradual shift from this extension to a 90° stance position follows. The latter provides for an adequate support of the body during the time that the other leg is in the swing phase. This sequence of changes in the position of the ankle (from flexed to extended to stance) should produce corresponding changes in the amount of stretch imposed on the ankle extensor and flexor muscles. Consequently, the stretch-sensitive primary endings of the muscle spindle receptors should send afferent volleys over the IA fibers and thus initiate a certain sequence of monosynaptic stretch reflex activity. Therefore, a question was raised as to how the interaction of the IA volleys at the motor neuron pools influences the contraction of those muscles which control the position of the ankle. This interaction was analyzed by studying the effects of a sequential activation of the gastrocnemius and the peroneal reflexes.

METHODS

These experiments employed eight cats as subjects. They were anesthetized with an intraperitoneal injection of Pentothal Sodium (35 mg per kg of body weight) to effect a spinal transection at Th 10. Following this transection, the

This study was aided in part by Grant NS-03266 from the National Institute of Neurological Diseases and Stroke.

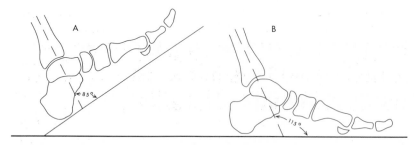

Figure 1. Skeletal view of the left foot in a position immediately before heel-strike (*A*) and after it (*B*). The horizontal line indicates the level of ground.

vertebral segments from L_1 to S_1 were subjected to laminectomy. The left ventral root corresponding to the fifth lumbar segment (L_5 VR) of the human spinal cord was dissected[1] and ligated near the dorsal root ganglion of that segment, and its central end was placed on bipolar Ag-AgCl electrodes. After the L_5 VR had made contact with the recording electrodes, their area was flooded with paraffin oil at the body temperature of the animal. The lateral gastrocnemius nerve (LGn) and the common peroneal nerve (CPn) of the left leg were prepared for stimulation in the same manner as the L_5 VR.

The experimental procedure was begun approximately 6 hr after the spinal transection, when the effect of the general anesthesia had, to a large extent, worn off. Single square wave pulses with a 0.1-msec duration were delivered once every 2 sec to each nerve separately. The stimulus intensity was adjusted to produce L_5 VR discharges in response to threshold activation of the IA afferents and some polysynaptic L_5 VR activity (Figure 2) in response to the stimulation of the skin afferents of the CPn (Figure 3). After these adjustments were completed, the interval between the stimulation of the two nerves was changed systematically. Initially, the stimulation of the CPn was delayed by approximately 20 msec following the stimulation of the LGn. Then the delay of the CPn stimulation was decreased progressively until both nerves were stimulated simultaneously. The stimulation of the CPn was advanced further in time to precede the stimulation of the LGn. The Grass C4 kymograph camera was used to obtain photographs of the oscilloscope traces showing reflex responses elicited at various interstimulus intervals. The amplitude of the L_5 VR monosynaptic reflex discharge was measured and compared with the amplitude of the appropriate control response in each case.

RESULTS

Changes in the amplitude of the L_5 VR reflex responses are shown in Figure 2. A decrease of the interstimulus interval between the LGn and the CPn

[1] Because the cat has a long lumbar region, it was actually the L_7 VR.

stimulation resulted in an increase in the amplitude of the monosynaptic component of the CPn reflex. This facilitation was observed on some occasions when the LGn stimulation preceded the CPn stimulation by 10 msec. A further reduction of the interstimulus interval to 8 msec, as shown in Figure 2 (*A2*), stabilized the occurrence of facilitation. It reached a magnitude of 140 percent of the control reflex, although on repeated trials this value was sometimes as low as 120 percent. Such a fluctuation of facilitation between 120 and 140 percent persisted at intervals shorter than 8 msec (*A3,4, B1,2*). When stimulation of the two nerves was nearly simultaneous, the monosynaptic components of the reflexes became partly fused (*B3,4*). Inasmuch as width of the compound potential increased, in addition to height, the measurement of height alone became insufficient for an adequate evaluation of facilitation in these cases. With a total fusion (*C1,2*), measurement of the response amplitude became a valid appraisal of facilitation once again. In these cases, however, facilitation was evaluated against that amplitude of the monosynaptic component which remained after the subtraction of the LGn control reflex. In records C1 and C2, this was 145 and 135 percent, respectively.

When stimulation of the CPn preceded the stimulation of the LGn, no facilitation of the LGn was observed. With the cases illustrated in Figure 2

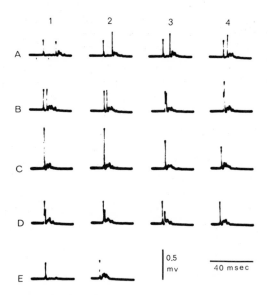

Figure 2. Oscilloscope recordings of L_5 VR discharges to stimulation of the LGn and the CPn. In all records, the LGn stimulation is delayed by 7 msec from the beginning of the oscilloscope trace sweeping from left to right (*A1, first dot*). The stimulation of the CPn is delayed by 18 msec (*A1, second dot*). With each successive recording (*A2,3,4, B1,2,* etc.), the delay of the CPn stimulation is decreased progressively. In C1 and C2, it is 7 msec (the interstimulus interval is 0). In D4, stimulation of the CPn is delayed by 1 msec. *E1*, LGn control reflex; *E2*, CPn control reflex.

(*C3,4,D1,2,3,4*), this was 90, 100, 90, 80, 70, and 0 percent of the control amplitude in the sequence of the illustration. Although the effect of the CPn volley on the LGn reflex was variable at relatively short interstimulus intervals (0.5 msec for C3, 1 msec for C4), at a 3-msec interstimulus interval (*D2*), the effect of inhibition became obvious; and, as the interstimulus separation was increased, the LGn reflex was abolished.

DISCUSSION

In view of the fact that the function of direct antagonists generally is governed by principles of reciprocal inhibition (Lloyd, 1960), results of these experiments are surprising. In fact, they indicate that the monosynaptic component of the CPn, which is known to be initiated by IA afferent volleys (Lloyd, 1960), undergoes facilitation when these volleys arrive at the motor neuron pool of the L_5 VR simultaneously or slightly later than the LGn volleys. It should be pointed out, however, that this effect can be observed only in case the IA volleys of the LGn are evoked by stimuli of threshold intensity and stimulation of the CPn is sufficiently strong to evoke the monosynaptic, as well as the polysynaptic, component of this reflex. A slight change in these conditions of stimulation abolishes facilitation at 0- to 2-msec interstimulus intervals. Facilitation at longer interstimulus intervals is less dependent on a critical adjustment of stimulus intensities. However, if the sequence of the afferent excitations is reversed, discharge of the L_5 VR motor neuron pool via the LGn afferents is impeded. The effectiveness of this inhibition is augmented when the intensity of stimulation of the CPn is increased to activate further the polysnaptic component of this reflex.

A delayed facilitation of the action of antagonists has been described previously (Lloyd, 1952) as being due to excitation via IB afferents which initiate the inverse myotatic reflex. Because in the present study such a facilitation also occurred with simultaneous stimulation of the CPn and LGn, it appears reasonable to think that, in addition to the action of the IB afferents, the subliminal fringe zones of the LGn and the CPn motor neuron pools should overlap (Figure 3). Neurons in the region of overlap, when excited by IA afferents by two sources, may discharge them and thus augment the L_5 VR activity (*dashed line with arrow*). When stimulation of the CPn precedes the stimulation of the LGn, however, the activity of the polysynaptic pathways may arrive via the inhibitory interneuron at the LGn motor neuron pool simultaneously or prior to the LGn afferent volley. This reduces or blocks the excitation of the LGn motor neuron pool. Such polysynaptic paths may be formed not only by skin afferents of the CPn, but also by collaterals of the Pn fibers (not shown in Figure 3).

Because of the sequence of changes in the position of the ankle, the gastrocnemius muscle is slightly stretched immediately before the heel-strike,

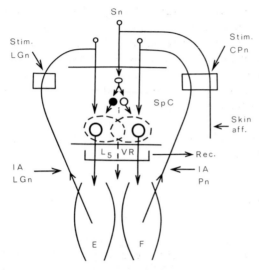

Figure 3. A diagram of reflex mechanisms involved in the control of muscles which move the foot. Circuits for reciprocal inhibition are not shown. Stretch imposed on muscles which function to extend the foot (*E*) activates IA afferents (*aff.*). Those of the lateral gastrocnemius nerve (*LGn*) will discharge monosynaptically within the spinal cord (*SpC*) a certain motor neuron pool (*closed circle*) of the fifth lumbar ventral root (*L_5 VR*) from which electrophysiological recordings (*Rec.*) were obtained. These motor neurons will activate the muscle which was stretched. Other neurons in the area limited by the *dashed line* surrounding the motor neuron discharge zone will be excited subliminally. According to the results of this study, the subliminal fringe zone of the LGn overlaps a similar zone of the common peroneal nerve (*CPn*). Stretch imposed on muscles which function to flex the foot (*F*) also activates IA afferents. They are found in the deep peroneal nerve (*Pn*) and are linked monosynaptically with those motorneurons of the L_5 VR which contract the stretched F muscles. Activation of receptors located in the skin, which covers the F muscles, initiates an afferent volley in certain small diameter neurons (*Sn*). They form polysynaptic connections (Lloyd, 1960). Some of the interneurons (*i*) can reexcite the Pn motorneuron pool, while others (*closed circle*) inhibit the LGn motor neuron pool.

and the muscles and skin innervated by the CPn are mechanically stimulated immediately after the heel-strike; thus, the activity of the corresponding motor neuron pools should be enhanced. A nearly simultaneous contraction of the E and F groups of muscles must ensue and thereby provide for an increased tension and stability at the ankle. Because the monosynaptic component of the CPn reflex is facilitated, the resulting contraction of the F muscles should lead to an active shift into the 90° stance position. The LGn and the CPn reflexes can be facilitated further by the excitation of certain subcortical structures of the brain (Emmers, 1961), an activity which should augment the stability of bodily support during walking. It is interesting to note that, usually, an intentional extension of the foot before the heel-strike greatly diminishes this stability of bodily support. Although in some cases of normal walking the absolute position of the ankle before the heel-strike may not differ from a stance position (Murray, 1967), it is the change toward a flexed position from a previous extension that is important in stretching of the E muscles. It is also interesting that, in case high-heeled shoes are worn, ankle flexion immediately prior to the heel-strike is slightly intensified (Murray, 1967).

It must be emphasized that the present study does not deny the existence of reciprocal inhibition between antagonists; it merely indicates that under some conditions it is obliterated by excitation. When a light, phasic type of stretch is imposed on muscles in a certain sequence, facilitation occurs that assists the walking cycle.

REFERENCES

Emmers, R. 1961. The role of limbic system in producing modifications of spinal reflexes. Arch. Ital. Biol. 99: 322–342.
Lloyd, D. P. C. 1952. On reflex actions of muscular origin. In: Patterns of Organization in the Central Nervous System. Vol. 30, pp. 48–67. Association for Research in Nervous and Mental Diseases.
Lloyd, D. P. C. 1960. Spinal mechanisms involved in somatic activities. In: J. Field (ed.), Handbook of Physiology. Vol. 2, pp. 929–951. Williams & Wilkins, Baltimore.
Murray, M. P. 1967. Gait as a total pattern of movement. Amer. J. Phys. Med. 46: 290–333.

Author's address: *Dr. Raimond Emmers, Department of Physiology, College of Physicians and Surgeons, Columbia University, 630 W. 168th Street, New York, New York, 10032 (USA)*

Mathematical modeling of the human gait: An application of the SELSPOT-system

K. Öberg

Een-Holmgren Ortopediska AB, Uppsala

In order to simulate a physical course of events, a well-defined model is very helpful. If the physical relations are known in a mathematical form, the model must be defined exactly in order to transform it to mathematical equations. A model in such a form is enhanced greatly by use of a computer in which a major part of the analyzing and calculating processes can be accomplished in a very short time. The disadvantage, however, is that a model is always a rough approximation of reality. However, depending on the specific problem that is to be solved, a model with approximations made with regard to that problem can be acceptable.

The initial aim of this project was to improve lower extremity prostheses by providing appropriate adaptations for different walking speeds and surfaces.

METHOD

As a first step, we chose to develop a method for calculating the forces and the moments working in the joints of a healthy person during normal walking. This has been obtained by means of a mathematical model which at present is undergoing further evaluation.

As a second step, it is our intention to simulate propositions to moment functions for prosthetic and, if possible, for orthotic, components on the basis of the moment functions in joints by normal walking. Results from this simulation will be compared with normal movement and then will be fed back to the simulating model again, thus providing changed moment functions.

It is also quite obvious that a gait model such as this can be used in the study of other problems that concern human locomotion. The aim of the project, therefore, has been extended to include analysis of how different types of gait can be characterized and how selected problems in the gait mechanism shall be evaluated.

The model consists of a two-dimensional mechanism with seven jointed segments corresponding to foot, shank, and thigh for both legs, as well as the trunk, including head and arms. Each segment is defined by a set of body segment parameters such as length, location of center of gravity, mass, and moment of inertia. At present, the model contains the following simplifying assumptions:

1. Trunk, arms, and head together are treated as a rigid body.

2. The legs are assumed to consist of three rigid bodies, thigh, shank, and foot, connected by joints. The foot is assumed to be weightless, i.e., its weight is included in the shank. This means that forces and moments in the ankle joint during the swing phase cannot be obtained from the model. During the stance phase, the action from the inertia of the foot is negligible because its movement is limited. This simplification, then, is from practical influence.

3. The movement is studied only in one plane. This is assumed to be

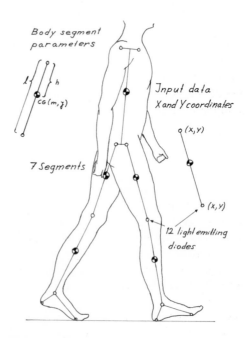

Figure 1. The mechanical model.

satisfactory for learning about the necessary moments. However, on the other hand, deviation of the movement from the plane is essential for maintenance of lateral balance.

The instantaneous position of the body is determined by the cartesian coordinates from 12 points on the body. These X and Y coordinates are the input data to the model and are obtained from experimental measurements (Figure 1). The forces and moments in each joint can be calculated from Newton's equations for each segment (Figure 2). A FORTRAN IV program is written to calculate the rest of the necessary coordinates and angles through the relations between the body segment parameters and the input data. It calculates velocities and accelerations through numerical derivation; thus, the moment and force function in the joints can be calculated by solution of the Newtonian equations (Figure 3).

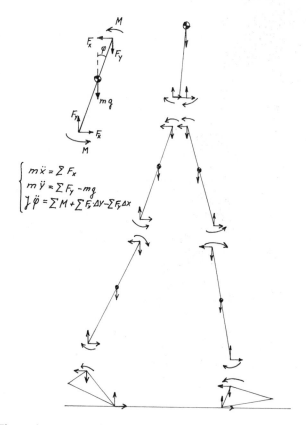

$$\begin{cases} m\ddot{x} = \sum F_x \\ m\ddot{y} = \sum F_y - mg \\ J\ddot{\varphi} = \sum M + \sum F_y \cdot \Delta y - \sum F_x \Delta x \end{cases}$$

Figure 2. The mathematical model.

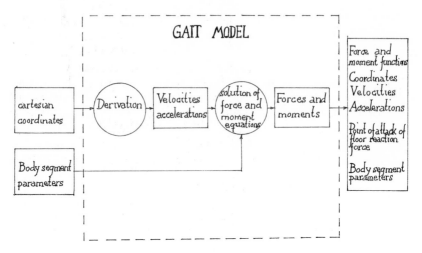

Figure 3. Gait model. Diagram of the calculation procedure.

The computer equipment currently in use includes:

Computer HP 2100	(16-K core memory)
Disc	(2.5 million word memory)
Graphical display	(Tektronix 4002A)
Hard copy unit	(Tektronix 4601)

A lot of information from input data and calculation is then available in the computer and can be obtained in varying forms.

The data which can be obtained are: (a) force and moment functions, (b) coordinates, (c) velocities and accelerations (d) point of application of floor reaction force, and (e) body segment parameters.

Some plotting routines were written, including a display of the stick model walking through a cycle, which has been very valuable for a quick evaluation of the kinematics of the model (Figure 4). Other plotting routines have been written in diagram form to present the variables as a function of time (Figure 5).

An integration model has been developed for testing the model. This means that a reverse calculation and the cartesian coordinates are calculated from the force and moment functions. The integration model then is going to be developed for use in simulation of hypothetical moment functions for prosthetic and, if possible, for orthotic, elements.

For routine use of the gait model in the future, a system for measuring the cartesian coordinates is necessary. During the development of the model, it has been difficult to test and evaluate it because of poor input data. The numerical derivation is particularly sensitive to errors in the input data. The

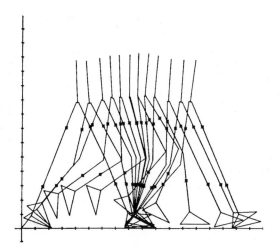

Figure 4. Graphical display of the stick model.

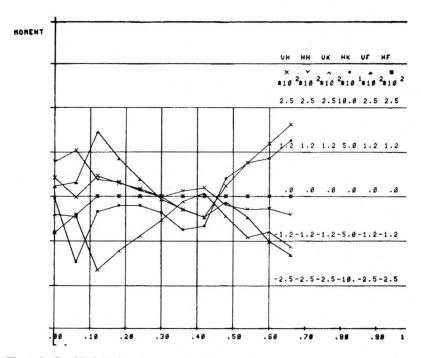

Figure 5. Graphical display of moment functions.

84   Öberg

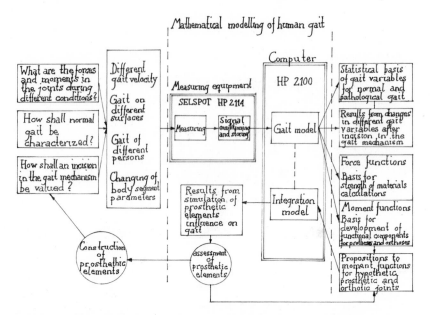

Figure 6. Mathematical modeling of human gait. General survey of the project.

best data that have been used so far are from cinematographical measurements. However, this technique requires too much time to evaluate the data from the flim, and the degree of accuracy is not adequate.

To continue the development of the model and to make it of practical value, the following requirements must be met: (a) high accuracy, (b) high measuring frequency, and (c) automated data reduction.

The SELSPOT-system seems to be the only one that can fulfill these demands. Two cameras on each side of the body measure the motion of 12 points (Figure 1). To process that much data, it is necessary to feed them into the core memory of a minicomputer (HP 2114). The data then can be punched on tape transferred to the computer (HP 2100), where the model calculation is performed (Figure 6).

Programs and equipment are made so that the model can be extended to a three-dimensional study and also to one involving more segments.

ACKNOWLEDGMENTS

This study was conducted in cooperation with H. Lanshammar, M.Sc., and L. Gustafsson, M.Sc. Teknikum, University of Uppsala, Uppsala, Sweden, and L. Lindholm, M.Sc., Chalmers University of Technology, Göteborg, Sweden.

Author's address: *Dr. Kurt Öberg, Een-Holmgren Ortopediska AB, Bergsbrunnagatan 1, 753 23 Uppsala (Sweden)*

Analysis of gait:
A method of measurement

E. N. Zuniga and L. A. Leavitt

Baylor College of Medicine, Houston

Over the years, locomotion has been explained in many terms. In the clinic, the principal motions of ambulation are determined by observable gait patterns. In order to understand these gait patterns, we must look at gait on a biomechanical level.

Foot-floor contact information has contributed much to the study of human gait. As a result, a variety of foot-floor contact systems have been developed and used, including those by Barnett (1956), Eberhart, Elftman, and Inman (1968), Elftman (1934), Finley, Cody, and Finizie (1969), Grieve (1969), Milner and Quanbury (1970), Schwartz et al. (1934), Smith, McDermid, and Shideman (1960), and Winter, Greenlaw, and Hobson (1972). This study introduces a method previously employed by Leavitt et al. (1970) for the analysis of gait and pressure parameters in amputees. The technique evaluates the time-related changes that occur in the gait cycle.

PURPOSE

The main purpose of this study was to examine a new method for evaluating the gait of normal subjects and amputees. Three groups of subjects were studied. The first was a group of 10 normal male subjects. The second group was composed of 10 normal female staff occupational therapists. The third group was made up of 10 male above-knee amputees who were judged clinically to have good gait patterns.

This research was supported in part by Grant No. 23-P-55230/6-03, by the Regional Research and Training Center Grant RT-4 from the Social and Rehabilitation Service, and by Grant No. RR-05425 from the U.S. Public Health Service, Department of Health, Education, and Welfare, Washington, D.C.

METHODS

A "universal harness," which consisted of a subject-cable assembly connected to footswitches and electronic goniometers, was used. Two footswitches were attached to the soles of each shoe and one to the heels. The electronic goniometers were attached to flexible knee cages and then secured to the subject's lower extremities (Figure 1). The subjects were instructed to proceed through an automatic start control and then through an automatic stop at the end of a walkway. The system measured the foot-floor contact at the heel, midfoot, and toe of both extremities, as well as the knee flexion-

Figure 1. Subject with attached foot-floor contact module and goniometrical module for measuring gait components.

extension angles. The signals obtained from this system were processed within a signal conditioning unit. An analog record identified the footswitch closures and knee angles of both lower extremities. The circuit which identified the floor contact of the heel, midfoot, and toe footswitches provided a binary weighted voltage of 4, 2, and 1 units, respectively. When all three footswitches were closed, the circuit produced a 2-volt output which represented seven units. Combination of footswitch closures also would appear in a gait pattern (Figure 2). The events also were recorded on an analog tape recorder.

RESULTS

Temporal Factors of Gait

The sequence of events during the stance phase was measured as footswitch closure times. In the male normal and amputee subjects, each successive sequence was noted as heel, heel-midfoot, midfoot-toe, and toe footswitch closures, respectively (Figure 3A,C). In the female subjects, a heel-midfoot-toe footswitch closure and the absence of the midfoot footswitch closure were observed in the majority during the stance phase (Figure 3B). The sequence of the successive footswitch closures and their combinations were considered as the characteristic gait pattern of each subject. The 10 female subjects also showed a shorter gait cycle in comparison with the temporal components of the gait cycle for the 10 normal male and 10 amputee subjects.

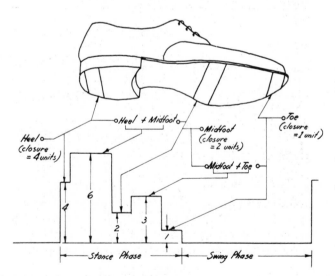

Figure 2. Footswitch voltage unit output pattern.

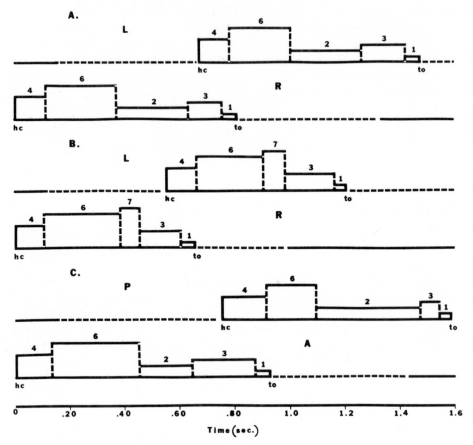

Figure 3. Mean footswitch closure pattern for 10 normal male subjects (*A*), 10 normal female staff occupational therapists (*B*), and 10 male above-knee amputees (*C*). *L* = left foot; *R* = right foot; *P* = prosthetic foot; *A* = anatomical foot; *hc* = heel contact; *to* = toe-off. Switch closure (units in parentheses): heel (4); heel-midfoot (6); heel-midfoot-toe (7); midfoot (2); midfoot-toe (3); and toe (1).

Knee Flexion Angles

In the normal males, the difference between the left and right knee angles was $3°$ ($51-54°$, respectively). In the females, the difference between left and right knee angles was $1°$ ($59-60°$, respectively). In the amputees, the difference between the prosthetic and anatomical knee angles was $4°$ ($45-49°$, respectively). The various types of knee units were not taken into consideration. Table 1 summarizes the results obtained from the gait patterns for the male, female, and amputee subjects.

Table 1. Mean values of gait parameters for male, female, and amputee subjects

Gait parameters	Normal male subjects (N = 10)		Female subjects (N = 10)		Amputee subjects (N = 10)	
	Left	Right	Left	Right	Prosthetic	Anatomical
Swing phase time (sec)	0.51	0.52	0.43	0.43	0.59	0.50
Stance phase time (sec)	0.81	0.81	0.66	0.66	0.84	0.93
Double stance time (sec)	0.14	0.16	0.12	0.12	0.18	0.16
Heel switch closure time (sec)	0.11	0.11	0.11	0.10	0.16	0.13
Heel-midfoot switch closure time (sec)	0.22	0.26	0.24	0.28	0.18	0.32
Heel-midfoot-toe switch closure time (sec)	0.00	0.00	0.08	0.07	0.00	0.00
Midfoot switch closure time (sec)	0.26	0.26	0.00	0.00	0.38	0.19
Midfoot-toe switch closure time (sec)	0.16	0.12	0.18	0.15	0.07	0.23
Toe switch closure time (sec)	0.06	0.06	0.06	0.05	0.05	0.06
Toe-off time to maximal knee flexion (sec)	0.14	0.15	0.13	0.13	0.19	0.15
Maximal knee flexion during swing phase ($^\circ$)	51	54	59	60	45	49

DISCUSSION AND CONCLUSIONS

The findings of this study provided some interesting differences between the gait patterns of normal men, normal women, and above-knee amputees. The most interesting difference was the finding of the foot-floor contact of the heel-midfoot-toe combination during stance phase between heel contact and toe-off and the absence of the midfoot contact in the gait cycle of the women. This could be an artifact possibly related to the short female foot, or it could be attributable to a short first metatarsal, notably prevalent in women, as found by Morton and Fuller (1952). There were also differences in the temporal components of the gait cycle, with the women having shorter swing phase, stance phase, and total gait cycle times than the men and amputees, as expected because of the shorter limb lengths. The swing-stance phase time ratios of both the men and women were found to be similar, indicating the structural symmetry of their gait. In the amputees, however, the decreased swing-stance ratio is indicative of a compensatory action for gait stability.

REFERENCES

Barnett, C. H. 1956. The phases of gait. Lancet 271: 617–621.

Eberhart, H. D., H. Elftman, and V. T. Inman. 1968. Locomotor mechanism of the amputee. *In:* P. E. Klopsteg, P. D. Wilson, et al. (eds.), Human Limbs and Their Substitutes, pp. 472–480. Hafner Publishing Co., New York.

Elftman, H. 1934. A cinematic study of the distribution of pressure in the human foot. Anat. Rec. 59: 481–486.

Finley, F. R., K. A. Cody, and R. V. Finizie. 1969. Locomotion patterns in elderly women. Arch. Phys. Med. Rehabil. 50: 140–146.

Grieve, D. W. 1969. The assessment of gait. Physiotherapy 55: 452–460.

Leavitt, L. A., C. R. Peterson, J. Canzoneri, R. Paz, A. Muilenburg, and V. T. Rhyne. 1970. A quantitative method to measure the relationship between prosthetic gait and the forces produced at the stump-socket interface. Amer. J. Phys. Med. 49: 192–203.

Milner, M., and A. O. Quanbury. 1970. Facet of control in human walking. Nature 227: 734–735.

Morton, D. J., and D. D. Fuller. 1952. Human Locomotion and Body Form, pp. 97–113. Williams & Wilkins, Baltimore.

Schwartz, R. P., A. L. Heath, W. Misiek, and J. N. Wright. 1934. Kinetics of human gait. J. Bone Joint Surg. 16: 343–350.

Smith, K. U., C. D. McDermid, and F. E. Shideman. 1960. Analysis of the temporal components of motion in human gait. Amer. J. Phys. Med. 39: 142–151.

Winter, D. A., R. W. Greenlaw, and D. A. Hobson. 1972. A microswitch shoe for use in locomotion studies. J. Biomech. 5: 553–554.

Author's address: *Dr. Efrain N. Zuniga, Department of Physical Medicine, Baylor College of Medicine, 1333 Moursund Avenue, Houston, Texas, 77025 (USA)*

An analysis of cinematographic and electromyographic recordings of human gait

B. R. Brandell and K. Williams
University of Saskatchewan, Saskatoon

This chapter describes graphical and computational techniques that are being used to analyze cinematographic and electromyographic records of human walking gait. An objective of this study was to define the characteristically constant features of angular motions and of muscle contractions in the lower limb. Floor and treadmill walking were compared in order to assess the validity of using the latter for gait studies.

METHODS

Five male subjects were photographed at 50 frames per second at the beginning, middle, and end of 320-ft walks made on a treadmill and on an 80-ft square floor track at speeds of approximately 2.5, 3.2, and 4.2 mph.

Action potentials were recorded simultaneously from the medial head of the gastrocnemius, tibialis anterior, vastus medialis, semimembranosus, gluteus medius, and adductor magnus muscles by means of indwelling wire electrodes (Basmajian and Stecko, 1962) and a portable electromyogram tape recorder (Brandell, 1973) that was carried by the experimenter.

The relative motions of marked points on the hip, knee, ankle, and foot were transferred from successive cinematic frames by means of an electronic x-y plotter directly onto digital tape, which was processed to give output on angular position, velocity, and acceleration of limb segments and joint motion plotted against frames as units of time. One- and two-way analyses of variance tests were made on the corresponding points of all curves to determine:

This study has been supported by MRC Grant No. MA-4748.

(a) the variability among the subjects and (b) the variation between the treadmill and floor modes of walking.

The curves of segment and joint motion were approximated by a two-stage harmonic analysis, and the resulting expressions were compared with the range of mean values obtained from the motion data of five normal people.

The raw electromyographic recordings were integrated and played back as uniform pulses of adjustable size which were plotted on the ordinate axis against the corresponding cinematic frames in register with the curves of limb motion.

RESULTS

Cadence Relationships

The cadence, stride length, and velocity for each of five subjects walking on treadmill and floor were determined and are shown in Figure 1A. Three of these subjects were tested on a separate occasion in which they walked for

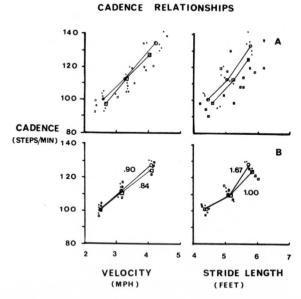

CADENCE RELATIONSHIPS

Figure 1. A: data taken from the motion pictures of five male subjects. Action potentials were recorded simultaneously with wire electrodes. Distance walked: 320 ft on treadmill and floor. B: data are averages taken by direct measurements from three of the same subjects as in A as they walked freely without encumbrance. Distance walked: .25 mi (1320 ft) on treadmill and cinder track. Small squares, individual floor data; large squares, mean positions of floor data; dots, individual treadmill data; open circles, mean positions of treadmill data.

Coefficient of slopes between 3.2 and 4.2 mph: Velocity = .90 and .84; Stride length = 1.67 and 1.00. Corresponding coefficients from a study by Murray et al. (1966) of 30 men: Velocity = 1.10 and Stride length = 1.62.

.25 mi at speeds of 2.5, 3.2, and 4.2 mph on the treadmill and on a firmly packed cinder track behind a tractor traveling at the same speeds (Figure 1*B*). In both experiments, the stride length and velocity tended to be greater for a given cadence on the floor than on the treadmill.

Analysis of Variance (ANOVA) Tests

One-way ANOVA tests based on the angular motion data determined the variation in motion data among pools of five strides for each of four subjects. Although variations among subjects were significant at the .05 level of probability for most of the data, the degrees of variation showed distinct profiles when plotted against the frames of the gait cycle (Figure 2). For

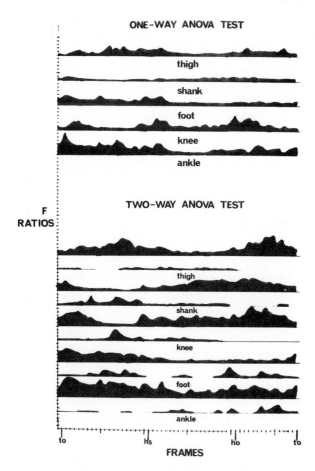

Figure 2. Analysis of variance (ANOVA) test for limb motion curves for one complete stride. *To,* toe-off; *hs,* heel-strike; *ho,* heel-off. In the two-way ANOVA test, the upper profile for each limb part represents variations among four subjects, and the lower profile represents the variation between treadmill walking and floor walking.

example, the least variability for foot and shank motions was in the stance phase, and for thigh motion it was in the swing phase.

Two-way ANOVA tests isolated the variations in limb motion that were attributable to floor walking as compared with treadmill walking from the variations among the subjects (Figure 2). Treadmill-floor differences were found to be nonsignificant at the .05 level for most of the limb motion data.

Harmonic Relationships

The angular displacement relationships for the thigh, shank, and foot were examined to determine the predominant harmonic components. Figure 3A shows the important fundamental harmonic components present in the motion of the thigh. For the shank and foot, in addition to the fundamental frequency term, significant second harmonic terms were found to exist. Figure 3B shows the resulting expression for the shank. The foot motion was found to be approximately $\phi_3{}^0 = -120 - 50 \cos 327T + 21 \sin (654T - 150)$. Differentiation of the displacement relationships with respect to time yielded angular velocity relationships; Figure 4A shows such a relationship for the shank. Further differentiation yielded angular acceleration relationships; Figure 4B shows such a relationship for the shank.

DISCUSSION

The small but consistent tendency for the velocity and stride length to be greater on the floor or track than on the treadmill at a given cadence, especially at 4.2 mph, was probably a consequence of the restrictiveness of the relatively short treadmill (5 ft, 10 in). The long-distance test represented a more valid comparison because velocity was more accurately controlled, and this test indicated almost identical slope coefficients in floor walking and treadmill walking between 2.5 and 3.2 mph.

Variations in joint and limb segment positions among different subjects appeared to occur at definite phases during the gait cycle, and it is suggested that such variation profiling may be useful to: (a) define the aspects of lower limb motion, which are relatively constant and stable in a normal gait, and (b) identify the phases of gait in which kinetic factors might be responsible for such inconsistencies.

The smaller variation that existed between floor and treadmill than among different subjects for most limb motions seemed to indicate the validity of using the treadmill for determining motion data. There appears to be a reciprocity, as yet unexplained, in which the treadmill variations tend to be greatest in phases in which the floor motions are the most consistent.

Preliminary harmonic analysis indicated that the fundamental stride frequency and double stride frequency were predominant in the motion of the

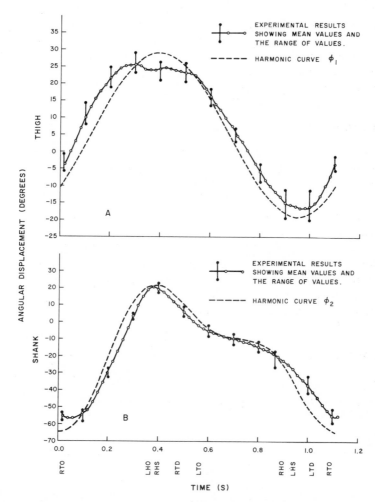

Figure 3. *A:* the angular displacements of the thigh as determined experimentally and a fundamental harmonic equation: $\phi_1{}^0 = 5 + 24 \sin (327T - 40)$, where T is the time from right toe-off (*RTO*), in sec (*S*). *B:* the angular displacement of the shank as determined experimentally and as a function of the first two harmonics as given by the equation: $\phi_2{}^0 = -18 - 35 \cos (327T + 16.5) - 15.75 \cos (654\ T - 33)$. *LHO*, left heel-off; *RHS*, right heel-strike; *RTD*, right toe down; *LTO*, left toe-off; *RHO*, right heel-off; *LHS*, left heel-strike; *LTD*, left toe down.

shank and foot, while the thigh was largely a function of the fundamental frequency. More careful selection of Fourier coefficients or the inclusion of one or more higher harmonic expressions should yield relationships which are representative of averaged data from a group of normal subjects. Such relationships would be particularly relevant to the dynamic analysis of normal walking. Further refinement eventually may permit the derivation of har-

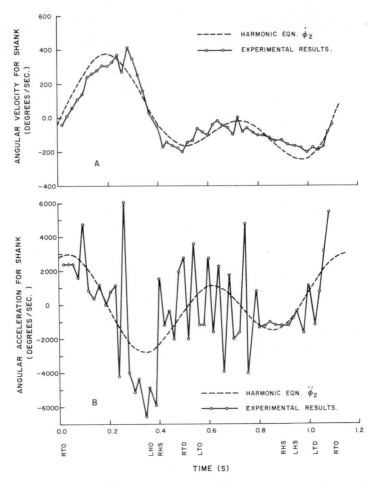

Figure 4. *A*: the angular velocity of the shank as determined from the experimental data and $\dot{\phi}_2$ as obtained by differentiating ϕ_2 with respect to time. *B*: the angular acceleration of the shank as determined from the experimental data and $\ddot{\phi}_2$ as obtained by differentiating $\dot{\phi}_2$ with respect to time. *EQN.*, equation.

monic equations for individuals, which will require the processing of considerably less data than are presently necessary. Differentiation to obtain both velocity and acceleration, the latter of which is essential for dynamic analysis, seems to agree with the rather unrefined differentiation of raw experimental data. Obviously, smoothing of the averaged experimental data is required.

REFERENCES

Basmajian, J. V., and G. A. Stecko. 1962. A new bipolar indwelling electrode for electromyography. J. Appl. Physiol. 17: 849.

Brandell, B.R. 1973. An analysis of muscle coordination in walking and running gaits. *In:* S. Cerquiglini, A. Venerando, and J. Wartenweiler (eds.), Biomechanics III, pp. 278–287. University Park Press, Baltimore, and S. Karger, Basel.

Murray, M. P., R. C. Kory, B. H. Clarkson, and S. B. Sepic. 1966. Comparison of free and fast speed walking patterns of normal men. Amer. J. Phys. Med. 45(1): 8–24.

Author's address: *Dr. Bruce Brandell, Department of Anatomy, University of Saskatchewan, Saskatoon, Saskatchewan (Canada)*

Effect of leg segmental movements on foot velocity during the recovery phase of running

C. J. Dillman
University of Illinois, Champaign-Urbana

Running is a form of human locomotion in which the body is projected over the ground by each leg alternately. After a runner has been thrust into the air, the driving foot most be moved forward and repositioned slightly in front of the torso in time for its subsequent period of support and propulsion. The recovery of the foot is an essential act in running, and the effectiveness of this movement becomes extremely important at higher speeds of running. At maximal velocity, a runner must recover the foot as quickly as possible and then, during the latter stages of recovery, decrease the forward velocity of the foot and effectively position it on the ground to minimize any possible retarding reaction.

The majority of research completed on the recovery of the leg and foot during running has been devoted to an analysis of the segmental angular displacement patterns and positions of the leg segments at certain times during the recovery cycle. Examples of this research can be found in studies by Deshon and Nelson (1964), Dillman (1971), Fenn (1930 a,b), Hopper (1969 a,b), Kroll and Morenz (1931), Murase et al. (1972), Saito et al. (1972), and Sinning and Forsyth (1970). Collectively, the results of these investigations concerning the movement patterns of the recovery leg have revealed that (a) the forward movement of the leg and foot is facilitated by the flexion of the lower leg behind the thigh, (b) the amplitude and rate of angular displacement for the leg segments increase as running speed increases, and (c) the segments of the leg move backward with respect to the body just prior to foot contact.

The findings of these studies illustrate the fact that the angular rotations of the leg segments during recovery influence, with respect to the body, the rate of foot recovery and the placement of the foot onto the ground. None of the previous investigations have ascertained specifically the influence of these leg segmental movements on the recovery of the foot. An understanding of the effect of these segmental movements on the forward velocity of the foot and its placement onto the ground would provide insight into those specific movement patterns of the leg which facilitate the efficient and effective recovery of the foot during running.

The purpose of this study was to examine the effect of leg segmental movement patterns on the horizontal velocity of the foot during the recovery phase of sprint running and the subsequent placement of the foot onto the ground.

METHODS[1]

The two highly skilled college cross-country runners employed in this study were filmed at a rate of 160 frames per second with a Locam 16-mm camera while they ran twice at maximal velocity on a rubberized asphalt track. The more consistent running trial from stride to stride, as indicated by relatively minor fluctuations in stride length and rate, was selected for analysis. Within this trial, one cycle of left-leg recovery was chosen for study because the left side of the body faced the camera and, consequently, all the segmental joints of this leg were in view throughout the movement cycle. The recovery cycle for the left leg began with the take-off of the left foot from the ground and concluded when the left foot was placed back onto the ground. During this period, the left leg was moved from behind the body to a forward position in front of the torso.

A Vanguard motion analyzer was used to obtain horizontal and vertical coordinates for the segmental joints of the left hip, knee, ankle, and meta-tarsophalangeal joint for every third frame of film throughout left-leg recovery. In addition to these frames, events such as touch-down and take-off of the foot were analyzed if they did not occur in the normal sequence of frames selected for analysis.

The horizontal coordinates for joint centers of the leg throughout recovery served as the basic data for this study. The horizontal displacement for each joint, measured in feet, was plotted as a function of time for the recovery period and was smoothed by hand. A polynomial equation of the fifth degree was fit mathematically to each displacement function by the method of least squares. Graphical comparisons were made between the actual displacement functions obtained from film and the predicted displacement patterns calculated from the polynomial equations. If a polynomial equation accurately described the original displacement pattern, then this

[1]The data for this study were obtained by the author while he was on the staff of the Biomechanics Laboratory of the Pennsylvania State University.

equation was differentiated with respect to time to obtain the corresponding velocity function. For those displacement functions or parts of displacement patterns which could not be described accurately by a polynomial equation, the methods of finite differences and graphical differentiation were employed to obtain the absolute velocity patterns.

Once the absolute horizontal velocity patterns for the joints of the leg were ascertained, a relative velocity analysis was conducted to determine the individual effect of the thigh, lower leg, and foot on the absolute velocity of the foot throughout recovery. Initially, the horizontal velocity of the knee with respect to the hip was determined by use of the following relative velocity relationship:

$$V_{k/h} = V_k - V_h$$

where $V_{k/h}$ is horizontal velocity of knee with respect to the hip, V_k is absolute horizontal velocity of knee, and V_h is absolute horizontal velocity of hip. The velocity pattern obtained from this equation for the recovery period represented the horizontal velocity of the knee caused solely by the absolute rotation of thigh about the hip.

In a similar manner, the relative velocity patterns for the ankle with respect to the knee and for the metatarsophalangeal joint (foot) with respect to the ankle were determined. The relative horizontal velocity functions of the ankle and foot indicated the individual contributions made by the lower leg and foot to the total absolute horizontal velocity of the foot. The relative velocity patterns for the segments of the leg were plotted as a function of time, and graphical interpretations were made concerning their effect on the absolute horizontal velocity of the foot during certain states of recovery. In addition, the contribution of each segment to the absolute horizontal velocity of the foot was investigated mathematically through the following relative velocity equation:

$$V_f = V_{mt/a} + V_{a/k} + V_{k/h} + V_h$$

where: V_f is absolute horizontal velocity of foot (metatarsophalangeal joint); $V_{mt/a}$ is horizontal velocity of metatarsophalangeal joint with respect to ankle (referred to as horizontal velocity of foot with respect to ankle); $V_{a/k}$ is horizontal velocity of ankle with respect to knee; $V_{k/h}$ is horizontal velocity to knee with respect to hip; and V_h is absolute velocity of hip.

RESULTS

An analysis of the running speeds during the analyzed cycle of leg recovery for each subject indicated that Subject A ran at an average velocity of 26.56

ft per sec (8.09 m per sec) and Subject B at an average velocity of 26.10 ft per sec (7.95 m per sec).

The horizontal velocity patterns for the left hip and foot during leg recovery for Subject A are depicted in Figure 1. The horizontal velocity of the hip remains relatively constant, whereas the foot velocity varies considerably. If the segments of the leg did not rotate about their respective joint centers during recovery, the linear horizontal velocity of the foot would be equal to the horizontal velocity of the hip. Thus, the differences between the horizontal velocities of the foot and the hip, as illustrated in Figure 1, represent the effect of the combined rotations of the leg segments to decrease and increase the velocity of the foot with respect to the velocity of the hip.

The horizontal velocity of the foot continually increased at varying rates during the initial two-thirds of the recovery cycle. Peak horizontal velocity of the foot was reached at the end of the support phase by the right foot. The peak horizontal foot velocity of Subject A was 62 ft per sec, while Subject B attained a slightly lower peak value of 56 ft per sec. These peak foot velocities were greater than twice the magnitude of the running velocities for both subjects.

Throughout the final flight phase of the recovery cycle, the horizontal velocity of the foot decreased. During the latter half of this period, the foot velocity decreased to well below the velocity of the hip. At foot contact, the horizontal foot velocity of Subject A was 2.5 ft per sec, while that for Subject B was 3.87 ft per sec.

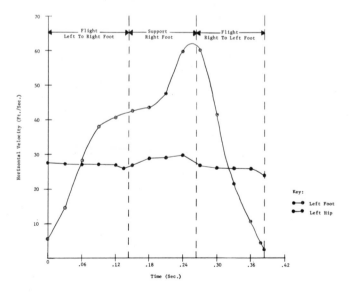

Figure 1. Horizontal velocity of left foot and hip during recovery of the left leg for Subject A (running velocity = 26.56 ft per sec).

The relative horizontal velocity patterns for the knee with respect to the hip and for the ankle with respect to the knee for Subject A are presented in Figure 2. A positive relative horizontal velocity indicated that the absolute rotation of the segment about its joint center was causing the horizontal velocity of the foot to be increased in a positive direction or in the direction of running. Conversely, a negative relative velocity indicated that a segment was acting to decrease the horizontal velocity of the foot in the direction of running.

Figure 2 illustrates that the relative horizontal velocities for the knee and ankle were not in phase with each other. The knee had high positive values during the initial half of recovery, while the ankle exhibited higher positive velocities during the later stages of recovery. These findings explain the sequential effect that the segments had in moving the foot forward. During the first half of the recovery cycle, the rotation of the thigh about the hip was primarily responsible for increasing the horizontal velocity of the foot and moving it forward. At a point halfway into the period of support by the opposite leg, the effect of the thigh decreased, and the lower leg and foot began to increase greatly the forward horizontal velocity of the foot until

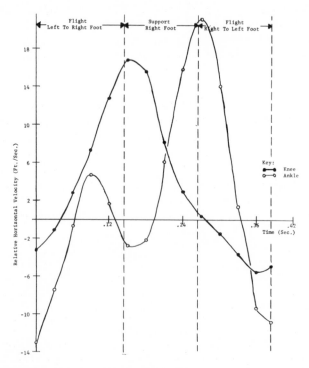

Figure 2. Relative horizontal velocity patterns for the left knee and ankle during leg recovery for Subject A (running velocity = 26.56 ft per sec).

after the support phase. The relative horizontal velocity patterns for the metatarsophalangeal joint with respect to the ankle (not shown in Figure 2) were very similar to the patterns for the ankle but of lower magnitudes.

The segmental velocity patterns that acted to reduce the horizontal velocity of the foot during the final third of the recovery cycle also are illustrated in Figure 2. The relative horizontal velocity of the knee became negative just after the support phase and remained negative until foot contact. These negative velocities of the knee about the hip reduced the forward horizontal velocity of the foot. At this point in the recovery cycle, the thigh was rotating backward about the hip joint, and the effect of this rotation was to produce a negative relative horizontal velocity.

The relative horizontal velocity of the foot (not depicted in Figure 2) became negative a short time after the thigh did so and had about the same effect as the thigh in decreasing the horizontal velocity of the foot prior to contact. About halfway through the final period of flight, the lower leg began to rotate backward about the knee joint and produced a negative relative horizontal velocity that continued to increase until foot contact. The relative motion of the lower leg had the greatest effect in reducing the horizontal velocity of the foot prior to contact.

The effect produced by each leg segment on the absolute horizontal velocity of the foot at four points during recovery for Subject B is illustrated mathematically in Table 1. The absolute horizontal velocity of the hip (V_h) remained relatively constant, whereas the horizontal velocity of the foot increased during the first two-thirds of recovery and decreased during the final third. At the beginning of the support period on the right foot, the relative horizontal velocity of the knee $(V_{k/h})$ contributed more than the other segments to the forward velocity of the foot. At take-off of the right foot, the relative velocities of the ankle $(V_{a/k})$ and foot $(V_{mt/a})$ contributed the most to produce the maximal forward velocity of the foot. When the foot contacted the ground, the relative velocities for all the segments were negative and thus acted to decrease the horizontal velocity of the foot. The lower leg had the greatest effect (−15.07 ft per sec) in reducing the horizontal velocity of the foot just prior to contact.

These last results indicated that the backward movements of the leg

Table 1. Relative horizontal velocity analysis for subject B*

Event	V_f	=	$V_{mt/a}$	+	$V_{a/k}$	+	$V_{k/h}$	+	V_h
Take-off left foot	11.18	=	(−3.10)	+	(−9.72)	+	(−1.70)	+	25.70
Touch-down right foot	43.50	=	+1.00	+	3.70	+	10.63	+	28.17
Take-off right foot	55.50	=	+11.30	+	14.87	+	2.91	+	26.42
Touch-down left foot	3.87	=	(−2.70)	+	(−15.07)	+	(−3.36)	+	25.00

*Units of velocity in ft per sec (positive velocity is in the direction of running).

segments with respect to the body served to decrease the horizontal velocity of the foot prior to contact. A further analysis was completed to determine the effect that these backward movements had on the positioning of the foot on the ground with respect to the body. Subject A had his foot (metatarso-phalangeal joint) 1.40 ft in front of the hip at initial contact with the ground. During the final flight period, the foot of Subject A was displaced a maximal distance of 2.20 ft in front of the hip. Thus, the backward rotation of the segments moved his foot backward with respect to the body, a distance 0.8 ft during the last half of the final flight period. The foot of Subject B was moved backward, in relation to his hip, a distance 0.55 ft before contact.

DISCUSSION

The most significant finding of this study was the segmental movement pattern employed to reduce the horizontal velocity of the foot prior to contact. The thigh began to rotate backward about the hip joint just after the support phase and continued to do so until the foot struck the ground. This backward rotation produced a negative relative horizontal velocity and conse-quently acted to reduce the horizontal velocity of the foot. About halfway through the final period of flight, the lower leg rotated backward about the knee and had a great effect in reducing the horizontal velocity of the foot. These facts indicate that the backward movement of the leg segments prior to foot contact are essential to reduction of the horizontal velocity of foot at contact in order to minimize any retarding ground reaction. In addition, it was found that these backward segmental rotations move the foot backward, with respect to the body, a distance greater than 6 in. prior to contact. Further research comparing trained and untrained runners by use of cinema-tography and a force platform might clarify the effectiveness of these back-ward segmental rotations.

CONCLUSIONS

Within the limitations of this study, the following conclusions are warranted.

First, a sequential segmental pattern that initially includes the forward rotation of the thigh about the hip and, subsequently, the forward rotations of the lower leg and foot about their respective joint centers effectively increases the horizontal velocity of the foot during the first two-thirds of the recovery cycle. Second, the backward rotations of the leg segments with respect to the body in the order of thigh, foot, and lower leg effectively reduce the horizontal velocity of the foot during the final third of the recovery cycle and position the foot more directly beneath the body.

REFERENCES

Deshon, D., and R. C. Nelson. 1964. A cinematographical analysis of sprint running. Res. Quart. 35: 451–455.

Dillman, C. J. 1971. A kinetic analysis of the recovery leg during sprint running. *In:* Biomechanics: Proceedings of C.I.C. Conference, pp. 137–165. Indiana University Press, Bloomington, Ind.

Fenn, W. O. 1930a. Frictional and kinetic factors in the work of sprint running. Amer. J. Physiol. 92: 583–611.

Fenn, W. O. 1930b. Work against gravity and work due to velocity changes in running. Amer. J. Physiol. 93: 433–462.

Hopper, B. J. 1969a. Characteristics of the running stride. Coach. Rev. Sept. 4–5.

Hopper, B. J. 1969b. Speed of driving foot. Coach. Rev. Dec. 3–4.

Kroll, W., and W. Morenz. 1931. The athletic run as quadruped motion. Arbeitsphysiologie 5: 227.

Murase, Y., S. Kamei, T. Hoshikawa, and H. Matsui. 1972. Differences in the foot speed during running between trained-runners and untrained-runners. Res. J. Phys. Ed. (Japan) 16: 273–279.

Saito, M., T. Hoshikawa, M. Miyashita, and H. Matsui. 1972. An analysis of motion patterns of the hip, the knee, and the ankle joint to running speed. Res. J. Phys. Ed. (Japan) 16: 265–271.

Sinning, W. E., and H. L. Forsyth. 1970. Lower-limb actions while running at different velocities. Med. Sci. Sports 2: 28–34.

Author's address: *Dr. Charles J. Dillman, Department of Physical Education, University of Illinois, Champaign-Urbana, Illinois 61820 (USA)*

Temporal patterns in running

M. Saito
Asahigaoka High School, Nagoya

K. Kobayashi and M. Miyashita
University of Nagoya, Nagoya

T. Hoshikawa
Aichi Perfectural University, Nagoya

Running is the end of a series of complex rotations of three leg levers. The purpose of this study was to reveal the temporal sequence in the rotary action of the leg in sprint running.

PROCEDURES

The subjects employed in this study were three trained runners and an untrained man who had no special conditioning in running. Their physical characteristics are shown in Table 1.

Electrogoniometers were used in order to measure the rotary actions in the hip, knee, and ankle joints. Foot-timing information was determined with microswitches secured to the ball and heel of a shoe. These data were recorded on the visigraph by means of a telemetry system.

All subjects were asked to run about 70 m on the regular track at as near a constant speed as possible. More than 10 trials were performed by each

Table 1. Characteristics of subject

Subject	Age (yr)	Height (m)	Weight (kg)
T. U.	18	1.74	64
S. U.	17	1.73	63
I. W.	20	1.70	55
M. I.	36	1.65	67

subject at various speeds from 4 m per sec to his maximum. The step length was determined by the interval of footprints on the ground. The progressive speed was calculated from the step length and the step frequency.

RESULTS

Relationship between the Step Length, Step Frequency and Progressive Speed

Figure 1 shows the relationship between the step length, step frequency, and running speed. The step is lengthened with an increase in running speed of up to 7 m per sec for the trained runner and 5.5 m per sec for the untrained, while the step frequency becomes greater over that speed. There is a distinct difference in length-frequency curve between three trained runners and the untrained subject. The difference is especially pronounced in the step length, with a maximal length of only 174 cm for the untrained and 204–233 cm for the trained runners.

Temporal Sequence in Rotary Action of Leg

Temporal sequences of rotary actions during running at five different speeds for the trained (T. U.) and the untrained runner (M. I.) are presented in

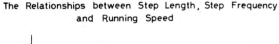

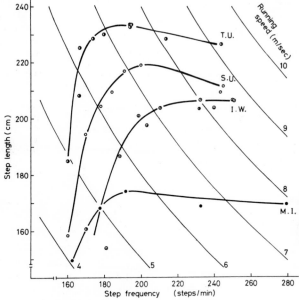

Figure 1. The relationships between step length, step frequency, and running speed.

Figure 2. The zero-reference position is the time when the foot makes contact with the ground, and the 100-reference is the time when the same foot makes contact again.

The hip joint is flexed so as to become a leading leg after the toe leaves the ground. This flexion is initiated at approximately 24 percent of one running cycle. At the maximal speed, the trained runner flexes the hip joint 3.4 percent earlier than the untrained runner.

Immediately after peak flexion, hip joint is extended at 75 percent of one cycle. At the maximal speed, the trained runner extends the hip joint 5.0 percent later than the untrained.

As soon as the foot strikes the ground, the knee joint begins to flex to absorb the shock after about 7 percent of the running cycle. It then begins to extend to produce the propulsive force. After the toe leaves the ground, the knee joint begins to flex again. The trained runner tends to flex it earlier than does the untrained runner. After peak flexion, the knee joint begins to extend. The trained runner extends it earlier than does the untrained runner. After the foot makes contact with the ground, the ankle joint is flexed at a point 10 percent into the running cycle. However, at the maximum, the plantar flexion in the trained runner commences earlier than that in the untrained. The dorsal flexion begins at 23 percent. However, between the trained and untrained runner, there is a large difference in the plantar flexion after peak dorsal flexion; the trained runner flexes his ankle earlier.

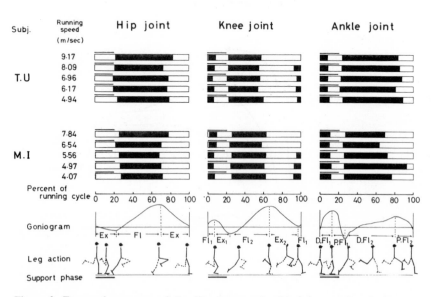

Figure 2. Temporal sequence of the hip, knee, and ankle joint rotation during one running cycle. *Ex*, extension; *Fl*, flexion; *Do*, dorsal; *P.*, plantar.

Rotary Actions of Leg during One Running Cycle

The flexion phase of the hip joint occupies a larger part of one cycle than does the extension phase as running speed is increased. The percentage for the trained runner is more than that for the untrained.

The durations of both flexion phase and extension phase of the knee joint during support diminish as running speed increases. Flexion phase during swing is almost constant and extension phase during swing is increased as running speed is increased. The difference between the trained runner and the untrained is that, at higher speed, the trained runner shows the longer extension phase and the untrained shows shorter extension phase during swing.

Dorsal flexion phase and plantar flexion phase of the ankle joint during support and those during swing vary with running speed. The changes in plantar flexion are especially noticeable.

DISCUSSION

Running speed is a product of the step length and step frequency. However, the step length has a greater effect on running speed than does the step frequency (Högberg, 1952; Matsui, and Miyashita, Hoshikawa, 1973; Hubbard, 1939). The results obtained in this study support this concept. Figure 2 indicates that the trained runner (T. U.) who can run at a speed of 9.1 m per sec demonstrates the longest step length for given speeds and frequencies.

A linear movement is produced through the rotary actions of leg levers during running. Therefore, the coordination of the leg levers is the most important factor in running faster. In this study, the temporal pattern of rotary actions of the leg was analyzed in relation to running speed.

The results show that the trained runner flexes his hip joint earlier after kicking than does the untrained runner, and also extends his knee joint earlier than the untrained runner in the swing phase, during which the other foot is in contact with the ground. Since the force produced by this extension adds to the kicking force exerted by the other leg, hip flexion and knee extension should be done earlier to produce faster running.

Moreover, the trained runner extends his hip joint later after peak flexion than does the untrained runner. This indicates that the trained runner flexes his hip joint earlier and extends it later. Consequently, the trained runner can keep the body in the air long enough to make full use of the forward component of the driving force to perform a longer step. Therefore, it is possible for the trained runner to extend his step to increase his running speed. These results coincide with the description by Hubbard (1939) that

Table 2. Duration of flexion and extension phase in each joint during one running cycle

Subject	Running speed (m/sec)	Step length (m)	Duration of one cycle (msec)	Hip joint		Knee joint				Ankle joint			
				Flexion msec (%)	Extension msec (%)	Flexion$_1$ msec (%)	Extension$_1$ msec (%)	Flexion$_2$ msec (%)	Extension$_2$ msec (%)	Dorsal flexion$_1$ msec (%)	Plantar flexion$_1$ msec (%)	Dorsal flexion$_2$ msec (%)	Plantar flexion$_2$ msec (%)
T.U.	9.17	2.25	492	299 (60.7)	193 (39.3)	38 (7.5)	69 (14.1)	170 (34.5)	215 (43.6)	37 (7.5)	76 (15.5)	296 (60.1)	83 (16.9)
	4.94	1.85	750	410 (54.7)	340 (45.3)	77 (10.2)	125 (16.7)	278 (37.0)	270 (36.0)	72 (9.6)	106 (14.2)	483 (63.3)	89 (12.9)
S.U.	8.62	2.11	490	271 (55.3)	219 (44.7)	85 (17.4)	65 (13.3)	167 (34.0)	173 (35.3)	39 (7.9)	72 (14.8)	324 (66.1)	55 (11.2)
	4.23	1.58	750	356 (48.7)	385 (51.3)	97 (12.9)	137 (18.2)	281 (37.5)	235 (31.3)	65 (8.7)	134 (17.8)	511 (68.1)	40 (5.3)
I.W.	8.50	2.03	478	250 (52.3)	228 (47.7)	45 (9.4)	77 (16.1)	159 (33.3)	197 (41.2)	33 (6.9)	87 (18.2)	239 (50.0)	119 (24.9)
	4.57	1.61	704	294 (41.8)	410 (58.2)	110 (15.7)	135 (19.1)	241 (34.3)	218 (30.8)	67 (8.7)	144 (20.7)	403 (57.2)	90 (12.8)
M.I.	7.84	1.68	430	225 (52.3)	205 (47.7)	34 (7.9)	70 (16.3)	158 (36.8)	168 (39.0)	46 (10.7)	70 (16.3)	183 (42.8)	130 (30.2)
	4.41	1.48	675	297 (44.0)	378 (56.0)	95 (14.1)	118 (17.5)	256 (37.9)	206 (30.5)	80 (11.8)	99 (14.4)	347 (50.7)	149 (23.5)

improvement in running is due to an increase in the step length rather than to an increase in rate of movement.

REFERENCES

Högberg, P. 1952. Length of stride, stride frequency, "flight" period and maximum distance between the feet during running with different speed. Arbeitphysiologie 14: 431–436.
Hoshikawa, T., H. Matsui, and M. Miyashita, 1973. An analysis of running pattern in relation to speed. *In:* S. Cerquiglini, A. Venerando, and J. Wartenweiler (eds.), Biomechanics III, pp. 342–348. University Park Press, Baltimore, and S. Karger, Basel.
Hubbard, A. W. 1939. An experimental analysis of running and of certain fundamental differences between trained and untrained runners. Res. Quart. 10: 28–38.

Author's address: *Dr. Mitsuru Saito, Asahigaoka High School, Kodekimachi Higashiku, Nagoya (Japan)*

Effects of a nondirective type of training program on the running patterns of male sedentary subjects

Y. Girardin and B. G. Roy
Université de Montréal, Montréal

This study was designed in an attempt to identify possible changes in the running pattern of male sedentary subjects after a 6-wk training program.

METHODS

The 22 subjects enrolled in the study received minimal instructions. They were to run 4,000 m around an oval indoor track three times a week for six weeks. In order to help the subjects find a convenient running pace, two practice sessions were held prior to the beginning of the experiment.

For the first testing session, the pertinent joint centers were identified on each subject's body according to the specifications of Dempster (1955). While the subjects were running their 20 laps, a 16-mm Locam camera that was set at a speed of 100 frames per sec and located 65 ft from the subject's path and perpendicular to the running track was used to register the appropriate cinematographic data. During this first testing session, two film sequences were obtained: the first during the fifth running lap, the second during the 20th lap. These will be referred to as the preinitial (PrI) and the postinitial (PoI) testing sessions. The same procedure was repeated at the end of the 6-wk training program, and the film sequences were designated as the prefinal (PrF) and the postfinal (PoF) testing sessions. A Vanguard motion analyzer was used to analyze the various film sequences.

The 12 kinematic factors retained for this study were as follows: angle of trunk lean (trunk and horizontal) at take-off and at touch-down; angle between thigh and horizontal at take-off and at touch-down; angle between leg and horizontal at take-off and at touch-down; horizontal distance between hip and toes at take-off and between hip and heel at touch-down; duration of periods of support and nonsupport; stride length; and average running velocity.

Numerous authors (Beck, 1965; Buchanan, 1971; Cavagna, Saibene, and Arcelli, 1965; Deshon and Nelson, 1964; Dittmer, 1962; Högberg, 1952; Nelson et al., 1972; Osterhoudt, 1968) have studied the human running pattern in relation to velocity, age, and level of ability. From these studies, it would appear that improvement in the running pattern is characterized by: (a) a decrease in the angle between the trunk and the horizontal at both take-off and touch-down, (b) a decrease of the angle between the thigh and the horizontal as well as of the angle between the lower leg and the horizontal at take-off, whereas the same angles show an increase at touch-down, (c) a larger horizontal distance between the hip and the heel at both take-off and touch-down, (d) a shorter period of support, (e) a longer period of nonsupport, and (f) a longer stride length. The discussion of the results is interpreted in the light of these findings.

Because the training program was nondirective in nature (because of the minimal amount of instructions given the subjects), an a posteriori approach was used to analyze and interpret the data.

For the testing and training sessions, the subjects were not restricted to a specific running velocity but were aksed to maintain a relatively steady pace. However, it was found that from the initial to the final testing sessions, many subjects showed an increase in average running velocity. Therefore, an attempt has been made to isolate the possible effects attributable to running velocity from those attributable to the training regimen.

It was decided that the subjects would be divided into two groups: (a) those whose average running velocity for the final testing session was maintained within ± 0.5 S.D. from the mean average running velocity of the initial session (referred to as the constant velocity group), and (b) those whose average running velocity for the final testing session increased by more than 0.5 S.D. from the mean average running velocity of the initial session (designated as the increased velocity group) (see Table 1).

Table 1. Mean values of running velocity of the two groups

Testing period	Constant velocity group (m/sec)	Increased velocity group (m/sec)
Preinitial	2.8 (N = 7)	2.4 (N = 15)
Prefinal	2.8 (N = 7)	3.2 (N = 15)
Postinitial	2.6 (N = 9)	2.5 (N = 13)
Postfinal	2.8 (N = 9)	3.4 (N = 13)

RESULTS

From the PrI to the PrF testing sessions, seven subjects qualified for the constant velocity group. It was found that, for these subjects, the results showed no significant differences for any of the factors studied when the final session was compared with the initial session. However, the slight changes observed when the data were considered in terms of absolute values seem to indicate a tendency toward improvement for most of the factors.

On the other hand, for the 15 subjects whose average running velocity increased (beyond 0.5 S.D.) from PrI to PrF testing sessions, a significant difference pointing toward improvement was observed for seven factors, namely, the angle between the trunk and the horizontal at both take-off and touch-down, the angle between the thigh and the horizontal at touch-down, the angle between the lower leg and the horizontal at touch-down, the horizontal distance between the hip and the heel at touch-down, the period of support, and the length of the stride (see Figures 1–7). As for the four

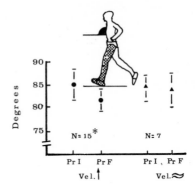

Figure 1. Angle of trunk lean at touch-down. A comparison of the mean values between the two velocity groups* ($p \leq 0.1$). *Vel.*, velocity.

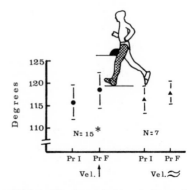

Figure 2. Angle between thigh and horizontal at touch-down. A comparison of the mean values between the two velocity groups* ($p \leq 0.1$).

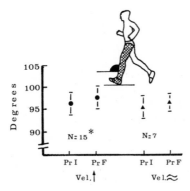

Figure 3. Angle between leg and horizontal at touch-down. A comparison of the mean values between the two velocity groups* ($p \leqslant 0.1$).

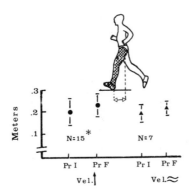

Figure 4. Horizontal distance between hip and heel. A comparison of the mean values between the two velocity groups* ($p \leqslant 0.1$).

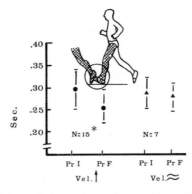

Figure 5. Duration of the period of support. A comparison of the mean values between the two velocity groups* ($p \leqslant 0.1$).

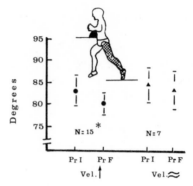

Figure 6. Angle of trunk lean at take-off. A comparison of the mean values between the two velocity groups* ($p \leqslant 0.1$).

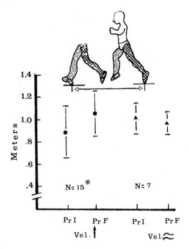

Figure 7. Stride length. A comparison of the mean values between the two velocity groups* ($p \leqslant 0.1$).

other factors, a tendency toward improvement as demonstrated by the changes observed in terms of absolute values could be detected.

When comparing the PoI with the PoF testing session, it was observed that nine subjects maintained an almost equivalent average running velocity. A close look at the data revealed that significant changes pointing toward improvement were observed in only two parameters: the angle between the thigh and the horizontal at touch-down and the angle between the lower leg and the horizontal at take-off (see Figures 8 and 10). On the other hand, it was noted that the period of nonsupport demonstrated a significant reversal of pattern of improvement observed in the other factors (see Figure 11). As for the eight remaining factors, very slight changes expressed in absolute terms could be observed.

Insofar as the 13 other subjects whose average running velocity increased (beyond 0.5 S.D.) from PoI to PoF testing session, significant changes indicative of improvement were found in five factors: the angle between the thigh and the horizontal at touch-down, the angle between the lower leg and the horizontal at take-off, the period of support, the period of nonsupport, and the stride length (see Figures 8–12). When considered in absolute terms, all the other factors indicated a slight improvement, except for the horizontal distance between the hip and the heel at touch-down, which remained unchanged.

DISCUSSION

These data subsequently were compared with those of other studies. For instance, the stride length of our subjects seems to follow a curve that

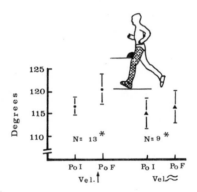

Figure 8. Angle between thigh and horizontal at touch-down. A comparison of the mean values between the two velocity groups* ($p \leqslant 0.1$).

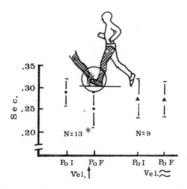

Figure 9. Duration of the period of support. A comparison of the mean values between the two velocity groups* ($p \leqslant 0.1$).

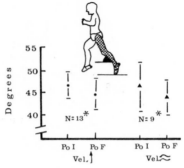

Figure 10. Angle between leg and horizontal at take-off. A comparison of the mean values between the two velocity groups* ($p \leqslant 0.1$).

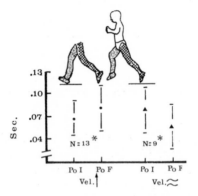

Figure 11. Duration of the period of nonsupport. A comparison of the mean values between the two velocity groups* ($p \leqslant 0.1$).

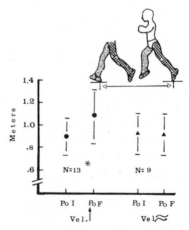

Figure 12. Stride length. A comparison of the mean values between the two velocity groups* ($p \leqslant 0.1$).

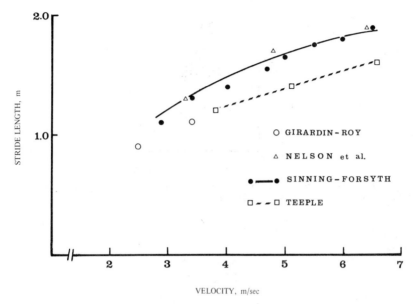

Figure 13. Changes in stride length with increases in running velocity.

approximates that reported by Nelson et al., Sinning and Forsyth (1970), and Teeple (1968) (see Figure 13). The low average running velocities exhibited by our subjects would suggest that they were effectively sedentary.

Some of the factors involved in this investigation also were studied by Osterhoudt (1968) and Teeple (1968). In most cases, the reported data proved to be quite similar, but they consistently pointed to the lack of proficiency exhibited in the running pattern of our subjects.

Another study involving training and various running velocities has been conducted by Buchanan (1971). His results seemed to corroborate our own, because they indicate that the observed modifications in the selected bio-mechanical factors were more affected by the changes in running velocity than by the training season.

From the results of the present study, it would seem that an increase in average running velocity would have more influence on the overall running pattern than would the training program per se. The data also seem to indicate that specific instructions in terms of how to improve various facets of the running pattern might be preferable to a nondirective approach or, at least, could bring about different results.

REFERENCES

Beck, M. C. 1965. The path of the center of gravity during running in boys grades one to six. Ph.D. dissertation, University of Wisconsin, Madison, Wis.

Buchanan, C. W. 1971. The effects of time and velocity on the strides of experienced middle distance runners. M.S. thesis, Pennsylvania State Univ., University Park, Pa.

Cavagna, G. A., F. P. Saibene, and E. Arcelli. 1965. A high speed motion picture analysis of the work performed in sprint running. Res. Film. 5,4: 309–319.

Dempster, W. T. 1955. Space requirements of the seated operator. WADC Technical Report, pp. 55–159. Wright-Patterson Air Force Base, Ohio.

Deshon, D., and R. C. Nelson. 1964. A cinematographical analysis of sprint running. Res. Quart. 34: 451–455.

Dittmer, J. 1962. A kinematic analysis of the development of the running pattern of grade school girls and certain factors which distinguish good from poor performance at the observed ages. M.S. thesis, University of Wisconsin, Madison, Wis.

Högberg, P. 1952. Length of stride, stride frequency, flight period and maximum distance between the feet during running with different speeds. Arbeitphysiologie 14: 431–436.

Nelson, R. C., C. J. Dillman, P. Lagasse, and P. Bickett. 1972. Biomechanics of overground versus treadmill running. Med. Sci. Sports 4,4: 233–240.

Osterhoudt, R. G. 1968. A cinematographic analysis of selected aspects of the running stride of experienced track athletes. M.S. thesis, Pennsylvania State Univ., University Park, Pa.

Sinning, W. E., and H. I. Forsyth. 1970. Lower limb actions while running at different velocities. Med. Sci. Sports. 2,1: 28–34.

Teeple, J. B. 1968. A biomechanical analysis of running patterns of college women. M.S. thesis, Pennsylvania State Univ., University Park, Pa.

Author's address: *Dr. Yvan Girardin, Department d'Education Physique, Université de Montréal, Montréal, 101 P.Q., (Canada)*

Effects of fatigue on the mechanical characteristics of highly skilled female runners

B. T. Bates and B. H. Haven

Indiana University, Bloomington

The purpose of this investigation was to identify and describe selected mechanical characteristics associated with fatigue in highly skilled female runners participating in a 440-yd run.

METHODS

The subjects for this investigation consisted of 12 females, nine of whom were members of the women's U. S. Olympic track team. The remaining three were women who had received special invitations to the training camp on the basis of previous performances. Best previous subject performances for the 400-m run ranged from 51.8 to 55.8 sec, with a mean performance of 53.8 sec.

In order to ensure a nearly maximal effort in performance, the filming was accomplished during the 4- x 440-yd relay of an international meet between Canada and the U. S. in August, 1972. Each runner was filmed twice for one complete running stride at points approximately 185 and 405 yd from the starting line. A camera speed of 100 frames per sec was used for the filming.

The film was analyzed by use of a stop-action projector and a digitizer. The film frames identified for analysis were those corresponding to the running cycle as defined by James and Brubaker (1972). The biomechanical parameters investigated were grouped as follows: (a) temporal analysis, and (b) stride and horizontal velocity. Means and standard deviations were computed and recorded for all sets of data.

The temporal analysis consisted of describing the six phases of the running cycle in terms of both absolute time and percentage of total stride time. In addition, various phases were combined to give (a) support and nonsupport phase times, (b) support and recovery phase times, and (c) step times.

Step and stride frequencies were determined. Step and stride lengths were calculated by the use of two methods: (a) measuring the horizontal distance between the toes in the corresponding take-off positions and (b) measuring the horizontal distance between the theoretical centers of gravity in the two appropriate take-off positions. Average horizontal velocities were computed for both right and left steps and total stride according to the corresponding step and stride lengths along with the times found in the temporal analysis. In addition, average horizontal velocities were determined for the support and nonsupport phases of the stride according to the distances as ascertained from the calculations of center of gravity.

RESULTS AND DISCUSSION

The results are presented in Tables 1, 2, and 3. Condition 1 data were taken from a previous study describing the running mechanics of highly skilled female runners (Bates and Haven, 1973). Mean stride lengths and stride frequencies decreased between Conditions 1 and 2. Average horizontal velocities also decreased for all stride phases, but decreases were greater for the nonsupport than for the support phases. The analysis indicated a definite change in the relationship between the support and nonsupport phases from Condition 1 to Condition 2. For the nonfatigued running condition, nonsupport phase times were greater than support phase times, whereas for the fatigued running condition, nonsupport phase times were less than support phase times. The absolute recovery times remained nearly constant for both running conditions, but the support times increased from Condition 1 to Condition 2.

Within the support phases, the percentage of time for the take-off phases decreased, whereas the percentage of time for the foot-strike and midsupport phases increased. Within the recovery phases, the percentage of time for the foot-descent phases increased, whereas the percentage of time for the forward-swing and follow-through phases decreased between Conditions 1 and 2. The increase in leg-descent time and the decrease in take-off time may indicate a less effective striding pattern.

In summary, stride lengths, stride frequencies, and average horizontal velocities decreased between the nonfatigued and fatigued running conditions. In addition, there were changes in the temporal patterns of the six defined phases of the running stride, as well as for the support, nonsupport, and recovery phases between Conditions 1 and 2.

Table 1. Temporal factors for support and recovery for phases of the running stride of 12 female runners

Phase	Condition 1*				Condition 2			
	Mean (sec)	S.D.	Percentage of support/recovery	Percentage of total stride	Mean (sec)	S.D.	Percentage of support/recovery	Percentage of total stride
Support Right	0.116	0.011	100.0	21.6	0.140	0.015	100.0	24.9
Support Left	0.118	0.009	100.0	22.0	0.145	0.017	100.0	25.8
Foot-strike plus midsupport	0.046	0.008	39.8	8.6	0.072	0.014	51.7	12.9
	0.049	0.007	41.5	9.1	0.075	0.019	51.7	13.3
Take-off	0.070	0.011	60.1	13.0	0.067	0.006	48.2	12.0
	0.069	0.007	58.4	12.8	0.070	0.008	48.2	12.4
Recovery	0.420	0.021	100.0	78.3	0.421	0.021	100.0	75.0
	0.419	0.022	100.0	78.00	0.418	0.016	100.0	74.4
Follow-through	0.031	0.010	7.5	5.9	0.025	0.009	5.9	4.4
	0.033	0.014	8.0	6.2	0.024	0.005	5.7	4.3
Forward swing	0.290	0.019	68.9	53.9	0.270	0.019	64.0	48.0
	0.286	0.017	68.3	53.3	0.274	0.014	65.5	48.8
Foot descent	0.099	0.012	23.5	18.4	0.126	0.021	30.0	22.5
	0.099	0.014	23.6	18.4	0.120	0.014	28.6	21.3
Total stride	0.537	0.023		100.0	0.561	0.022		100.0

*Condition 1 data taken from Bates and Haven (1973).

Table 2. Temporal factors for support and nonsupport phases of the running stride of 12 female runners

Phase	Condition 1*			Condition 2		
	Mean (sec)	S.D.	Percentage of total stride	Mean (sec)	S.D.	Percentage of total stride
Support	0.234	0.020	43.6	0.285		50.8
Right step	0.116	0.011	21.7	0.140	0.0154	24.9
Left step	0.118	0.009	22.00	0.145	0.0178	25.8
Nonsupport	0.302	0.024	56.4	0.276		49.2
Right step	0.149	0.010	27.8	0.138	0.0137	24.6
Left step	0.153	0.016	28.6	0.138	0.0123	24.6
Total stride	0.537	0.023	100.0	0.561	0.0228	100.0
Right step	0.267	0.012	49.8	0.283		50.4
Left step	0.270	0.014	50.2	0.278		49.5

*Condition 1 data taken from Bates and Haven (1973).

Table 3. Stride frequency; step, and length, and horizontal velocity of 12 female runners

	Condition 1*		Condition 2	
	Mean	S.D.	Mean	S.D.
Frequency (steps/sec)				
Right step	3.74	0.17	3.60	0.20
Left step	3.70	0.26	3.54	0.20
Stride	1.86	0.15	1.78	0.08
Length (ft)				
Right step (toe)	6.48	0.41	6.16	0.24
Right step (COG)†	6.46	0.49	6.25	0.34
Left step (toe)	6.53	0.48	6.15	0.32
Left step (COG)	6.47	0.39	6.10	0.37
Stride (toe)	13.01	0.72	12.26	0.63
Stride (COG)	12.93	0.83	12.35	0.58
Horizontal Velocity (ft/sec)				
Right step (toe)	24.22	1.43	21.65	1.44
Right step (COG)	24.18	1.25	22.10	1.58
Left step (toe)	24.22	1.11	22.20	1.58
Left step, (COG)	23.98	1.16	22.04	1.39
Stride (toe)	24.22	1.11	21.92	1.43
Stride (COG)	24.08	1.08	22.07	1.49
Phase (ft/sec)				
Right support	23.44	1.42	21.57	1.31
Right nonsupport	24.61	1.37	22.15	1.55
Left support	23.93	1.14	21.83	1.45
Right nonsupport	24.89	1.32	22.32	1.52

*Condition 1 data taken from Bates and Haven (1973).
†COG, center of gravity.

REFERENCES

Bates, B. T., and B. H. Haven. 1973. An analysis of the mechanics of highly skilled female runners. *In:* J. L. Blaustein (ed.), Mechanics and Sport. Vol. 4, pp. 237–245.
James, S. L., and C. E. Brubaker. 1972. Running mechanics. JAMA 221: 1014–1016.

Author's address: *Dr. Barry T. Bates, Department of Physical Education, University of Oregon, Eugene, Oregon 97403 (USA)*

Biomechanics
of
sport

Jumping
and
vaulting

Determination of biomechanical parameters of athletic movements utilizing a noncontact measurement system

W. Gutewort

Friederich-Schiller University, Jena

The development of biomechanical analysis procedures is directed not only toward an improvement in accuracy and a better evaluation of space and time but also to better practical application of these procedures toward improvement of athletic practice. New fields of application are opened by automation of measurement procedures, excluding laboratory conditions, and by minimizing the size of transmitters and transducers, as well as by more complex usage of these methods.

Very short evaluation periods and well-processed data make it possible to use such measurement devices not only for biomechanical analysis but also for informational routines describing the biomechanical propotions of performance patterns (Farfel, 1962; Ratov, 1972). The further development of measurement signals as a basis for improving sport performance is of basic importance. If correctly applied, they can make a contribution to the physical perfection of human beings. Thus, biomechanical procedures are used to reduce learning time, to stabilize techniques, and to intensify training while causing as little discomfort as possible to the athlete.

METHODS

Functional Principle of the Transmitting System

In this connection, high-frequency probes serve as electronic approach triggers which provide for a noncontact method. A capacitance probe was inserted into the oscillating circuit of an oscillator. Approaching the body parts of an athlete, with their high percentage of water, causes changes in the dielelectrics, thus altering the capacitance of the condenser. The circuit thereby is detuned and the oscillations are interrupted. This simple device, however, is not adequate for biomechanical measurements because of hysteresis and drifting, which make it unstable for long-term experiments.

Possibilities for Improving Electronic Approach Triggers

The first possibility for adaptation of this procedure to biomechanical or practical sport conditions lies in the construction of the probe itself. A probe consists of thin, plastic-covered copper wire or metal-coated foil. The field structure is determined by the arrangement, size, and distance of the electrode elements. The probe can be constructed in such a way that only the steep decrease of flux density following direct skin contact (sole of the foot, hand) initiates the trigger signal. For most athletic purposes, however, a noncontact process seems desirable, so that the athlete may wear any kind of shoes, and the probe can be installed under plastic covers, gym floors, gym mats, and even in sport equipment. It can be modified by a capacity-measuring bridge and by a phantom (foot model) that is optimally adjusted to the actual conditions, and a measurable signal can be received without contact within a defined distance from the probe.

Further improvement of the transmitting system can be achieved by suitable dimensioning. An oscillator that has just been stimulated to oscillate reaches a critical point as a result of attenuation and/or changes in capacitance and stops functioning. Switching the oscillator does not provide a solution to this problem because of the hysteresis between oscillation start and breakdown of a transistor as an active element and because of the shifting of the working point caused by temperature changes during longer periods of work.

The best results to date have been obtained with a device that responds to detuning of a high-frequency bridge. This approach produces no harmful reactions to oscillation generators by the passive elements and requires that only the voltage level be changed.

A variable D. C. source operates a simplified Schmitt-trigger. More recently, positive feedback operation amplifiers have been used successfully for this purpose. They enable the investigator to minimize hysteresis and to employ a moveable trigger threshold so that the transmitting system can be adjusted easily.

It is recommended that the steepest possible trigger characteristics be used to ensure maximal trigger-point stability and reproducibility at moderate probe sensitivity and minimal interference.

Measurement and Storage System

The measurement signals, thus having been received at contact or take-off from the ground or an apparatus, are used simultaneously for electronic short-time measurements and for automatic photographic phase recording.

Time measurements are made with a multichannel analyzer into which time impulses of 10^{-3} and 10^{-4} sec have been stored. The impulses arriving from the transmitters induce an address-on-switch so that the support period or flight time may be stored on many channels in succession. These results can be made directly visible on the screen of a storage oscilloscope as an electronically written graph. At the same time, the channels can be checked automatically, and this result can be printed or punched onto tape. The wealth of information can be processed adequately only with an electronic data-processing system.

The same measurement signals cause the disengagement electronics to release one electronic flash each, so that the body positions of the athlete can be recorded photographically—exactly at these moments of touching or leaving the ground. From these Strobophotograms, the motor phases can be defined according to place and time; this cannot be done with chronocyclo-photograms or other such series processes.

Since these important phases of motion are recorded at the same time and space, they can be compared directly, thus permitting a quantitative evaluation of single frames.

The limitations of the procedure include the troublesome problem of reproducibility of the trigger threshold. Further, the system does not function properly if moisture or wet conditions are present.

APPLICATION OF THE MEASUREMENT AND INFORMATION SYSTEM

The procedure can be used for noncontact measurements of periods of support or nonsupport, for objective analysis of motor rhythms, and for storing and quantitative evaluation of selected body positions and body angles during movements. It should be emphasized that these parameters are of special importance in the biomechanical analysis of sports skills. Two examples are presented.

Jumping Ability Test

One method for measuring vertical jumping distance involves a pointer connected by a string to a part of the body that moves as the subject jumps. The maximal displacement of the center of gravity, which is of essential interest in biomechanical respects, resulted in a measurement error because of the change in the body position during the period of nonsupport. The actual distance moved by the center of gravity, however, can be approximated from the flight time. If air resistance is neglected because of low velocity, and ascending and descending times are considered to be equal, then the jumping height can be calculated from the flight time according to formula:

$$h = \frac{g}{2}\left(\frac{T}{2}\right)^2$$

where *h* is vertical displacement of the center of gravity, *g* is acceleration caused by gravity, and *T* is total flight time.

Time measurement must be in msec because about 1.5 cm of displacement occurs in 0.01 sec (see Figure 1).

Real Time Data Recording in Side Horse Vaulting

High-frequency probes were installed at the lower side of the jumping board (10), in the horse (11), and under the gym mat (12) (see Figure 2). Impulses for time measurement and electronic flashes were given according to the motor phases in Figure 2. Three more body positions were registered photographically for further information. Additional flashes were released by electronic delay elements 0.25 sec after the subject left the board (3) and at 0.2 and 0.4 sec after the subject left the horse (6 and 7). The quality of vaults, including velocity parameters and especially the rotations occurring during the flight phases, could be determined accurately. A vault thus is characterized by eight body positions and their corresponding periods (see figures 3 and 4).

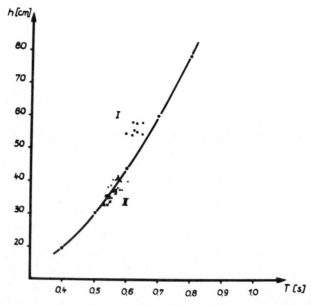

Figure 1. Relation between flight time and jumping height. T = flight time in sec;, h = lifting height of the center of gravity in cm (difference between maximally reached height of center of gravity and height of center of gravity at the moment when the effect of the jumping force comes to an end); --- = theoretically calculated values; · = empirically measured values with swinging of arms; x = empirically measured values without arm swing; I = attempts with a zero-point fixation in standing position as we did it up to now; II = attempts with a zero-point fixation in maximal toe-stand as the force effect comes to an end.

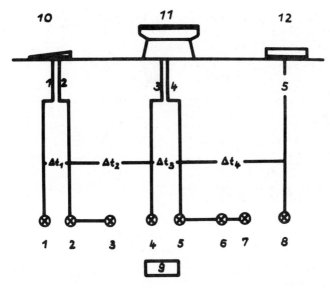

Figure 2. Experimental outlay for on-the-spot information during horse vaulting. High-frequency probes: *1–8* Electronic flashes. *1*, jumping onto the board; *2*, leaving the board; *4*, approaching the horse; approach flight phase; *5*, leaving the horse; push-off flight phase; *8*, landing phase; *3, 6, 7*, electronic flashes controlled by electronic delay elements; *9*, camera; *10*, jumping board; *11*, vaulting horse; *12*, gym mat; Δt_1 = period of touching the board, take-off phase; Δt_2 = period of the first flight phase; Δt_3 = period of touching the horse = support phase; Δt_4 = period of the second flight phase.

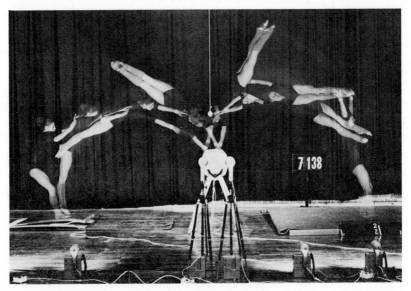

Figure 3. Strobocinegram of a horse vault (compulsory routine at the 1972 Munich Olympics). Jumping-off time, 0.095 sec; flight time, 0.325 sec (first flight phase); support time, 0.257 sec; flight time, 0.501 sec (second flight phase).

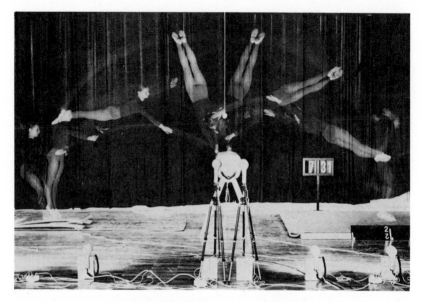

Figure 4. Strobocinegram of a horse vault (executed by Karin Janz).

The strobocinegrams achieved with this procedure depict parameters of a biomechanically sound and effective technique related to the morphology of vaulting. This system provides for immediate evaluation of a gymnast's vault and permits immediate corrections of improper techniques during training sessions.

The efficiency of the method can be seen clearly by comparing Figures 3 and 4. The same vault was executed by world champion and Olympic winner Karin Janz and by another young gymnast. It shows that the trajectory of the Olympic winner is higher and the rotation of the body is far more progressed after leaving the jumping board, so that she has reached almost handstand position at the moment of touching the vaulting horse. Unlike the other young gymnast, Karin Janz does no rotation about the longitudinal axis of the body during the support phase and her rotation is finished 0.4 sec after pushing-off from the horse.

Author's address: *Dr. Wolfgang Gutewort, Friedrich-Schiller-Universität, Institut für Korpererziehung, Seidelstrasse 20, 69 Jena (German Democratic Republic)*

The mechanics of the use of the Reuther board during side horse vaulting

E. Kreighbaum

Montana State University, Bozeman

The purpose of this study was to describe the interaction of women side horse vaulters with the Reuther board during performance of the handspring vault. Specific parameters to be identified were the forces and impulses of the board as a function of the vaulter's contact position and the sequential pattern of segmental motion and force parameters during board contact.

METHODS

Eight women side horse vaulters from the 1972 Northwest District Gymnastic Meet were subjects. The following anthropometric measurements were taken for each subject: height and weight, and lengths of the toes, foot, leg, thigh, trunk, arm, forearm, and hand. The segmental centers of gravity, radii of gyration, and weights were not obtained directly but were calculated from percentages taken from Kjeldsen (1969) and Plagenhoef (1971). During performance, the vaulter used a Reuther board to which eight strain gauges were attached in a Wheatstone bridge circuit and attached in turn to a galvanometrical type of oscillograph that recorded deflections in the board by means of a light tracing on photosensitive paper. The board was calibrated with vertical static loading for nine positions, 4 in apart. During the vaulter's contact, forces from the board were determined by using the deflection trace and the calibration curve that corresponded to the position of the vaulter's metatarsophalangeal joint on the board. The Reuther board forces were considered external forces and were used in the kinetic analysis as upward vertical forces applied to the vaulters' feet. A planimeter was used to determine the area under each force deflection curve, from which the total impulse of the board could be determined.

Each subject was filmed with a 16-mm Locam camera at 180 frames per sec from hurdle step through preflight. The sides of the Reuther board were marked along the edge to identify the landing positions used by the vaulters. Cinematographical data were collected from the film through the use of a computerized graphic tablet system that fed coordinate points into an IBM 360 computer. Programs were written or adapted for the collection of the data by a member of the computer science department at Montana State University. Segmental angular velocities, accelerations, and summation of forces and torques about each joint center were calculated.

RESULTS AND DISCUSSION:

Reuther Board Kinetics

Each deflection trace consisted of a maximal deflection followed by a steep decrease and then by a submaximal deflection before take-off (Figure 1). The limits of the maximal and submaximal forces were 2,160–1,180 lbs. and 1,050–550 lbs., respectively. The magnitude of the maximal deflection showed no significant correlation with the size of the submaximal deflection, according to the Spearman rank difference correlation coefficient. The maximal deflection showed a significant correlation with the weight of the vaulter ($p < 0.05$); however, submaximal deflection did not. None of the deflection traces between the maximal and submaximal deflections dropped below the level that corresponded to the weight of the vaulter. Temporal interaction of the vaulter and the board showed that the vaulter reached her maximal downward displacement at the peak of the maximal deflection, continued to roll onto the ball of the foot while the deflection of the board was decreasing, and extended the hip, knee, and ankle joints simultaneously with the submaximal deflection.

The limits of the total impulses calculated from the deflection traces were 72.86–130.26 lb per sec. The total impulse showed a significant rank correlation ($p < 0.05$) with the magnitude of force produced in the submaximal deflection and, to a lesser extent, ($p < 0.10$) with the vaulter's weight. The total impulse showed no relation to the force of the maximal deflection, the vaulter's take-off speed, or the change of her resultant speed. Although the change in momentum of the vaulter would be expected to change in relation to the impulse applied to the vaulter, the following factors must be considered. First, although the assumption was made that the human body is rigid, part of the impulse from the board would be lost in the damping effects of the articulations and the soft tissues of the body. Second, some of the impulse could be absorbed by the partial relaxation of the musculature surrounding the articulations, so that the joints were not rigid connections and thus flexed in reaction to the upward force. The changes in the speed

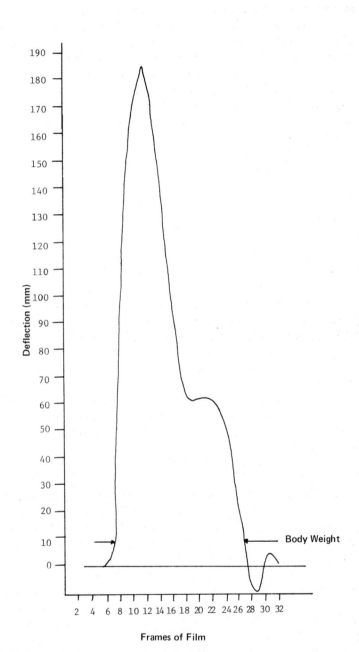

Figure 1. Typical deflection of the Reuther board of a female vaulter while performing a handspring vault.

patterns of the vaulters while in contact with the board, the sequence of the joint extensions, and the Reuther board deflections seemed to indicate that the impulse calculated from the beginning of the second deflection until take-off may have been a better indication of the effective impulse from the board than was the total impulse.

Kinematics of the Center of Gravity

The mean and range, respectively, for the kinematic parameters of the total body center of gravity were as follows: angle of contact, 64 and 6°; contact speed, 20.12 and 9.49 ft per sec; contact time, 0.114 and 0.027 sec.; angle of take-off, 96 and 7°; speed of take-off, 22.61 and 6.99 ft per sec; and change in speed, 2.52 and 10.02 ft per sec. No significant correlations in the rankings were found between any two of the variables according to the Spearman rank difference correlation coefficient. The limits of the landing positions were from 10 to 22 in behind the front edge, with a mean of 18 in. The resultant speed increased most during the extension phase of board contact that coincided with the second deflection of the board.

Kinematics of the Body Segments

The sign and the direction of the velocity and acceleration vectors for each segment specified the direction of motion of that segment. A positive angular velocity indicated that the segment was moving in a clockwise direction, and a negative sign indicated that the segment was moving in a counter-clockwise direction. Acceleration signs indicated whether the segment was accelerating (+) or decelerating (−) in the indicated direction of motion; thus, a kinematic description of the body segments could be made. The velocity and acceleration values of the arm, forearm, and hand segments were negligible when compared with the other four segments and consequently were omitted from analysis. As the vaulter contacted the board, the trunk was extending and continued to extend in a blocking motion through the first phase of contact. The knee and ankle joints controlled the descent and absorbed the force of the board by accelerated flexion. The vaulter's body moved forward over the ball of the foot, thus bringing the leg forward in a clockwise direction and contributing to the dorsiflexion of the ankle. The counter-clockwise movement of the foot brought the heel down to the board. The foot was the first segment to begin accelerated plantar flexion and appeared to present a resistive reaction to the upward force of the board, inasmuch as the body was still settling onto the board.

While the entire foot was in contact with the board, the trunk began accelerated extension, thus indicating that it continued to absorb the force of the board for the longest period of time and possibly allowed time for the body to progress forward over the ball of the foot. As the vaulter began to roll onto the ball of the foot, all of the joints were extending in a motion that contributed to take-off. Two of the subjects showed periods of deceleration and/or accelerated flexion of the knee and ankle joints during the push-off phase. It appeared that they attempted to reduce the magnitude of a linear velocity that was too great for them to control. One of these subjects was the only one who decreased her speed in the period from hurdle to preflight.

Kinetics of Body Segments

The muscular torques contributing to the push-off were those which indicated the predominant action of the hip extensors and ankle plantar flexors, while the torques at the knee indicated that the knee flexors were predominant at that joint. Even though the knee extensors probably were contracting, the knee extension appeared to be brought about by the large torque of the hip extension on the proximal end of the thigh and the torques produced by the ankle plantar flexion on the distal end of the leg. It appeared that the hip and ankle extensions are the greatest contributors to the take-off of the vaulters, while the knee flexors appear to work to control the forces of the knee extension and to absorb the force of the board when needed. A training program for vaulters should include work on the speed and force of hip and ankle extension as well as an attempt to increase the range of extension of these joints.

Kinematic and Kinetic Interaction

According to Plagenhoef (1971), a moving body will decrease in the sum of the vertical forces on a body, making it easier to lift. Proper timing of the deceleration of the body segments during the push-off from the board, in order to utilize this decrease in vertical force, would decelerate the segments from the trunk to the feet, with the deflection of the board occurring at the point of deceleration of the plantar flexion of the foot.

Although the sequential deceleration for these subjects occurred in an efficient order, the submaximal deflection of the board occurred prior to the deceleration of any of the segments and thus was not occurring at the time when it could have been most effective. In order to facilitate this sequence, the vaulter should attempt to accomplish the joint extension before the greatest deflection of the board.

Although contact time on the board generally is considered inversely related to the quality of the vault, maximizing the deflection of the board would maximize the time that the vaulter could use the board effectively and hence would give the vaulter more time to accomplish the extensions of the hip, knee, and ankle joints. It appears that the correlation between the total contact time and the quality of the vault may not be a reliable comparison unless one considers the temporal and sequential patterns of the vaulter while she is in contact with the board, as well as the position of contact and the sequence of muscular and board forces prior to take-off. Because the duration of contact is so short, it may be possible that the body cannot accomplish an absorption of force, a rolling onto the ball of the foot, and a sequential accelerated extension and deceleration of the joints prior to the final impulse from the board. By decreasing the vertical component coming onto the board, the vaulter could contact the board as a more rigid body with hips, knee, and ankles only slightly flexed. This reduction of the vertical component could reduce the vertical impact of landing and thus reduce the necessity of as much energy absorption by the joints. The extension sequence should begin immediately at contact, with the hips extending first, the knees second, and the ankles last. With conditioning of the plantar flexors, the dorsiflexion of the ankle could be minimized and thus enable the vaulter to begin the plantar flexion during the deceleration of knee extension. In this way, the foot possibly would reach its deceleration just at the maximal deflection of the board.

Although the arm contributions appeared to be negligible, if the vaulter prefers to hold her arms down at her side during the hurdle, she should attempt to accelerate the arms upward just prior to the extension of the trunk. In this way, the subsequent deceleration of the arms upward may occur as the trunk is accelerating. This timing is not suggested as contributing to take-off but rather so that the arm action does not increase the sum of the vertical forces downward on the trunk at the time when the trunk is attempting to accelerate upward.

CONCLUSIONS

A descriptive analysis of the Reuther board contact during side horse vaulting indicated the following.

1. All subjects showed two deflections of the board with a partial unweighting between deflections, the first being associated with the weight of the vaulter and the second being associated with the vaulter's push from the board.

2. The total impulse from the board did not appear to be an indication of the contribution of the board to the take-off.

3. The acceleration and deceleration sequences of the segments should proceed from the top segment downward, with the deceleration of one segment occurring just prior to the accelerated extension of the next lower segment and with the deflection of the board occurring at maximal deceleration of the last segment.

4. The predominant muscular torques during push-off were the hip and ankle extensors and the knee flexors.

5. Visual observations and kinematic descriptions do not give a true indica-

tion of the contributions of the segments to the quality of Reuther board contact, because muscular torques about a joint are not necessarily consistent with the motion of the segments about that joint.

REFERENCES

Kjeldsen, K. 1971. Body segment weights of college women. M.S. thesis, University of Massachusetts, Amherst, Mass.

Plagenhoef, S. 1971. Patterns of Human Motion. Prentice-Hall, Englewood Cliffs, N.J.

Author's address: *Dr. Ellen Kreighbaum, Department of Physical Education, Montana State University, Bozeman, Montana 59715 (USA)*

The use of angular momentum in the study of long-jump take-offs

M. R. Ramey
University of California, Davis

It is well known that most competitive long jumps are executed in such a way that the body has a tendency to rotate in a forward direction during the flight phase. This fact has been observed many times, and texts on the subject clearly note that the rotation is caused by angular momentum developed during the take-off phase (Dyson, 1963). In a recent study of the flight phase that used a computer simulation, it was shown that each of the three common styles of long jumping—"hitch kick," "hang," and "sail"—require significantly different magnitudes of angular momentum to produce equally successful performances (Ramey, 1974). This latter result indicates that the angular momentum required for the success of one style of jumping will not necessarily be compatible with that of another style.

This study illustrates the manner by which the angular momentum at take-off can be calculated from film and force-plate data. Further, it is shown that by varying the parameters in a governing equation, one can estimate the changes produced by altering the take-off configuration.

GOVERNING EQUATION

Recognizing that angular momentum, or the moment of the linear momentum, is not a physical quantity that may be measured directly, it is necessary to compute the angular momentum on the basis of its definition from mechanics. Textbooks on analytical mechanics show that the resultant moment on a system of rigid bodies is equal to the time rate of change of the angular momentum (Beer and Johnson, 1972). Thus, we have

$$\underline{M}_o = \underline{\dot{H}}_o \tag{1}$$

where $\underline{M}_o$ is the resultant moment about a point, o, in a reference frame attached to the mass center, $\dot{H}_o$ is the angular momentum vector about the point o, and the *dot* ($\cdot$) signifies differentiation with respect to time. Integration of the above equation from time $t = 0$ to time $t = t_1$ yields the angular momentum at time $t = t_1$. That is,

$$\underline{H}_o = \int_0^{t_1} \underline{M}_o \, dt \qquad (2)$$

Thus, for long-jump studies, the resultant angular momentum experienced by the jumper during the flight phase can be determined from an integration of Equation 2 for the total time of contact with the take-off surface.

Assuming the long jump to be a planer activity, one can determine the resultant moment, $\underline{M}_o$, about an axis through the mass center of the body by using force-plate and motion picture records of a particular jump. The force-plate data consist of the vertical and horizontal forces represented as time varying functions $V(t)$ and $H(t)$, respectively. The motion picture records can be used to estimate the variation of the body mass center relative to the foot in contact with the take-off surface. With reference to Figure 1, the moment, $\underline{M}_o$, about the system mass center is given as

$$\underline{M}_o = V(t)\,x(t) - H(t)\,y(t) \qquad (3)$$

Here $x(t)$ and $y(t)$ are the horizontal and vertical functions representing the displacement of the center of mass relative to the take-off foot. Substitution of Equation 3 into Equation 2 and integration for the time of contact with the take-off surface yields the angular momentum occurring in the flight phase of the jump.

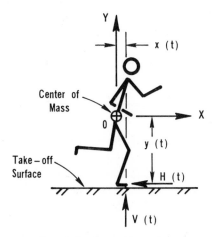

Figure 1. Coordinate system for angular momentum calculations.

DETERMINATION OF ANGULAR MOMENTUM FOR A SPECIFIC JUMP

The force-plate records for a specific jump are shown in Figure 2,*a* and *b*. The details of the experimental set-up were described by Ramey (1970). The motion picture records for this particular jump provided the configuration of the body at various time intervals from which the location of the mass center at each interval was calculated on the basis of the mass distribution presented by Williams and Lissner (1962). Figure 2,*c* and *d* shows the variation of horizontal and vertical distances of the mass center, x(t) and y(t), measured with respect to the foot in contact with the take-off surface. The film record consisted of 17 frames at nominal 0.01-sec intervals.

The variation of the moment $\underline{M}_o$ based on Figure 2 and Equation 3 is shown by a solid line in Figure 3. Positive values of $\underline{M}_o$ signify rotations in the counter-clockwise direction in accordance with a right-hand coordinate system (to the jumper it would appear that he is rotating backward). It is seen that the initial contact produces a relatively large resultant backward rotation, and subsequent time intervals tend to diminish this effect with the exception of a small portion at the very end of the take-off. Numerical integration of Figure 3 results in a final angular momentum value of

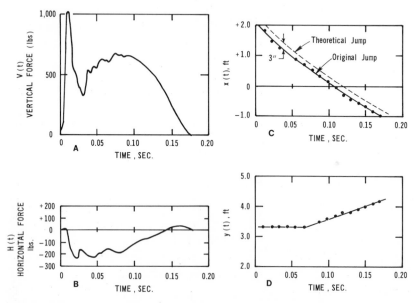

Figure 2. Force and mass center displacement during a long jump. (*a*), vertical force; (*b*), horizontal force; (*c*), horizontal displacement of the mass center relative to the take-off foot; (*d*), vertical displacement of the mass center relative to the take-off foot.

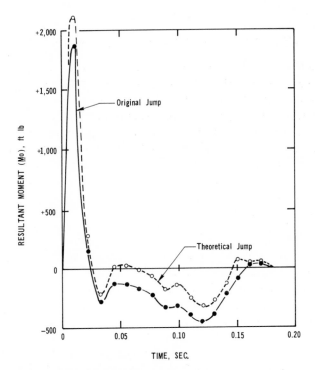

Figure 3. Resultant moment during a long jump.

$$\underline{H}_o = -9.6 \text{ ft-lb-sec}$$

The negative sign indicates that the resultant rotation is clockwise or forward. The magnitude and direction of the resultant angular momentum closely approximates those values reported by Ramey (1974) for simulated long jumps.

DISCUSSION AND CONCLUSIONS

The foregoing discussion outlines the manner by which the angular momentum can be calculated. Such data may be used in the study of long-jump take-off in various ways. For example, it is known that some styles of jumping require less forward rotation than others, and it may be desired to alter the rotation of a jumper to suit his technique better. Equations 2 and 3 indicate that any of the terms could be changed to result in a different value of the resultant angular momentum. For a serious effort, one cannot easily change $V(t)$ and $H(t)$, but it is within the capabilities of the jumper easily to change $y(t)$ and $x(t)$. The vertical variation, $y(t)$, from one jump to another often is changed by crouching on the stride just preceding contact with the take-off surface. The horizontal variation of the mass center, x(t), is varied, not usually on purpose, by taking an elongated stride before contact with the take-off surface.

Assume, for the moment, that a jumper changes x(t) by increasing the last stride by 3 in and that he maintains the other variables, y(t), V(t), and H(t) as in the jump previously discussed. This raises the curve of Figure 2,c, as shown by the dotted line. The effect of this change on the moment, $\underline{M}_o$, is shown by the dotted line in Figure 3, and the resulting numerical integration of $\underline{M}_o$ for the length of contact yields a value for the angular momentum

$$\underline{H}_o = + 15 \text{ ft-lb-sec}$$

Note now that the direction of the resultant rotation is backward, whereas the previous calculation indicated a forward rotation. It is recognized that it would be difficult actually to vary only x(t) without causing a change in the other variables, but this calculation does indicate how a small change in one of the parameters can alter drastically the functioning of the jump.

Although angular momentum is a difficult quantity to obtain, its influence on successful jumps is very important. The calculations presented were computed to illustrate how the angular momentum could be determined and how one might vary the parameters to obtain an estimate of the effect on another performance. With the proper equipment, one can study the take-off phase from this point of view and begin to estimate the changes caused by different take-off foot locations and variations in the take-off forces. These estimates could be made before any changes are suggested in the technique of a jumper and could form the basis for any suggested alteration in his style.

REFERENCES

Beer, F. P., and R. Johnston, Jr. 1972. Vector Mechanics for Engineers: Statics and Dynamics. 2nd Ed. McGraw-Hill, New York.

Dyson, G. H. G. 1963. The Mechanics of Athletics. University of London Press, London.

Ramey, M. R. 1970. Force relationships of the running long jump. Med. Sci. Sports 2: 146–151.

Ramey, M. R. 1973. Significance of angular momentum in long jumping. Res. Quart. 44: 488–497.

Williams, M., and H. R. Lissner. 1962. Biomechanics and Human Motion, Apprendix A. W. B. Saunders, Philadelphia.

Author's address: *Dr. Melvin R. Ramey, Department of Civil Engineering, College of Engineering, University of California, Davis, California, 95616 (USA)*

Investigation of the take-off technique in the triple jump

T. Bober

College of Physical Education, Gdańsk

The authors dealing with the technique of triple jump, including Gundlach (1961), Hoffmann (1966), Nett (1964), Starzyński (1969), and Wierchoszań-ski (1963), were interested mainly in the proportions of the distance of the particular steps and kinematic parameters of the take-off.

The aim of this study was the analysis of the course of the performance of the particular steps, of the deviation from the approaching axis of the triple jump, and of interrelations between these deviations and other features of the movement. This chapter also contains some data characterizing the technique of the triple jump performed by Polish top-class athletes.

METHOD

Seventeen triple jumps, as performed by eight Polish athletes, were studied. The jumps, which ranged from 15.05 to 16.16 m, constituted 90–96 percent of the best performances of the competitors. All jumps were filmed simultaneously with two cameras at 64 frames per sec. One of them was set up perpendicularly to the direction of movement and the other filmed the movement from the front.

RESULTS

The distance of the particular steps varies and is measured from the marks, toe, toe, and heel (Hoffmann, 1966).

The greatest percentage of the total distance jumped was achieved in the first step (Figure 1). In the light of the investigation by Wierchoszański (1963), the proportions achieved by the jumpers under analysis are optimal

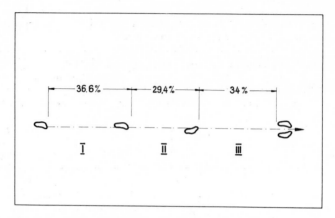

Figure 1. The measurement of the distance covered by each step and the middle range of the distance of the entire triple jump (in percentages).

for distances of approximately 16 m. However, he draws attention to the fact than an unduly prolonged first step may shorten the triple jump. The proportions among the steps depend on, among other things, general perfection, technique, and the type of competitor. For example, Szmidt, the well-known record breaker and Olympic champion, was a high-speed type of competitor, and his average proportions for the best jumps were 35.3 : 29.2 : 35.5 (Hoffmann, 1966).

The time of the take-off increased with each step, but the main difference was found between the first take-off and the next two (Table 1).

The horizontal velocity decreased as the time of the take-off lengthened: it was 8.2 m per sec after the first take-off, 7.2 m per sec after the second, and 7 m per sec after the third. At the last step, the competitors compensated for the drop in horizontal velocity (v_x) by increasing the effectiveness of take-off, thus resulting in comparatively high vertical velocity at the end of the take-off and an increase in the projection angle (see Table 2).

Because of these parameters, the center of gravity during the flight reaches the same height in the first and in the last step, and in the second step

Table 1. The time of take-offs in seconds and in percentages of the particular steps, and the time of the take-off of Szmidt's record triple jump in percentages

Steps	Time in sec			Time in percentages	
	$\bar{x}$	Minimum	Maximum		Szmidt
I	0.14	0.11	0.16	27.4	24.1
II	0.18	0.16	0.2	35.3	36
III	0.19	0.18	0.2	37.3	39.9

Table 2. Vertical velocity at the end of the take-off and projection angle in the particular steps of the triple jump

Steps	Vertical velocity in m/sec			Projection angle in degrees		
	x̄	Minimum	Maximum	x̄	Minimum	Maximum
I	2.65	2.1	2.82	18	17	20
II	2.06	1.61	2.54	16	13	20
III	2.57	2.3	2.75	20	18	23

it is about 12 percent lower. At the end of the take-off phase of the second step, the center of gravity is also lower than in the others by 3–4 percent.

Lateral Deviations in Triple Jump

In the triple jump, some competitors find it difficult to maintain the whole jump in a straight line, resulting in lateral deviations. Attention was drawn to this fact by Starzyński (1969) after the Olympic games in Mexico. In this study, the following points have been reconsidered:

1. The lateral deviation happens when the foot at a given take-off is placed beside the axis running through the foot-step left by the preceding take-off, and the magnitude of this deviation amounts to more than 2 percent of the step length.
2. The inside deviation happens when the take-off of the right foot is directed to the left or when the left foot is directed to the right.
3. The outside deviation happens when the take-off of the right foot is directed to the right or when the left foot is directed to the left.

After all 17 jumps were analyzed, the following types of lateral deviation were defined (see Figure 2): (a) total, or lateral deviation of the whole triple jump; (b) crossing, or particular steps that form a zigzag line; and (c) curved, or steps made on a curved line.

These deviations from the triple jump axis are composed of deviations of the particular steps. On the basis of criteria for the total of 51 take-offs, previously described, 31 were performed in a straight line as measured in relation to the preceding take-off. There were 10 take-offs each to the outside and the inside of the axis. The highest deviation was 8 percent of the distance of the jump (Table 3), which in absolute figures amounts to 40 cm.

The lateral deviations in the triple jump are related to the position of the whole body and projection of the center of gravity in relation to the supporting surface at the take-off phase. The body position was described by the inclination of the supporting leg and trunk in the frontal plane in the middle phase of the take-off (see Figure 3). The inclination of the leg

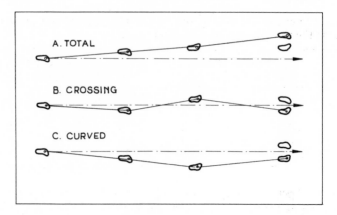

Figure 2. Typical deviations in triple jump.

amounts to an average of 5.90° with the dispersion of 1.96°. The average deviation of the trunk is greater than that of the leg. However, in some cases it was found that the trunk was in a vertical position or was inclined to the opposite side. There is a statistically significant correlation (r = 0.62) between the inclination of the trunk and that of leg.

This analysis of the inclination of trunk and leg has served to describe the features of the structure of body movements and their effect on the deviations from the approaching axis in the particular steps. Results of the statistical analysis revealed significant differences in leg and trunk inclination as a result of the deviations (see Table 4).

In relation to a more or less inclining position of the leg in frontal plane, there may be a deviation of the step to the inside or the outside of the approaching axis. Similarly, there is a statistically significant correlation between the inclination of the trunk and the deviation of the step to the inside.

SUMMARY

In the triple jump, the first step, in most cases, is longer than the third, and the second is the shortest, consisting of about 30 percent of the triple jump.

Table 3. Number of deviations in both directions of particular steps, with a division into deviation values

Deviations in percentages in relation to the jump length	Steps		
	I	II	III
≤2	12	12	7
3–4	2	4	4
5–8	3	1	6

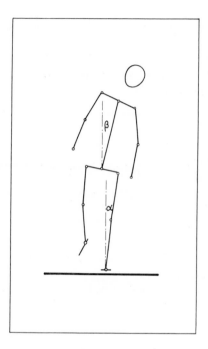

Figure 3. The inclination of the body in frontal plane in the middle phase of the take-off.

Of the three steps, the flight path of the second is the lowest. The projection angle of only the third step is similar to that of the long jump; that of the other two is smaller.

In the triple jump, three basic types of lateral deviations have been found: total, crossing, and curved. Some competitors perform the triple jump in a straight line.

Table 4. Inclination of leg and trunk in degrees in relation to the direction of the deviation of the jump

Direction of the jump deviation	Inclination of							
	leg				trunk			
	$\bar{x}$	s	d	t*	$\bar{x}$	s	d	t
Inside	3.89	1.2			2.67	2.88		
			2.24	3.8			5.44	4.8
Normal	6.13	1.6			8.13	2.64		
			2.37	4.2			1.77	1.8
Outside	8.5	1.14			9.9	2.38		

*Differences are significant when t > 2.75.

The greatest lateral deviations found in this study amounted to 8 percent of the total length of the step.

Finally, statistically significant correlations were found between the inclination of the trunk and the supporting leg, on the one hand, and the lateral deviation of the steps on the other.

REFERENCES

H. Gundlach. 1961. Zur Technik der weltbesten Dreispringen. Der Leichtathletik. Trainer, no. 33.

K. Hoffmann. 1966. Analiza rytmu i wybranych elementów techniki niektórych czołowych trójskoczków świata. Kult. Fizyczna 10: 358—363.

T. Nett. 1964. Immer mehr Flachsprunge im Dreisprung. Die Lehre der Leitchathletik, no. 33.

T. Starzyński. 1969. Na linii Rzym—Tokio—Meksyk. Lek. Atlet. 6: 5—7.

J. Wierchoszański. 1963. Proporcje skoków w trójskoku. Lek. Atlet. 2: 4—6.

Author's address: *Dr. Tadeusz Bober, College of Physical Education, Biomechanics Laboratory, Wiejska 1, 80-336, Gdańsk (Poland)*

Kicking and throwing

Kinetic parameters of kicking

E. M. Roberts, R. F. Zernicke, Y. Youm, and T. C. Huang
University of Wisconsin, Madison

A concern of many investigators in biomechanics has been obtaining accurate information on the kinetics of rapid human motion. The history of forces and moments at the joints of accelerating limb segments undoubtedly will tell us something about the muscular forces involved and also about the neural control. Technically, however, it has proven very difficult to quantify both the magnitude and time course of limb segment accelerations with reasonable confidence.

In the present investigation, an attempt was made to compare vertical ground reaction forces calculated from film displacement measures with vertical reaction forces recorded by a force platform. A planar, simulated kick in which the trunk and supporting leg were assumed to be essentially stationary was chosen for initial treatment prior to extension of the equations of motion to more than three segments and to spatial motion.

The particular kinetic analysis used has been described in detail elsewhere (Youm et al., 1973). It differs from those used by others (Dillman, 1971; Plagenhoef, 1971) in that calculations are made in terms of segmental member coordinates and then transformed to inertial coordinates. The method gives forces along and normal to the individual segments as well as in inertial coordinate terms and considerably simplifies calculations as the method is extended to spatial motion.

METHODS

An experienced, 25-year-old male kicker was the subject of the study. Anthropometrical data for the subject are presented in Table 1.

This research was supported in part by the Wisconsin Alumni Research Foundation and in part by the National Institutes of Health Biomedical Support grants administered by the Research Committee of the Graduate School, the University of Wisconsin, Madison, Wis.

158   Roberts et al.

Table 1. Anthropometric data of the subject

Weight (Dempster, 1955)	
Total body	87.32 kg*
Segment 1, foot	1.14 kg
Segment 2, leg	3.84 kg
Segment 3, thigh	11.26 kg
Length	
Height	1.78 m
Length of foot (ankle joint to first metatarsophalangeal joint)	0.16 m
Length of leg	0.41 m
Length of thigh	0.40 m
Mass moments of inertia (Dempster, 1955)	
Segment 1, foot	6.50 g cm sec^2
Segment 2, leg	58.21 g cm sec^2
Segment 3, thigh	195.22 g cm sec^2

*Values have been rounded to the nearest hundredth.

Cinematographic data of a simulated kick and a two-step approach soccer toe kick (Figure 1) consisted of 16-mm film taken under the following conditions. A Milliken 16-mm pin registered camera with a 25-mm Comat lens was positioned perpendicularly to the object plane of motion at a lens-to-ball distance of 11.08 m. Lens-to-ground distance was 1.20 m. A 160° shutter opening was used with a film exposure time of 0.0044 sec. The filming rate of 100 frames per sec was verified by a clock in the field of view and by internal timing lights that were set at 100 pulses per sec.

The force platform utilized was the property of the Biomechanics Laboratory of the Department of Physical Education for Men at the University of Iowa. The platform surface was 0.635 m square. An inkwriter Offner polygraph with paper speed set at 100 mm per sec was used to record the vertical reaction forces of the kicking activities on four channels, with gain at 27.216 kg per cm. Force records were digitized with a Hewlett-Packard 9864A digitizer, accurate to within 127 μ, in conjunction with a Hewlett-Packard 9810A calculator. The mean values of three independent digitizings of the four components were summed to obtain total vertical force.

A Recordak microfilm reader (Model P4O) with 40 X magnification was used to project the film image onto the digitizing platen. The X-Y coordinates for two vertical reference points, for markers placed at the iliac crest, hip, knee, and ankle joints and at the foot, and for the center of gravity of the kicking lower extremity, were digitized for each frame and transferred to punched paper tape. Computer programs were written to determine the center of gravity of the thigh and leg from the hip, knee, and ankle joint coordinates, as well as the angular segmental inclinations of the thigh, leg, and foot. The digitized film coordinates and calculated angles were curve-fitted to provide continuous time function of displacement estimates. Polynomial

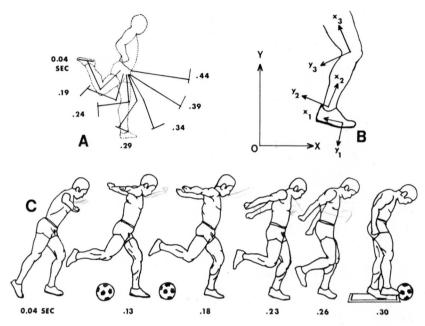

Figure 1. Experimental activities and coordinate systems. *a*, serial frames of the simulated kick. Maximal foot velocity reached during the swing at 12.58 m per sec. *b.*, illustration of the local coordinate systems and the inertial coordinate system. *c.*, serial frames of the two-step soccer toe kick. Foot velocity at contact 17.73 m per sec, with resultant ball velocity reaching 24.09 m per sec.

equations were applied, and the first and second derivatives were obtained to provide velocity-time and acceleration-time estimates. Following the kinematic analyses, kinetic analyses of joint forces and joint moments were made by use of segmental mass estimates derived from Dempster (1955). Free-body diagrams in which the unknown muscle forces were replaced by equivalent forces and moments at the joints were drawn, and the equations of motion for each segment were formulated in their respective local coordinate system (Youm et al., 1973). Magnitudes of the resultant forces obtained from the program by the local and inertial coordinate systems were checked against a program utilizing only the inertial coordinate system, and identical resultants were obtained.

RESULTS

A comparison between the vertical force recorded by the force platform and the calculated vertical ground reaction force for the simulated kick is shown in Figure 2,*a*. The overall trend of the two curves is quite similar, with an average deviation of the calculated from the directly recorded values of only 6 percent. The agreement is least where the force platform record deviates from a smooth polynomial-shaped curve. It is likely that the polynomial

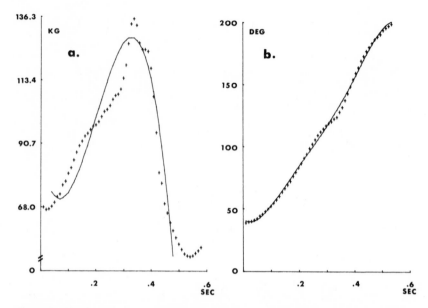

Figure 2. Simulated kick. *a.*, comparison between vertical reactive forces recorded by force platform and those calculated using a fifth-degree polynomial fitted curve of the displacement data; ————, computed; + + +, force platform. *b.*, thigh segmental inclination; ————, curve-fitted values; + + +, values from film. *DEG*, degrees.

curve-fitting techniques applied to the raw data were responsible for much of the difference between the force curves. The assumption that limb segment motions are smooth in an action such as that shown here is not entirely accurate. The displacement of the thigh, especially (Figure 2,*b*), deviates from the fitted curve between approximately 0.26 and 0.35 sec while the knee is extending. Thus, a single polynomial curve does not completely describe the whole action. The particular polynomial selected as giving the best fit to the data also has a bearing on the deviation of the calculated data from directly recorded data. A five-degree polynomial was selected in this case; it fitted the force platform better than did a six-degree equation. Curve-fitting procedures need further study. Spline-function curve-fitting techniques that are now under study show promise of describing some complex human motions more accurately than do polynomials.

Some of the deviation between the two vertical force curves probably is due to the fact that the trunk was not entirely stationary, as it had been assumed to be. In the soccer toe kick, where the inertial forces of the trunk are considerable, most of the difference between the calculated and recorded forces is likely to be accounted for by this factor (Figure 3,*d*). The calculations soon will be extended to include the trunk.

Figure 3,*a* and *b* shows the experimental and fitted angular displacement and the fitted angular acceleration of the thigh and leg up to the time of ball

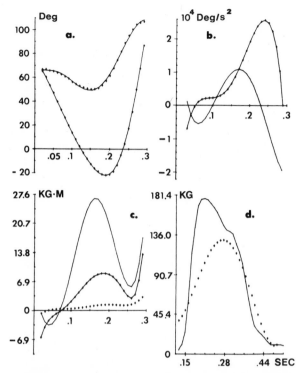

Figure 3. Two-step soccer toe kick. *a.*, comparison of sixth-degree polynomial fitted curve and experimental displacement data for the inclination angles of the kicking leg and thigh; + + +, experimental; —————, fitted (upper curve, thigh; lower curve, leg). Maximal difference between experimental and computed values was 1.5°. *b.*, derived angular accelerations of the thigh (—————) and leg (+ + +). *c.*, resultant moments of the hip, knee, and ankle joints. Counter-clockwise is positive; —————, hip joint; —————————, knee joint; + + +, ankle joint. Values in a, b, and c are illustrated for that point in time from when the toe of the kicking leg leaves the ground until ball contact. *d.*, comparison between vertical reactive forces recorded by force platform (—————), and the computed forces (+ + +). Values are illustrated from the time of landing on the supporting foot through the follow-through.

impact. It can be seen that the fitted six-degree polynomial curves closely approximated the experimental data and that the displacements and accelerations of the two segments were out of phase with one another, as noted by Roberts and Metcalfe (1968).

The resultant moments at the hip, knee, and ankle are illustrated in Figure 3,*c.* The trends seem reasonable in that all three values were negative as the kicking foot left the ground during the initial backward rotation of the segments. They then became positive, with the hip joint moment reaching maximum just before peak thigh angular acceleration and as the direction of rotation reversed (between 0.16 and 0.18 sec). Maximal knee-joint moment followed that of the hip joint, and its relationship to peak leg acceleration and reversal of the direction of the knee-joint action (deepest flexion) was similar to that for the hip (from 0.20 to 0.25 sec).

The potential of kinetic analyses of human motion, particularly high-speed human motion, is far from realization at present. The approaches discussed here may contribute to that potential, provided that they are used with great caution. Cross-comparison and verification of results obtained by different methods seem to be almost essential. Electromyographic recording, for example, is an additional tool, and preliminary recording of the toe kick (Roberts, Anderson, and Zernicke, 1974) showed that electromyograms of quadriceps and hamstrings were compatible with the film and force platform data.

ACKNOWLEDGEMENTS

We are grateful to Dr. James G. Hay for making available the facilities of the Biomechanics Laboratory of the Department of Physical Education for Men at the University of Iowa and for generously giving of his time. We are grateful too for the invaluable assistance of T. W. Roberts and Margaret B. Anderson.

REFERENCES

Dempster, W. T. 1955. Space Requirements of the Seated Operator. WADC Technical Report. Aerospace Medical Research Laboratory, Wright Air Development Center, Wright-Patterson Air Force Base, Ohio.

Dillman, C. J. 1971. A kinetic analysis of the recovery leg in sprint running. In: J. M. Cooper (ed.), Selected Topics on Biomechanics, pp. 137–165. The Athletic Institute, Chicago.

Plagenhoef, S. C. 1971. Patterns of Human Motion: A Cinematographic Analysis. Prentice-Hall, Englewood Cliffs, N. J.

Roberts, E. M., and A. Metcalfe. 1968. Mechanical analysis of kicking. In: J. Wartenweiler, E. Jokl, and M. Hebbelinck (eds.), Biomechanics I: Medicine and Sport, Vol. 2, pp. 228–237. University Park Press, Baltimore.

Roberts, T. W., M. A. Anderson, and R. F. Zernicke. 1974. Electromyography using MOS circuitry. In: R. C. Nelson and C. A. Morehouse (eds.), Biomechanics IV: University Park Press, Baltimore, and S. Karger, Basel.

Youm, Y., T. C. Huang, R. F. Zernicke, and E. M. Roberts. 1973. Mechanics of a Simulated Kick. ASME Symposium on Mechanics and Sport, November.

Author's address: *Dr. Elizabeth M. Roberts, Department of Physical Education for Women, Lathrop Hall, 1050 University Avenue, Madison, Wisconsin 53706 (USA)*

Effects of ball mass and movement pattern on release velocity in throwing

H. Kunz

Swiss Federal Institute of Technology, Zürich

In ball and javelin throwing, the velocity that is imparted to the object is the most important factor for maximal distance. At a given angle of release and a given attitude angle of the javelin, the distance of the throw depends on the speed of the javelin at the moment of its release.

Various authors have described the relationship between maximal static strength and speed of movement (Bergmaier and Neukomm, 1973; Larson and Nelson, 1969; Nelson and Fahrney, 1965). Hochmuth (1967) and Marhold (1963) have completed fundamental studies of the influence of the preliminary wind-up movement on throwing velocity. Toyoshima and Miyashita (1973) investigated the relationship between strength and speed of throwing. In this study, the influence of the type of movement and mass of the ball on the speed of release for the ball throw was investigated.

METHODS

The path and velocity of the hand and the hip were measured with two rubber-band goniometers (Neukomm, 1973). The goniometers were attached to a wooden board corresponding to the median direction of movement at an angle of 30° for the hand and 10° for the hip (angle was measured from the horizontal plane). The rubber band for the hand was attached to the forefinger of the right hand, and that for the hip was attached to a belt around the hips (at the crest of the ilium). The right foot was on a mark 4 m from the turning axis of the goniometer.

Figure 1 shows the research methods, and the test conditions and symbols are shown in Table 1. Every test condition was repeated three times, and the best result was evaluated. The test subject was told to throw the ball with

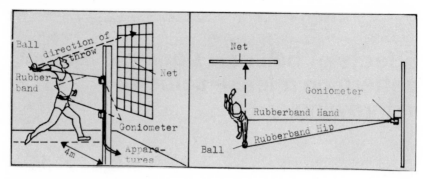

Figure 1. Research methods.

maximal effort. Twenty subjects volunteered to participate in the investigation. The subjects were divided into two groups on the basis of previously recorded best performances in javelin throwing. Group 1 was composed of good throwers, with personal best performances of between 55 and 65 m, and Group 2 was composed of poorer throwers, with personal best performances of between 30 and 52 m. The signals from the release angles α and the signal from the angular velocities ψ (from electronic differentiation) were recorded on ultraviolet paper. The path of the hand and hip and the velocity of the hand and hip were calculated from these data.

RESULTS AND DISCUSSION

Figure 2 shows that a linear relationship existed between the personal best performance in the javelin throw and the velocity of the hand at the moment of release. The good throwers (Group 1) achieved higher velocities than the poorer throwers (Group 2). The linear correlation coefficient between V_{Ha} and the personal best performance in the javeline throw was 0.84. Figure 2 also indicates that the good javelin throwers were able to take advantage of

Table 1. Test conditions

	Mass of the ball (M_{Ba})			
Type of movement	0	80 g	400 g	800 g
Throw from standing position with straight arm	A_0	A_{80}	A_{400}	A_{800}
Throw with a preliminary wind-up movement	B_0	B_{80}	B_{400}	B_{800}
Symbols*	○	●	△	▲

*Refer to figures.

best result
javelin throw

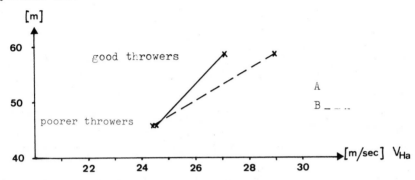

Figure 2. Relationship between the personal best performance for the javelin throw and the velocity of the hand at the moment of release.

the preliminary wind-up movement. The velocity of release of the hand (V_{Ha}) decreased with increasing ball mass, as would be expected.

It is an interesting observation, however, that the results achieved by the poorer throwers were not as good without a ball as they were with a ball weighing 80 g (see Figure 3). For the good throwers, the velocity of the hand at release was also less without a ball for Movement A. This discrepancy may be attributable to lack of effort or to poorer technique (*shorter path*). The good throwers achieved higher velocities than did the poorer throwers. Their results in the preliminary wind-up movement were better than in the movement with a straight arm. The poorer throwers achieved almost the same velocities under both throwing conditions.

Figure 4 shows that the velocity of release V_{Ha} tended to increase with increasing length of the path of the hand. The subjects of Group 2 (poorer

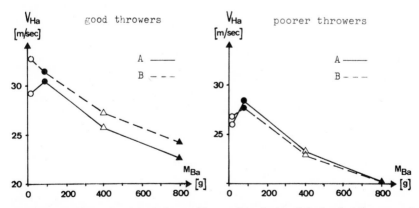

Figure 3. The influence of the ball mass M_{Ba} on the velocity of the hand at the moment of release V_{Ha}.

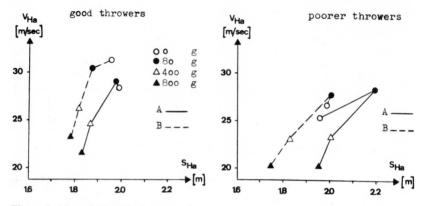

Figure 4. The relationship between the path of the hand S_{Ha} and the velocity of the hand at the moment of release V_{Ha}.

throwers) achieved nearly the same velocities with Movement B over a shorter path as with Movement A.

The velocity of release of the good throwers using Movement B was even higher (*shorter path, higher velocity*). This indicates the advantage of the preliminary wind-up movement. Figure 4 also shows that the length of the path decreased with the increase in the ball mass. The poorer throwers accelerated the ball over a greater distance than the good throwers (especially for Movement A). This is discussed in detail below. The difference in the length of the path between the two test movements was greater for Group 2 than for Group 1. The poorer throwers did not perform the preliminary wind-up movement as well; therefore, they did not achieve better results with Movement B.

Figure 4 also shows that the subjects from Group 1 achieved higher velocities V_{Ha} than did the subjects from Group 2. These higher velocities

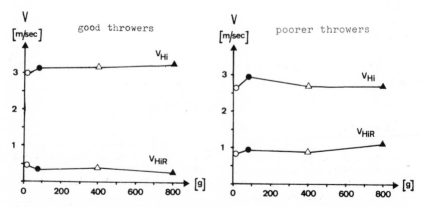

Figure 5. Differences in movements between the two groups (Movement A, V_{Hi}; velocity of hip at release, V_{HiR}).

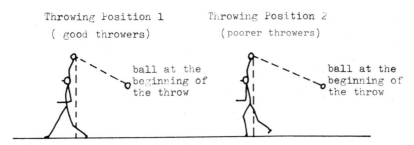

Figure 6. Differences in the throwing position at the point of release.

could have been the result of better training conditions (power, etc.) or better techniques, but these factors were not considered in this investigation.

Figure 5 shows that the good throwers achieved higher maximal velocities of the hip (V_{Hi}) than did the poorer throwers. In contrast, the velocities of the hip at the moment of the release (V_{HiR}) were essentially less for good throwers in comparison with the poorer throwers. The movement of the hip was stopped at the moment of the release by an intensive pressing of the left foot against the ground, making the position at the point of release different for the two groups. This shortened the movement paths of the hand and the hip (Group 1).

In throwing position 2 (see Figure 6), the path of the hand was longer than in Throwing Position 1. However, it was not possible to have a good back arch tension because the posterior foot lost contact with the ground too early. The path of the hand for the throw does not have to be of maximal length but should be of optimal length for each individual.

For both groups, the maximal velocity of the hip with and without the preliminary wind-up movement was almost the same. That indicates that the advantage of the preliminary wind-up movement to the velocity at release is the result of a greater tension in the upper body and the arm.

CONCLUSION

By using a preliminary wind-up movement, better results in throwing movements such as javelin and ball-throwing should be possible.

REFERENCES

Bergmaier, G., and P. Neukomm. 1973. The correlation between static muscular force and speed of movement. *In:* S. Cerquiglini, A. Venerando, and J. Wartenweiler (eds.), Biomechanics III, pp. 235–238. S. Karger, Basel, and University Park Press, Baltimore.
Hochmuth, G. 1967. Biomechanik Sportlicher Bewegungen. Sportverlag, Berlin.

Kunz, H. 1973. The influence of various weight-loads and types of movements upon the velocity of movement of forearm flexion. Paper presented at the international symposium, The Problems of Motion-Biology in Track and Field, Budapest, April 17.

Larson, C., and R. Nelson. 1969. An analysis of strength, speed and acceleration of elbow flexion. Arch. Phys. Med. Rehabil. 50: 274–278.

Marhold, G. 1963. Biomechanische Untersuchungen Sportlicher Hochsprünge. Inauguraldissertation Sporthochschule, Leipzig.

Nelson, R., and R. Fahrney. 1965. Relationship between strength and speed of elbow flexion. Res. Quart. 36: 455–463.

Neukomm, P. A. 1973. Some modern methods and equipment for the measurement of angles and position in biomechanics research. Paper presented at the international symposium, The Problems of Motion-Biology in Track and Field, Budapest, April 18.

Toyoshima, S., and M. Miyashita. 1973. Force-velocity relation in throwing. Res. Quart. 44: 86–95.

Author's address: *Hansruedi Kunz, Biomechanics Laboratory, Swiss Federal Institute of Technology, Plattenstrasse 26, 8032 Zürich (Switzerland)*

Contribution of the body parts to throwing performance

S. Toyoshima and T. Hoshikawa
Aichi Prefectural University, Nagoya

M. Miyashita
University of Nagoya, Nagoya

T. Oguri
Toyota Technical College, Nagoya

The human body, mechanically speaking, comprises a system of levers capable only of rotary motions. Most human movement consists of a series of rotations of many segments of the body. The more rotations that are brought into the action, if they are timed in sequence, the higher the final speed of the action.

A throwing movement consists of many rotations, including hip, trunk, and shoulder, as well as elbow extension and wrist flexion. In this study, we analyzed the contribution of each segment to the ball velocity in throwing of balls weighing 100, 200, 300, 400, and 500 g. In addition, using electromyography and electrogoniometry, we investigated the successive transformations of force to the ball.

PROCEDURES

Equipment for Measurement of Ball Speed

In this experiment, an instrument system was developed in order to measure thrown ball velocity by means of two sets of phototransistors connected to a digital counter, which in turn was connected to a minicomputer, as shown in Figure 1. Fifteen tubes with 15 phototransistors were placed equidistant at intervals of 5 cm in a metal frame, 97 cm long, 15 cm wide, and 6 cm thick. In order to concentrate the light, a lens was placed in each tube at the end opposite to the phototransistor.

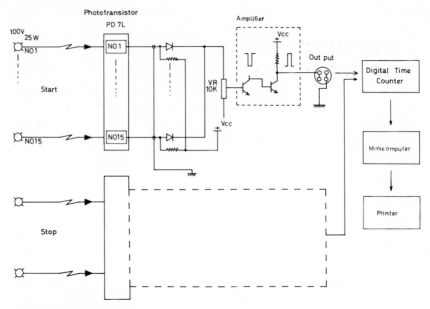

Figure 1. Block diagram.

   The first frame of phototransistors was placed just in front of the position at which the ball was released from the hand. The distance from the first frame to the second frame was 1 m. The thrown ball passes in front of the first frame and intercepts the light that enters one of the phototransistors. This change in light intensity causes a slight change in voltage within the photoelectric circuit. Furthermore, this change is enlarged through an amplifier that is in the bottom of the metal frame. The change in voltage of the first set of phototransistors started the digital time counter, whereas the change in voltage of the second set of phototransistors stopped the digital time counter. The time was measured in milliseconds. The computer calculated the velocity from the elapsed time and the known distance between phototransistor frames. These calculations of velocity were printed automatically on paper.

Subjects and Conditions

Seven male adults participated in this experiment. The subjects were asked to throw hard rubber balls with a constant diameter of 7 cm and weights of 100, 200, 300, 400, and 500 g, respectively. Each subject threw each ball 25 times, using five different patterns of throwing (five throws for each pattern).
   The throwing patterns were as follows: (Pattern 1) overhand with step (normal), (2) overhand without step, (3) overhand with lower body immobilized, (4) overhand with upper body immobilized, and (5) overhand with the upper arm placed on the arm of chair and immobilized. For each

condition (ball weight and throwing pattern), only the throw with the highest velocity was used for analysis.

Electromyographic and Electrogoniometric Techniques

Electrogoniometers were attached to the upper extremity to obtain changes in angles at the wrist and elbow joints. Bipolar surface electrodes were used to obtain electromyograms from wrist flexors, wrist extensors, biceps brachii, triceps brachii, deltoideus, pectoralis major, gluteus maximus, and adductor magnus.

RESULTS AND DISCUSSION

Average ball velocities relating to the ball weight in five different throwing patterns are shown in Figure 2. The highest velocity of 29.4 m per sec was obtained when Subject S. T. threw the lightest ball with Pattern 1. The lowest velocity of 26.3 m per sec was obtained when Subject M. M. threw the heaviest ball with Pattern 5..

The correlation equation between ball velocity (V) and ball weight (P) in the normal overhand throw was $(P + 1.749)(V + 25.689) = 100.700$ for five

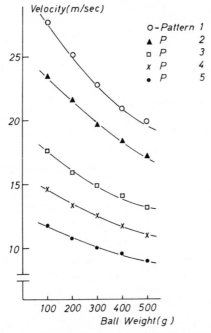

Figure 2. Average ball velocities relating to the ball weight in five different throwings.

male adults, as reported previously (Toyoshima and Miyashita, 1973). In the present study, we found that when the body parts participating in the throw are immobilized successively, the following equations were obtained:

1. $(P + .480)(V - 8.59) = 110.83$
2. $(P + .669)(V - 4.99) = 14.17$
3. $(P + .580)(V - 5.72) = 8.08$
4. $(P + 1.18)(V + 0.84) = 19.89$
5. $(P + .633)(V - 3.87) = 5.81$

Although the equations were not identical, the average velocity decreased as the ball weight increased. The actual decrease may be viewed as a percentage of the respective maximum for the Pattern 1 condition.

For example, the mean ball velocity for the seven subjects using Pattern 1 was 27.7 m per sec (100 percent). For Pattern 2, it was 23.4 m per sec (84 percent); for Pattern 3, 17.6 m per sec (63. 5 percent); for Pattern 4, 14.7 m per sec (53.1 percent); and for Pattern 5, 11.8 m per sec (42.6 percent). On the other hand, when throwing the heaviest ball using Pattern 1, these subjects produced a mean velocity of 19.9 m per sec (100 percent). Using Patterns 2, 3, 4, and 5, the mean velocities were 17.1 m per sec (85.9 percent), 13.2 m per sec (66.3 percent), 11.0 m per sec (55.3 percent), and 9.0 per sec (45.2 percent), respectively.

Therefore, the decreasing ratio was approximately the same regardless of the ball weight. Consequently, the contribution of the body parts to the velocity of the ball was independent of ball weight as used in this study.

Broer (1969) has reported the contribution of the body parts in two women throwing a tennis ball. Approximately 50 percent of the velocity of the overhand throw resulted from the step and the body rotation, and the remainder came from the shoulder, elbow, wrist, and finger action. The values in this study were 46.9 percent for the step and body rotation and 53.1 percent for arm action in throwing of the lightest ball (100 g). These figures approximate the results obtained by Broer.

For a better understanding of the mechanisms causing the percentage of contributions of the body parts, electromyographic and goniometric data were analyzed.

Intensity of quantitative muscle contraction was estimated on the basis of 5-, 10-, 20-kg calibration signals for comparison of the muscle activity among ball weights and among patterns.

Generally, there were strong action potentials for the gluteus maximus, adductor magnus, and pectoralis major in Pattern 1. This tendency substantiated the large contribution produced by trunk (pelvis and shoulder) rotation during this pattern.

Figure 3 shows the electromyogram, wrist flexor, wrist extensor, biceps brachii, triceps brachii, and goniograms of the wrist joint and elbow joint action for Patterns 1 and 5. The left-hand side of the figure shows the goniogram and the electromyogram of the forearm throw (Pattern 5). The

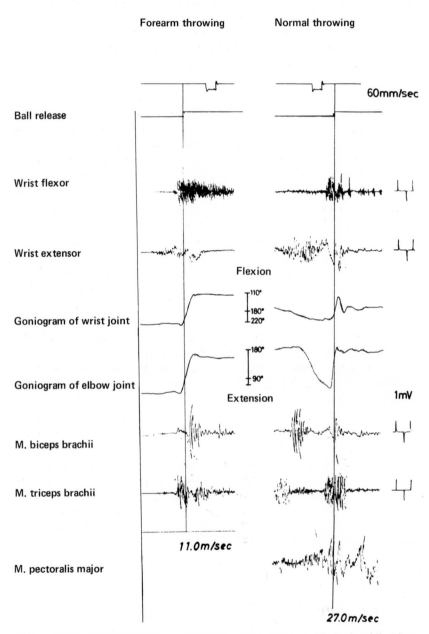

Figure 3. The electromyogram, wrist flexor, wrist extensor, biceps brachii, triceps brachii, and goniograms of the wrist and elbow joints.

174   Toyoshima et al.

right-hand side shows a case of normal throwing (Pattern 1). The velocity of the ball when thrown with only the forearm was 11.0 m per sec, whereas, in normal throwing, the velocity was 27.0 m per sec. Comparison of the two cases reveals differences and similarities in muscle activity. Pattern 5 shows greater wrist flexor and less wrist extensor activity.

Although the intensities of the triceps brachii were approximately the same, the duration of muscle contraction differed. This longer duration during Pattern 1 may be explained by comparison of goniogram results. In this study, it was noted that there was a considerable difference between the goniogram results in the two throwing patterns. That is to say, with regard to elbow joint and within the same time span, approximately twice the range of motion was observed in elbow extension prior to release during normal throwing in comparison with forearm throwing. Therefore, in order to obtain more exact numerical values, the Visicorder was used to measure the angular velocities of the forearms of two subjects, T. H. and S. T., at a paper speed of 500 mm per sec. In normal throwing, the results were 31.14 rad per sec; in forearm throwing, they were 15.57 rad per sec. It was verified, therefore, that the extension of the elbow joint was performed at a higher speed in normal throwing than in the activity of maximal voluntary effort using the elbow joint. This indicated that the forearm was being swung like a whip by the rotary actions of other parts of the body, such as hip, trunk, and shoulder. Moreover, the greater the radius of rotation, the greater will be the production of speed in these ball-throwing situations.

In summary, it appeared that the contribution of the extension of the elbow joint to the speed of the ball did not result only from the power caused by voluntary muscular contraction of the triceps brachii, but also from the torque produced by rotation of the body. The radius of rotation of the forearm in the flipping motion was increased by body rotation, thus resulting in a greater speed. It therefore would seem that the contribution of the elbow joint in normal throwing performance may be less than the 42.6 percent calculated in this study; as a matter of fact, a larger percentage of the velocity of the thrown ball resulted from the body rotation. Furthermore, it is interesting to note that the rapid arm action acts not on the basis of conscious muscle contraction, but from the physical phenomenon and reflection of the neuromuscular system.

REFERENCES

Broer, M. R. 1969. Efficiency of Human Movement. W. B. Saunders, Philadelphia.
Toyoshima, S., and M. Miyashita. 1973. Force-velocity relation in throwing. Res. Quart. 44: 86–95.

Author's address: *Dr. S. Toyoshima, Aichi Prefectural University, 3–28, Takado-cho, Mizuho-ku, Nagoya City (Japan)*

Biomechanical analysis of the shotput

G. Marhold

Research Institute for Physical Culture and Sports, Leipzig

It is essential that clear-cut target ideas for athletic techniques be formulated if we are to establish correct methods for both teaching and training. That aim, which is a rational approach to the solution of problems with an existing motor task, must be established and objectively represented as effectively for a long-term orientation as for an actual comparison. The athletic task, namely, to put a shot of a certain mass as far as possible under given conditions, is apparently an uncomplicated problem of coordination. Motor coordination, however, is implemented on the basis of a variable level of specific energetic capacities, thus necessitating new adaptations. Therefore, special energetic presuppositions are required to implement the progressive features of O'Brien's technique. A higher level of power production, increased by systematic training, demands more highly developed technical qualities, as well as better coordination. Greater force potentials are transferred into the shot-put performance only in accordance with the adaptation of the specific movement patterns.

Athletic techniques are reflected in so-called biomechanical curves in an abstract manner (Hochmuth, 1967). Generalizing the individual somatic proportions and the individual level of motor abilities of world-class athletes, we can summarize their biomechanical characteristics. Further, specific values of movements can be transformed into data and curves which represent the most rational solution to problems with top-class performances. They can be used as general orientations of a first approximation for the desired technique in athletic events.

Furthermore, it is interesting to evaluate typical changes of biomechanical functions and data with top-class athletes on a long-term basis in order to deduce general features of changes related to the progressive development of special motor abilities.

The different functions and measurements are based on mechanical laws and biological conditions. There are some problems in training and competition concerning validity and accuracy, thus necessitating a special selection and restriction.

PROCEDURES

Electromyography

Biopotentials of eight selected muscles have been recorded directly by means of superficial, bipolar, and silver-wire electrodes.

Cinematography

Film sequences of the shot put have been taken with a camera in a fixed position from the side and from above at 64 and 96 frames per sec, respectively. Coordination of frames and other mechanical sources of information was possible by frequency proportional impulse sequences generated by a light beam within the camera. The information was evaluated with special analyzers (fixed focus, rear projection, and reading via measure spindles in coordinates).

Dynamography

The reactive forces of the athletes were taken at the points of support by three dynamographic platforms, transformed according to the strain-gauge principle, and were recorded with an oscillograph in x- and y-directions.

Speedography

The shot-put movement was transformed by a cord connection between hand and transducer (tachogenerator) into a velocity proportional to alternate current with constant frequency, rectified, and recorded on an oscillograph. Necessary corrections of the push-off velocity in comparison with the curve velocity were made according to the trajectory of the shot.

Following well-known procedures (Grigalka, 1967; Kutov, 1966; Lindner, 1960; Net, 1961; Schmolinsky, 1973; Tutevitch, 1955), five characteristic phases of the shot-put movement were distinguished (see Table 1). These phases, which deviated from normal dynamic considerations, were limited in photographic detail by the characteristic body positions of the performer. They form the basis for cinematic measurements and comparisons (Figure 1).

RESULTS

There was a discrepancy between the amount of investigational efforts and the amount of information actually obtained in some procedures. That was

Table 1. Shot put. Biomechanical functional phases

No.	Phase	Function	Visible event	
			Start	Finish
1	Starting	First acceleration of the shot and athlete	First movement from starting stance	Rear foot leaves the ground
2	Gliding	Fast change of support of the rear leg to increase the total distance of acceleration; the whole system is without ground contact	Rear foot leaves the ground	Rear foot touches the ground again
3	Transitional	Preserving the "landing" energy and quick acceptance of a favorable starting position for the final acceleration	Rear foot touches the ground again	Front foot touches the ground
4	Push-off	Maximal final acceleration	Front foot touches the ground	Shot leaves the hand
5	Amortization	The athlete's impulse is being amortized	Shot leaves the hand	Athlete is in rest again. Return-point of the movement

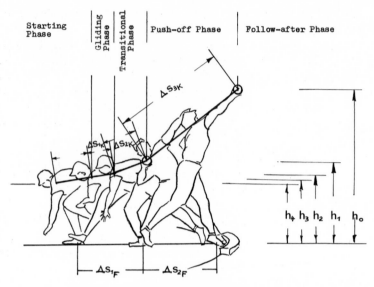

Figure 1. Shot put. Characteristic positions and measured data.

true of both electromyography and dynamography. The electromyographic characteristic reflects a coordinative pattern which is subjected to more or important variations from one movement to another with the same individual, and which can be reported only statistically. Thus, gross differences existing between beginners and advanced athletes become visible. Differentiated variations in different shotputs of one world-class athlete are neglected within the range of deviations.

Interpretations of dynamographic characteristics in detail meet with two essential difficulties. Irrespective of the problem of sufficiently high lower limits of resonance frequency of such platforms, an extensive amount of work resulted from the necessary vector division between the torque and acceleration impulses; this had to be carried out with cinematographic aids. Further efforts are required by the division of gravity, depending on the position in support on two platforms (push-off phase). Speedography, in combination with cinematography, provided useful information concerning the aspects previously mentioned.

The projection of the spatial shot curve with respect to the vertical level shows a tendency toward a straight-up movement with top-class athletes. A small difference in angle between the inclination preceding the push-off phase and that during the push-off phase presents a high technical level with respect to the spatial characteristics (so-called opening angle = 150–165°). After a longitudinal analysis, the tendency for a lower starting point on the basis of improved motor abilities was identified. The opening angle increased, e.g., during 3 yr with German top-class athletes, from about 140 to 165°. In this connection, the distance between shot and ground at the beginning of the gliding phase (h_3) decreased from 1.04 to 0.80 m.

Technically perfect athletes were superior to others in their ability to increase the velocity of the shot prior to the beginning of the push-off phase. Another important characteristic of technical perfection is a relatively high level of velocity at the beginning of the push-off phase (v_1) while a maximal possible final acceleration is reached. In shotputs of over 21 m, values between 3.5 and 3.7 ms^{-1} have been reached, i.e. 25–27 percent of the take-off velocity $(v_o.)$.

German top-class athletes also have reached values in that range. This level of velocity represents an individual optimum. It can be increased only on the basis of advanced coordinative abilities. There will be coordinative troubles and reduced final accelerations if this level of velocity is increased beyond the optimal range without first achieving lower levels.

The length of the gliding path, Δs_{F_1}, is in equal proportion to the length of the step during the push-off phase, Δs_{F_2}, with top-class athletes. Contrary to what is found in discus-throwing (Marhold, 1973), we see here a "short-long" relationship.

From first calculations of measurement correlations of several single factors and their relationship to the shot-put performance, we found significant interrelations for the length of the push-off path, Δs_{3K}, for the shot path during the transitional phase, Δs_{2K}, as well as for the time of the transitional phase, Δt_2. From longitudinal analyses of German top-class athletes, we found the tendency to perform better with increasingly greater accelerative distances during the push-off phase. These distances are accomplished by reduced shot paths during the transitional phase and by shorter times during the same period. These tendencies should be investigated further.

REFERENCES

Grigalka, O. 1967. Grundlagen der modernen Kugelstoßtechnik. Der Leichtath. 39: 40.

Hochmuth, G. 1967. Biomechanik Sportlicher Übungen. Sportverlag, Berlin.

Kutov, M. 1966. Kugelstoß von Spitzenathleten. Teori. Prak. Moskau 11: 8.

Lindner, E. 1960. Der Kugelstoß. Die Leibeserzie. 6: 7.

Marhold, G. 1973. Zur biomechanischen Analyse der Technik des Diskuswurfs. Referat auf dem 1. Internationalen Symposium zu Problemen der Bewegungsbiologie/Biomechanik in der Leichtathletik, Budapest.

Nett, T. 1961. Das Übungs- und Trainingsbuch der Leichtathletik. Verlag Barthels und Wernitz Berlin-W. oder Band 3 Wurf and Stoß.

Schmolinsky, G. 1973. Leichtathletik. Sportverlag, Berlin.

Tutevitch, V. N. 1955. Tolkanije jadra. Verlag Körperkultur u.Sport, Moskau.

Author's address: *Dr. Gert Marhold, Deutsche Hochschule für Korperkultur, Abt. Biomechanik, Friedrich-Ludwig-Jahn-Allee 59, 701 Leipzig (German Democratic Republic)*

Optimal angle of release for the competition javelin as determined by its aerodynamic and ballistic characteristics

J. Terauds

University of Texas of the Permian Basin, Odessa

The purpose of this study was to determine the optimal angle of release for the competition javelin as determined by its aerodynamic and ballistic characteristics. In particular, aerodynamic measurements included lift, drag, and the pitching moment, while ballistic tests consisted of horizontal flight distances of the javelin at various angles of release combined with various velocities.

METHODS

The aerodynamic and ballistic tests were conducted on 19 models of competition javelins at the Department of the Army Ballistic Research Laboratories at the Aberdeen Proving Grounds and at the Glenn L. Martin Institute of Technology Low Speed Wind Tunnel at the University of Maryland. The aerodynamic measurements of lift, drag, and the pitching moment were conducted in the wind tunnel at angles of attack ranging from 0.0 to 45.0°, and in the Transonic Free Flight Range at Aberdeen, where the flight aerodynamic and ballistic measurements were recorded by spark photographic stations in the path of the flying javelin. Further ballistic data on the javelin were obtained as the javelin was launched by a pneumatic gun at velocities of 60.0–140.0 ft per sec and at angles of release of 0.0–50.0° (see Figure 1).

RESULTS

The aerodynamic data reveal that lift increases with an increase in the angle of attack of up to 45.0°. Of course, lift also increases linearly with an increase

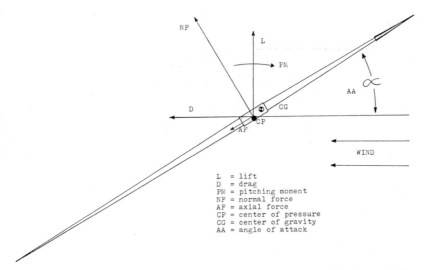

Figure 1. Aerodynamics of the javelin are measured in terms of lift, drag, pitching moment, axial force, and normal force. The center of pressure is found by dividing the pitching moment about the center of gravity by the normal force.

in velocity. For example, at a velocity of 90.0 ft per sec, the lift force on the Held Mark Two javelin is 0.0, 0.246, 0.722, 1.924, and 2.929 lbs. at an angle of attack of 0.0, 10.0, 20.0, 30.0, and 40.0°, respectively. For the same javelin, the lift force at an angle of attack of 35.0° is 1.673, 2.231, and 2.789 lbs. at velocities of 60.0, 80.0, and 100.0 ft per sec, respectively. The lift pattern for all javelins tested was similar, but, on the whole, javelins with the largest planiform area generated the highest lift force. At 100.0 ft per sec, lift equals the weight of the javelin at angles of attack of 27.5–29.5°.

Drag is present at all angles of attack and increases as the angle of attack increases. Again, the javelins with the largest planiform area develop the greatest drag force. The lift force equals the drag force at an angle of attack of between 42.0 and 46.0°, while the maximal lift-drag ratio is found at an angle of attack of between 10.0 and 16.0°.

The pitching moment of the javelin changes as the center of pressure meanders along the long axis of the shaft. Javelins on which the pressure remains behind the center of gravity are always stable. A javelin becomes unstable and develops a positive pitching moment as the center of pressure moves ahead of the center of gravity. All javelins are unstable from an angle of attack of 0.0–10.0°. Some javelins become unstable at an angle of attack of somewhere between 10.0 and 32.0°, but all javelins become stable after an angle of attack of 32.0°.

Tests with the pneumatic javelin launcher reveal that the javelins that are unstable at certain angles of attack fly further than the perpetually stable javelins because of the gliding effect achieved through a larger angle of attack. Generally, the optimal angle of release for a javelin decreases as the initial

Table 1. Distance of javelin flight, in ft, with initial velocity of 85 ft per sec

Javelin model	Angle of release in degrees					
	20	25	30	35	40	45
A - 1	200.2 F	224.8 F	245.5	253.3	250.6 F	233.0
B - 2	208.2 F	228.0 F	246.0 F	253.0	254.4 F	244.5
C - 3	213.7	227.8	241.0	252.0	255.5	243.4
D - 4	196.1	215.0	231.0	246.0	248.0	233.9
E - 5	198.9	215.2	231.0	248.5	252.7	241.6
F - 6	201.8	211.0	220.6	230.6	235.5	233.5
I - 9	227.2	233.6	241.4	253.7	255.4	250.0
J - 10	203.5	226.4	251.0	253.6	248.5	223.5
K - 11	201.9	217.5	233.6	250.0	251.6	217.8
L - 12	200.5	215.4	229.6	244.7	247.0	237.8
M - 13	181.7	240.0	217.7	230.5	234.8	229.2
G - 7	194.1	210.2	222.2	228.5	225.0	216.6
P - 16	201.3	215.4	232.1	247.6	250.0	232.1
Q - 17	197.3	215.5	234.7	254.2	263.0	262.8
R - 18	201.0	217.0	234.9	247.7	250.5	244.0
S - 19	202.0	215.6	230.8	245.6	250.5	241.0
T - 20	200.9	219.0	237.5	261.4	258.0	236.0

velocity increases. With most javelins, a gain in distance is realized if the angle of release is reduced by $1.5°$ while the velocity is increased 1.0 ft per sec. It is possible that, on lowering of the angle of release, the velocity can be increased sufficiently actually to gain horizontal distance. Novice javelin throwers have demonstrated that velocity can be increased with a decrease in the angle of release. Although data are not available, it seems reasonable to expect similar results from the expert thrower. If, for example, a thrower uses the Sandvik Super Elite (which has an optimal angle of release of $35.0°$ at 85.0 ft per sec), he may increase his throw from 253.6 to 270.0 ft by lowering the angle of release from 35.0 to $20.0°$ while he increases the initial velocity from 85.0 to 95.0 ft per sec.

REFERENCES

Terauds, J. 1970. Aerodynamics of the Held javelins at low angles of attack. Unpublished paper, University of Maryland, College Park, Md.

Terauds, J. 1972. A comparative analysis of the aerodynamic and ballistic characteristics of competition javelins. Ph.D. dissertation, University of Maryland, College Park, Md.

Terauds, J. 1972. Low speed wind tunnel tests on the competition javelin. Unpublished paper, University of Maryland, College Park, Md.

Terauds, J. 1973. Free flight aerodynamic tests of the javelin. Unpublished paper, Department of Army, Ballistic Research Laboratories, Aberdeen, Md.

Terauds, J. 1973. Ballistic tests of the competition javelin. Unpublished paper, University of Maryland, College Park, Md.

Terauds, J. 1973. Initial velocity of the javelin at various angles of release. Unpublished paper, University of Maryland, College Park, Md.

Author's address: Dr. Juris Terauds, Department of Physical Education, Center of Research in Sports, University of Texas, Odessa, Texas 79762 (USA)

Swimming and diving

Total resistance in water and its relation to body form

J. P. Clarys
Vrije Universiteit Brussel, Brussels

J. Jiskoot
Academy of Physical Education, Amsterdam

H. Rijken
Netherlands Ship Model Basin, Wageningen

P. J. Brouwer
Vrije Universiteit Brussel, Brussels

Using the analogy of calculation and comparison of the resistance of ship models, it appears that, in the research of water resistance on human beings, one also must strive toward an effective choice of form parameters.

In shipbuilding research, the total resistance is divided into frictional resistance and rest resistance. The latter can be subdivided into eddy and wave resistance (Lap, 1954). In this study, it appeared that it is as yet impossible to use such a division for the human body because a number of coefficients (such as the frictional resistance coefficient of human skin in water) cannot be calculated with mathematical accuracy. Therefore, we used only the concept of "total resistance" (W_o) in the biomechanical relationship drawn between the human body and water. A similar statement has been made by Karpovich (1933) in his attempt to calculate water resistance by using more accurate formulas than those used in connection with ship models. It is our aim to find out how form phenomena can influence the created resistance of a moving body in water.

CHOICE OF THE FORM PARAMETERS

Although it is a fact that the form of the human body is irregular and not "harmonically streamlined" in comparison with the geometrical forms of a ship, it still is useful to analogize with existing form parameters of ship models. Thus far, the anthropometric and constitutional measurements can be used in such circumstances. In the following synopsis, the form parameters used for ship models are plotted against the corresponding anthropometric relations of the human subject (see Figure 1).

CHOICE OF THE POPULATION AND THE TESTS

It was hypothesized that the human shape influences total resistance and that this total resistance can be stated best by using extreme body types, e.g., three heavy and three thin types. The entire population has been selected

SHIP	Influenced Resistance Component	HUMAN BEING	Influenced Resistance Component
Coefficient of Slenderness $\dfrac{L}{\Delta^{1/3}}$		$\dfrac{\text{Body Height}}{\text{Body Volume}^{1/3}}$	
Length - Breadth relationship $\dfrac{L}{B}$	WAVE RESISTANCE W_g	$\dfrac{\text{Body Height}[1]}{\text{bi acromial Breadth}}$	
Breadth - Draught relationship $\dfrac{B}{T}$		$\dfrac{\text{bi-acromial Breadth}}{\text{Sag. Thoracal diam.}}$	
Midship - Section Area $\alpha = \beta$ B.T.		greatest Body Cross Section	TOTAL RESISTANCE W_o
Ship Length l		Body Height	
Wetted Area $\dfrac{L}{B}$	FRICTIONAL RESISTANCE W_r	(wetted) Body Surface cfr. (1)	
Length - Draught relationship $\dfrac{L}{T}$		$\dfrac{\text{Body Height}}{\text{Sag. Thor. diam.}}$	
Square of Length Midship Section Area relationship $\dfrac{L^2}{\alpha}$		$\dfrac{\text{square of Body Height}}{\text{Body Surface}}$	

Figure 1. Descriptive comparison of human and ship form parameters. *Sag.*, sagittal.

according to the Heath-Carter Somatotyping method (Carter, 1972). These extreme types are plotted on a somatochart to illustrate their divergence (Figure 2). The anthropometrical characteristics are listed in Table 1.

The tests were carried out at the water surface and 60 cm below the surface. They were performed using the initial hypothesis that total resistance is less under the water than at its surface (according to the results of Schramm, 1959). In this manner, wave-making resistance could be defined by a simple subtraction of the underwater resistance from the undersurface results, i.e., wave resistance = surface resistance − underwater resistance.

In order to limit the study to the influence of the human form on resistance only, all form variations had to be omitted from this basic study. The body is towed in a horizontally stretched position with the arms extended forward. The head is in the water between both arms, and the feet are in plantar flexion. At low velocities, the feet were supported in order to maintain the same test conditions. The subjects were towed through the water at velocities ranging from 1.50 m per sec^{-1} to 2.00 m per sec^{-1}, with

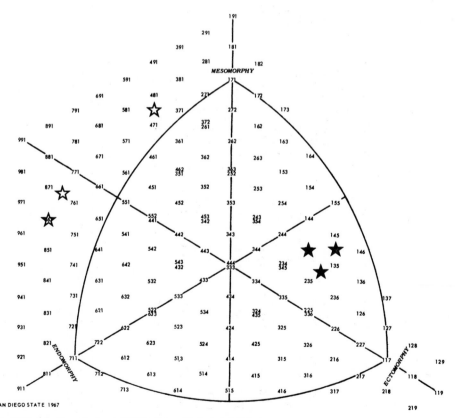

Figure 2. Somatochart distribution of heavy and thin types.

Table 1. Anthropometrical characteristics of 3 heavy and 3 thin types

Subjects	Ht. (cm)	Wt. (kg)	Vol. (L)	Surface (cm^2)	β B.T.* ∞ (cm^2)	B (cm)	T (cm)	L. Factor (cm)	Br. Factor (cm)	Muscle factor (cm)	% Fat
A. G. (2)	196.0	129.0	121.5	2.600	1239.9	44.8	28.5	108.7	15.6	42.1	15.7
R. S. (1)	164.3	95.0	90.0	2.009	831.7	37.1	21.4	91.3	14.0	37.1	19.0
J. W. (3)	180.1	115.0	110.0	2.241	988.8	46.8	25.6	103.4	15.3	42.4	23.5
$\bar{X}_1$	180.1	113.0	107.2	2.283	1020.1	42.9	25.2	101.1	15.0	40.5	19.4
C. S. (5)	189.5	69.5	64.1	1.949	782.6	41.5	19.9	106.3	13.7	32.9	7.4
D. A. (6)	191.0	75.0	68.9	2.016	792.1	43.5	18.5	107.3	14.2	33.7	5.8
H. V. H. (4)	183.1	70.0	64.8	1.905	570.2	42.5	18.1	106.3	12.9	31.1	8.6
$\bar{X}_e$	187.4	71.5	65.9	1.957	715.0	42.5	18.8	106.63	13.6	32.6	7.2
t_{1-e}	0.77	3.60	3.87†	1.7195	2.29‡	0.12	3.63†	1.006	1.72	3.37†	6.13†
$\overline{SE}_d$	9.39	11.54	10.66	0.19	133.25	3.41	1.75	5.468	0.80	2.36	1.98

*For clarification of abbreviations, refer to Figure 1.
†Significant at 0.05 level.
‡Significant at 0.1 level.

the same resistance-measuring points for surface and underwater tests. The speed was registered automatically on a direct-developing photo strip chart.

APPARATUS

The apparatus used belongs to the ship model test station of Wageningen (The Netherlands) and has been adapted for the specific research circumstances of human subjects.

The procedure for the measurement of the forces acting on the moving body, the towing device, and the definition of the resistance velocity points all have been described in previous studies (Clarys, Jiskoot, and Lewillie, 1973; De Goede, Jiskoot, and Van Der Sluis, 1971).

RESULTS AND DISCUSSION

We are dealing with individually different models (body forms). A comparison of the resistance among all the subjects and between both the heavy and the thin types has only a biomechanical meaning when evaluated as a function of their water displacements and velocity coefficients and on the basis of nondimensional relationships. The dimensionless resistances of the extreme types are plotted in Figure 3, and the dimensionless relationships are listed in Table 1. The absolute results of surface and underwater resistances are brought together in Figure 4.

With an increase of velocity, there is an increase of resistance (Karpovich, 1933). For the same speed, the increase of absolute resistance is greater in the group of heavy subjects than in the group of thin ones.

By using a nondimensional presentation of the resistance-velocity relationships, it is possible to approach the influence of form under the same control circumstances and in function of the water displacement (volume) and coefficient of speed (Froude number). It was noted (Figure 3) that resistance relates more positively to the water displacement (mass) [expressed in volume (L) = mass (kg or dm^3)], among the heavy types. In other words, this form relationship creates better possibilities to overcome the absolute resistance. They develop less resistance per kg of mass or develop greater power per kg of mass (see "muscle factor," Table 1) in relation to their coefficient of speed ($\frac{V}{\sqrt{gL}}$).

Between both extreme populations, we have found significant differences in the resistance-water displacement relationships at the 0.1 level, together with significant differences in the breadth–draught relationship, the length–draught relationship, the square of length–greatest cross-section relationship, and in the coefficient of slenderness at the 0.05 level. The smaller these

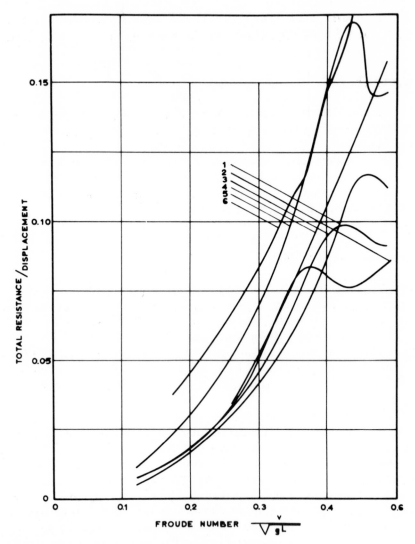

Figure 3. Nondimensional presentation of total resistance.

relationships, the more favorable the circumstances created for the resistance-velocity relationship.

When subjects were towed at both the water surface and at 60 cm under the water, a greater resistance was found for all subjects under the water. (This result also has been found in 60 physical education majors.) This result disproves the hypothesis that less resistance is measured while the subject is underwater, as was stated by Schramm (1959). The reason for this discrepancy may be the use of a different towing position; the heads of Schramm's

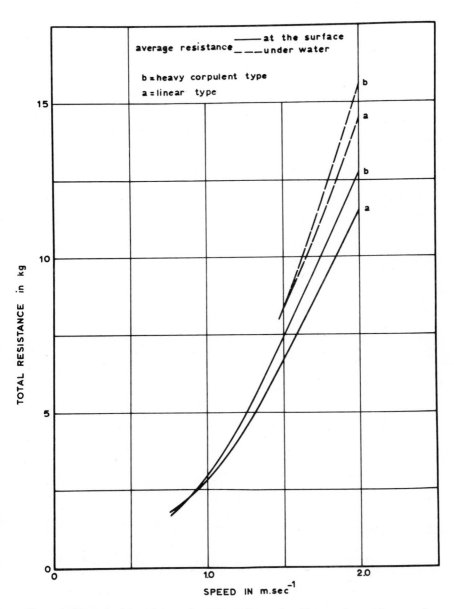

Figure 4. Mean absolute resistance (at water surface and at 60 cm underwater). $m.sec.^{-1}$, m per sec^{-1}.

Table 2. Nondimensional form parameters

	$\frac{W}{\Delta}$	Froude number* mgH	H	T	H	Greatest cross-section coefficient $c = B \cdot T$	H2	Y1/3
AG (2)	0.083	0.374	4.375	1.572	6.877	0.782	30.983	0.396
RS (1)	0.097	0.405	4.434	1.734	7.678	0.819	32.457	0.367
JW (3)	0.094	0.387	3.848	1.828	7.035	0.790	32.803	0.376
$\bar{X}$	0.091	0.389	4.219	1.711	7.197	0.797	32.081	0.380
CS (5)	0.135	0.380	4.566	2.085	9.523	0.772	45.886	0.431
DA (6)	0.136	0.380	4.361	2.351	10.254	0.784	45.431	0.465
HVH (4)	0.097	0.387	4.339	2.348	10.116	0.817	58.796	0.456
$\bar{X}$	0.123	0.382	4.422	2.261	9.964	0.791	50.038	0.451
t	2.2199†	0.630	1.2477	16.6163+	17.5504+	0.3468	4.4266	4.104+
$SE_{\bar{d}}$	0.0141	0.01	0.1627	0.033	0.1577	0.0173	4.0565	0.0173

*V = 1.64 m/sec
†significant at 0.1 level
+significant at 0.05 level
Δwater displacement (kg or dm³) = volume

subjects were lifted out of the water, creating ± 30 percent more resistance (De Goede et al., 1971).

The important resistance during submergence probably is due to the different relationships between frictional and rest resistances. Under water, the frictional resistance is greater because of the complete immersion of the body. Thus, the potential water stream has a greater contact because of the larger wetted surface area, whereas the rest resistance becomes smaller.

At the surface of the water, the situation is reversed because of the smaller wetted surface area in contact with the water and the greater effect of wave-making resistance. We presume that the differences between the frictional resistances at the surface and at the 60-cm depth are greater (the underwater resistance being higher) than the differences between the rest resistances (the surface resistance being higher). In other words, the disturbance of the potential stream at the water surface has a less negative influence on the increase of resistance than does the increase of frictional resistance underwater.

Because of a lack of precise relationships between both resistance components, one cannot evaluate friction and rest resistance in absolute values for human subjects. It is realized that the rest resistance also is indicated as form resistance because the three-dimensional pattern of the body is the source of wave and eddy resistance. Also, the form phenomena and the way they influence total resistance cannot be determined mathematically. On the other hand, we can deduce from the nondimensional relationships (Table 2) a few components that might cause an increase or decrease of total resistance.

The body shape influences resistance, as seen by study of extreme populations. However, one can suppose that by experimenting with a more uniform population, the form influence will be less pronounced and the individual differences will become very small.

CONCLUSIONS

For the same increase in velocity, we have observed a different resistance increase attributable to different (extreme) body forms.

Under similar nondimensional circumstances ($\frac{W}{\Delta}$ and $\sqrt{\frac{V}{gL}}$), the heavy types show a more positive resistance-velocity relationship. Important influences on resistance are the sagittal thorax diameter as it is used in the greatest body cross-section and the volume (water displacement) as it is used in the coefficient of slenderness.

The towed submerged body creates a greater resistance than does the body towed at the water surface; consequently, the respective resistance components can be only approximated.

The human form influences total resistance, but absolute value relationships cannot be described.

In order to obtain a complete view of the biomechanical aspects of movement in the water, it is indispensable to have a complete knowledge of the factors that can influence the resistance created by human motion.

ACKNOWLEDGEMENT

The authors are grateful to Prof. Dr. J. E. L. Carter for revision of the somatotype ratings.

REFERENCES

Carter, J. E. L. 1972. The Heath-Carter somatotype method. San Diego, Calif.
Clarys, J. P., J. Jiskoot, and L. Lewillie. 1973. A kinematographical, electro-myographical, and resistance study of waterpolo and competition front crawl. *In:* S. Cerquiglini, A. Venerando, and J. Wartenweiler (eds.), Biomechanics III, pp. 446–452. S. Karger, Basel, and University Park Press, Baltimore.
De Goede, H., J. Jiskoot, and A. Van der Sluis. 1971. Over de stuwkracht bij zwemmers. Zwemkroniek (The Netherlands) 48: 77–90.
Karpovich, P. V. 1933. Water resistance in swimming. Res. Quart. 4: 21–28.
Lap, A. J. W. 1954. Fundamentals of ship resistance and propulsion. International shipbuilding ˜rogress. Shipbuilding and Marine Engineering Monthly. Publication No. 129a of the N.S.M.B., Rotterdam.
Schramm, E. 1959. Untersuchungsmethoden zur bestimmung des wieder-standes, der kraft und der ausdauer bei schwimmsportlern. Wiss. Z. Deutsch. Hochsch. Körp. Kult. (Leipzig) 2: 161–180.

Author's address: *Mr. Jan P. Clarys, Instituut voor Morfologie, Vrije Universiteit Brussel, Eversstraat 2, B-1000 Brussels (Belgium)*

Impulse and work output curves for swimmers

R. K. Jensen
Laurentian University, Sudbury

D. G. Bellow
University of Alberta, Edmonton

The purpose of this study was to determine whether impulse and work output curves for shoulder extension against an appropriate resistance (a) are different for different levels of front-crawl swimming ability and (b) change as a result of swimming training.

The motion of a body segment underwater produces a drag force that is proportional to the square of the velocity of the segment moving through the water (Gallenstein and Huston, 1973, Seireg and Baz, 1971; and Seireg, Baz, and Patel, 1971). Body segment motion is due to the rotational forces created by muscle contraction and external forces. Differences in swimming ability between individuals can be explained partially in terms of differences in the availability of the rotational force caused by muscle contraction. If the swimming action is attributed wholly to shoulder extension in the sagittal plane, with the body segment considered as a rigid body rotating about a fixed, noncentroidal axis (the shoulder joint), then the equilibrium equation taken about the joint axis is:

$$M_a = I_o\alpha + F_g (\cos \phi)r_g - F_b (\cos \phi) r_b + M_d \qquad (1)$$

where M_a is moment caused by muscle contraction; I_o is moment of inertia of segment about joint axis; α is angular acceleration; F_g is weight of the segment; ϕ is segment angular displacement; r_g is radius to center of gravity of body segment; F_b is buoyancy force on the segment; r_b is radius to center of

This research was supported by General Research Grant 55-32285 from the University of Alberta.

buoyancy; and M_d is drag moment. This model is a gross simplification corresponding approximately to tethered swimming, but it does provide insight into the rotational forces involved.

Swimming speed does not depend on the availability of rotational force at a particular instant or angle.

Sinusoidal variations of horizontal momentum are evident (Miyashita, 1971), and the forms and the magnitudes of this curve determine the time taken to cover the swimming distance. Variations in horizontal momentum can be attributed to the impulses for the body segments. Impulse for the contraction moment for the simple model postulated is given by:

$$Imp_{1 \to 2} = \int_{t_1}^{t_2} M_a dt \qquad (2)$$

where *Imp* is impulse and t_1 and t_2 are limits of time argument.

The total work that the swimmer expends in propelling himself over the distance is also of considerable importance and is the sum of the work outputs for each segment. Work output for a single action can be calculated from the contraction moment and is the sum of the work $(w_{1 \to 2})$ required for drag and for moving the body segment:

$$W_{1 \to 2} = \int_{\theta_1}^{\theta_2} M_a d\theta \qquad (3)$$

where θ_1 and θ_2 are limits of the displacement argument.

EXPERIMENTAL APPARATUS AND TEST

The conditions envisaged in the above model can be simulated in the laboratory. The problem was to build a testing device that was compact and that permitted accurate and precise measures of the experimental parameters under conditions that were uniform for all subjects. A rotary torque actuator (Roto Actuator, Model DS4-4) was fitted with hydraulic tubing and valves to bypass the stator and with a lever arm and handle to propel the vane (Jensen, 1973). Rotation of the actuator caused fluid to flow through a valve. Application of the momentum equation to the constriction caused by the valve showed that the resistance moment was proportional to the square of the angular velocity. It was assumed that the action of the body segment paralleled that of the lever arm, and the joint axis was aligned at all times with the fixed instrument axis. The equilibrium equation is given by:

$$M_a = I_o' \alpha + F_z (\cos \theta) r_z + M_r \qquad (4)$$

where I_o' is moment of inertia of segment and instrument about joint axis; F_z is weight of segment/lever/handle; r_z is radius to center of gravity of segment/lever/handle; and M_r is resistance moment caused by instrument. A comparison with Equation 1 shows that there are differences between the two models, but the instrument developed was compact and could be used conveniently to test a large number of subjects and movements.

Bending strain for the lever arm was calibrated in terms of moment of force, and angular displacement was measured with a LVDT. The results were digitized electronically, and the analysis was fully computerized. Accuracy and precision were found to be high for the instrument. Impulse (Equation 2) and work output (Equation 3) were calculated for efforts started at $150°$ from the anatomical position, with the subject supine and stabilized. A uniform valve setting was used, and test reliability across trials was judged to be adequate.

A sample composed of 42 boys, 8–11 years of age, was divided into three ability levels on the basis of 50-m front-crawl speed (A, higher, B, medium, C, lower ability). The sample then was paired randomly into an experimental group that underwent 6 wk of training, 3 sessions per wk, and a control group. Test 1 was for 3 wk and Test 2 was for 6 wk after the start of training. A further testing session was conducted at the start of the training period to familiarize the subjects with the test. Four trials per test were conducted, and the best trial was selected.

RESULTS AND DISCUSSIONS

The impulse and work output curves calculated were of the means of the best trials for each of the 12 experimental subgroups (Tests 1 and 2, experimental, control, Abilities A, B, and C). Comparisons between curves were made using a one-tailed t test and a probability of 0.05.

Comparisons between levels of ability were made using the results of Test 2 for the control group. The three mean curves (Figure 1) were distinct for both impulse and work output. Comparisons of the differences between mean points indicated that Curve A was greater than Curve C and that, for the initial segments of the curves, Curve A was greater than Curve B (degrees of freedom = 12). The remaining differences were not significant. It can be said, then, that the test does distinguish between levels of swimming ability.

Comparisons of the ability level curves for the experimental group indicated that the Group B curves were probably greater than the corresponding control group curves. In Figure 2, a comparison has been made between experimental and control curves for Tests 1 and 2. The curves for the experimental group showed that, during the initial 0.28-sec period, the second test was significantly greater than the first test. Also, for the second half of the work output curves, the experimental curve was greater than the control curve on Test 1. All other differences were not significant. These

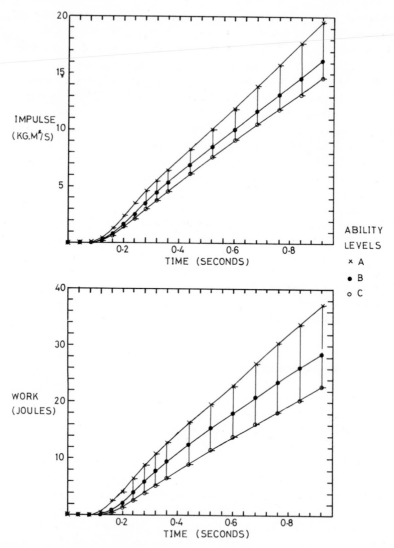

Figure 1. Control group ability level curves for Test 2. Bars between data points indicate a nonsignificant difference at the 0.05 level.

results were not supported by results from Ability Groups A and C, but suggest that the B group may have had a greater capacity for change.

CONCLUSIONS

An apparatus has been presented that provides a resistance moment similar to that for a simple underwater arm movement. It has been shown that differ-

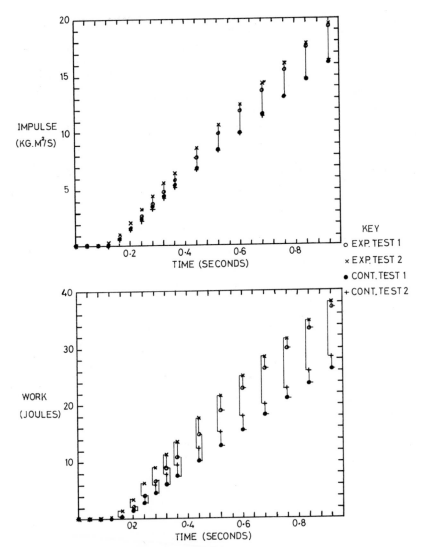

Figure 2. Comparison of curves for the B Subgroups. Bars between data points indicate a nonsignificant difference at the 0.05 level.

ences between individuals can be detected, and some changes that could be attributable to training have been demonstrated. Further investigations with more subjects and more intensive training are warranted.

REFERENCES

Gallenstein, J., and R. L. Huston. 1973. Analysis of swimming motions. Hum. Factors 15: 91–98.

Jensen, R. K. 1973. Dynametric measurement of static and dynamic responses. Paper presented at CAHPER, Calgary, July.

Miyashita, M. 1971. An analysis of fluctuations of swimming apeed. *In:* L. Lewille and J. P. Clarys (eds.). First International Symposium on Biomechanics of Swimming. pp. 53–58. Université Libre de Bruxelles, Brussels.

Seireg, A., and A. Baz. 1971. A mathematical model for swimming mechanics. *In:* First International Symposium on Biomechanics of Swimming. pp. 81–103. Université Libre de Bruxelles, Brussels.

Seireg, A., A. Baz, and D. Patel. 1971. Supportive forces on the human body during underwater activities. J. Biomech. 4: 23–30.

Author's address: *Dr. Robert K. Jensen, School of Physical and Health Education, Laurentian University, Sudbury, Ontario (Canada)*

Telemetry of electromyographic and electrogoniometric signals in swimming

L. Lewillie
Université Libre de Bruxelles

Biomechanics of sport and especially of swimming are being studied increasingly in the normal environment as opposed to an artificial, experimental situation. This trend has been stimulated by the availability of miniaturized electronic and telemetric instruments. For example, this author previously has studied the electrical activity of superficial muscles during swimming (Clarys, Jiskoot, Lewillie, 1973; Lewillie, 1968a, b, 1971a, b, 1973). It was deemed essential that simultaneous assessment be made of muscle activity and its effect on human motion. To accomplish this, the authors have superimposed electrogoniographic recordings of the elbow joint on the integrated electromyogram (EMG) of the biceps and triceps brachii muscles.

METHOD

The EMG telemetry system that has been described previously (Lewillie, 1968a,b) utilized antennas supported by balloons. The telemetered signals were recorded on magnetic tape and transcribed directly on a 700-counts per sec polygraph by intervention of an analog-to-digital converter. Surface electrodes were fixed on the shin by waterproof tape. Interpretation of the EMG recordings was made on the basis of percentage of the isometric maximum.

It is recognized that no simple relationship exists between integrated EMG and the force exerted by a muscle. However, recent research by Pertuzon (1972) indicated that, in practical terms, the relationship between external force and EMG is linear. Further, EMG demonstrates a linear relationship to the maximal acceleration of the movement, a quadratic relationship to the maximal velocity, and a linear relationship to the external work (Goubel, 1970; Bouisset, 1973). Consequently, the use of maximal isometric contraction as a reference level is justified. This assumption is especially tenable in the case of movement in water, which is characterized by its regularity and relative slowness.

Previous research utilizing goniometry to investigate swimming strokes has been reported by Filcak (1971) and by Ringer and Adrian (1969), who studied the elbow action in the crawl stroke. Also, Barthels and Adrian (1971) combined goniometry and EMG of the lower limb to investigate variability of the dolphin kick. The present study represents an extension of this line of investigation.

The electrogoniometer was made of two arms articulated around a water-proofed potentiometer. The axis of the apparatus was centered on the axis of the elbow joint, at the level of the external epicondyle. The arms of the goniometer were fixed on the arm and forearm of the subject by means of tape; in the limits of swimming movements, the pronation-supination movements did not induce any modification of the position of the apparatus. Calibration was made by means of an anthropometric protractor while the apparatus was attached to the subject.

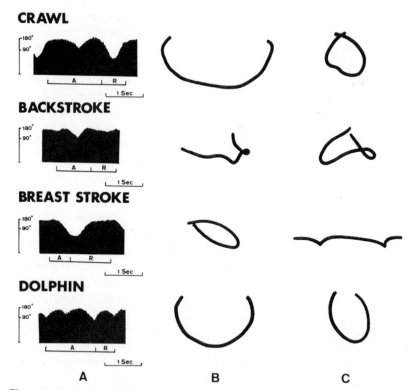

Figure 1. Comparison of electrogoniometry and light traces. *A*, electrogoniometry of the four swimming strokes (same subject); *B*, light trace of the hand (on termination); *C*, light trace of the hand (actual swimming).

RESULTS AND DISCUSSION

The evolution of the goniometrical trace appears to be very characteristic of the stroke. The individual differences do not allow separation of the different levels of ability of our subjects. These conclusions confirm those of Ringer and Adrian (1969).

Figure 1 depicts electrogoniograms and light tracing patterns of four swim strokes. The variability of the complete cycle including its different parts is very low and diminishes when the swimming speed increases. The motions of active and recovery phases must be blended, because of a gliding phase that is different for each stroke.

A comparison of the movements of the arm, as described by goniometry, with the light traces often used to describe the movement shows the complexity of the relationship among the segments of the limb and between them and the body. The comparison of the four strokes provides a good view of the differences of the state of flexion-extension of the arm and of the angular speed reached. For the crawl, our results were similar to those of Ringer and Adrian (1969), with the exception of a slight difference (maximum 150° per sec).

Our subjects, who were not of international caliber, clearly showed a greater angular speed during flexion than during extension. For the four strokes, the arm stays relatively extended, with the greatest flexion appearing during the recovery phase.

Comparison of EMG's and electrogoniograms presented in Figures 2 and 3 provides accurate identification of the moment of activity of each muscle. Figure 3 shows a recording obtained during actual swimming by a qualified subject that emphasizes the relationship of flexion-extension during the active phase in crawl. No relationship was present between the speed of the

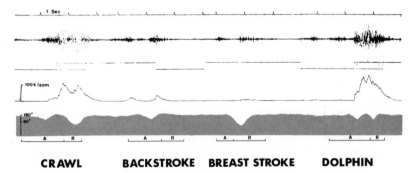

CRAWL   BACKSTROKE   BREAST STROKE   DOLPHIN

Figure 2. Recording of EMG and integrated EMG of triceps brachii muscles and electrogoniometry of the elbow joint during one single movement of each of the four swimming strokes (A, active phase; R, recovery phase).

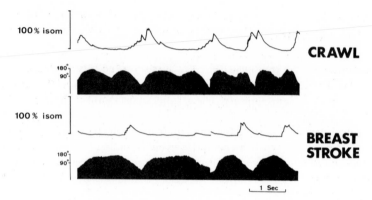

Figure 3. Integrated EMG of biceps brachii muscles and electrogoniometry of the elbow joint during swimming at slow and sprint speed.

flexion-extension movement and the level of activity of biceps and triceps brachii muscles.

CONCLUSIONS

The following conclusions are made. First, the electrogoniogram is very characteristic of the stroke and much less so of the individual swimmer. Second, the electrogoniogram is very useful for correlation with the EMG activity in the movement. Finally, no simple relationship seems to link the speed of flexion-extension of the elbow joint and the EMG activity level of biceps and triceps brachii muscles.

REFERENCES

Barthels, K. M., and M. J. Adrian. 1971. Variability in the dolphin kick under four conditions. *In:* L. Lewillie and J. P. Clarys (eds.), First International Symposium on Biomechanics in Swimming, pp. 105–118.
Bouisset, S. 1973. EMG and muscle force in normal activities. *In:* J. D. Desmedt (ed.), New Developments in Electromyography and Clinical Neurophysiology. Vol. 1, pp. 547–583.
Bouisset, S., and B. Maton. 1973. Comparison between surface and intramuscular EMG during voluntary movement. *In:* J. D. Desmedt (ed.), New Developments in Electromyography and Clinical Neurophysiology. Vol. 1, pp. 533–539.
Clarys, J. P., J. Jiskoot, and L. Lewillie. 1973. A kinematographical electromyographical and resistance study of water-polo and competition front crawl. *In:* S. Cerquiglini, A. Venerando, and J. Wartenweiler (eds.), Biomechanics III, pp. 446–452. University Park Press, Baltimore.
Filcak, M. 1971. Application possibilities of the electrogoniographic method in swimming technique evaluation. *In:* L. Lewillie and J. P. Clarys (eds.), First International Symposium on Biomechanics in Swimming, pp. 73–80.

Goubel, F. 1970. Etude et signification des relations entre l'activitié électromyographique intégrée et diverses grandeurs biomécaniques en contraction anisométrique. Thesis, Lille.

Lewillie, L. 1968a. Telemetrical analysis of the electromyogram. In: J. Wartenweiler, E. Jokl, and M. Hebbelinck (eds.), Biomechanics I, pp. 147–149. University Park Press, Baltimore, and S. Karger, Basel.

Lewillie, L. 1968b. Analyse télémétrique de l'électromyogramme du nageur. Trav. Soc. Med. B. Ed. Phys. 20: 174–177.

Lewillie, L. 1971a. Quantitative comparison of the electromyogram of the swimmer. In: L. Lewillie and J. P. Clarys (eds.), First International Symposium on Biomechanics in Swimming, pp. 155–159.

Lewillie, L. 1971b. Graphic and electromyographic analysis of various styles of swimming. In: J. Vredenbregt and J. Wartenweiler (eds.), Biomechanics II, pp. 253–257. University Park Press, Baltimore, and S. Karger, Basel.

Lewillie, L. 1973. Muscular activity in swimming. In: S. Cerquiglini, A. Venerando, and J. Wartenweiler (eds.), Biomechanics III, pp. 440–445. University Park Press, Baltimore, and S. Karger, Basel.

Pertuzon, E. 1972. La contraction musculaire dans le mouvement volontaire. Thesis, Lille.

Ringer, L. B., and M. J. Adrian. 1969. An electrogoniometric study of the wrist and elbow in the crawl arm stroke. Res. Quart. 40(2): 353–363.

Author's address: Dr. Leon Lewillie, Laboratoire de l'Effort, Université Libre, Avenue Paul Heger 28, 1050 Bruxelles (Belgium)

Analysis of kinematic parameters during competitive backstroke swimming

F. Krüger

Johann Wolfgang Goethe Universität, Frankfurt

The motivation for this article stems from the interest of the author in the biomechanics of competitive backstroke swimming. The results presented are concerned with three aspects of this sports skill: (a) description of the results of an analysis of 100-m backstroke performance and the methods used, (b) techniques for maintaining the competitive environment while undertaking biomechanical research, and (c) a proposed system for on-line recording and its use in the training program of swimmers.

ANALYSIS OF BACKSTROKE PERFORMANCE

The following kinematic parameters were measured during competitive backstroke swimming (see Table 1): (a) swimming performance [(SP) (Subject no.) (sec 10^{-2})], defined as the time between an electrical-acoustical starting signal and the finish touch of the swimmer; (b) motor arm reaction time [(ART) (Subject no.) (sec 10^{-2})], defined as the time between the starting signal and the release from the hand grip bar (range, 0.70–0.85 sec); (c) complex starting performance [(CSP) (Subject no.) (sec 10^{-2})], defined as the time between the starting signal and the moment that the swimmer reaches a point 5 m from the starting block (range, 2.28–2.36 sec); (d) pull phase (PP) of right (R)/left (L) arm (sec 10^{-2}), defined as the time between the changes—air-water, water-air—by the right/left hands (range, 0.80–1.30 sec); (e) recovery phase (RP) of the right (R)/left (L) arm (sec 10^{-2}), defined as the time between the changes—water-air, air-water—by the right/left hand (range, 0.34–0.44 sec); (f) time for the right (R)/left (L) arm-stroke (AS) (sec 10^{-2}), defined as the time between two successive changes—water-air—of the right/

left hand (range, 1.17—1.73 sec); (g) combined pull phase (CPP) of both arms, right (R)/left (L) arm remains in water (sec 10^{-2}), defined as the time between the change—air-water—of the right/left hand and the change—water-air—of the left/right hand (range, 0.15—0.45 sec); and (h) mean swimming speed ($\overline{V}$) (m per sec), defined as the quotient of swimming sections of different lengths and the corresponding swimming times (range, 1.30—1.54 m/sec).

The methods used for measuring most of these parameters were developed especially for this study. The techniques used for their measurement are described in the following sections.

Measuring Method for SP

A battery-powered buzzer was activated at the start and finish of the race. The signal voltage released a square-wave pulse that triggered a timer accurate to within 30 msec.

Method for Measuring CSP

The procedure of Maglischo and Maglischo (1968) was altered in the following manner. An alligator clip with a short-circuited phone jack was attached to the suit of the swimmer. One end of a 5-m control cable was affixed to the upper edge of the starting block. The circuit was broken at the swimmer's suit when he reached the 5-m distance, thus stopping the counter.

Measuring Method for Arm-Stroke Phases and ART

A signal voltage was transmitted by two two-conductor control cables from the head to the hands of the swimmer. Each cable ended in a copper sheet ($1.5\ \text{cm}^2$) on a plastic ring. The subjects wore one ring with an electrode on the palm, forefinger, and ring finger of each hand. The signal cables had to be led away from a water-polo cap to the pool side by means of a pole. By the immersion of each hand into the water, a circuit was closed and the signal voltage was triggered. Exceeding the trigger level limited the measuring quantities. In order to equalize a certain surface coating of electrolysis, the polarity was changed in each pair of electrodes by means of a four-pole switch.

Method of Measuring $\overline{V}$

Six reflection photoelectric cells (PEC) were aligned 30 cm above the water surface along the poolside at selected distances (Table 2), with beams parallel to the water surface and perpendicular to the pool side. A light metal sheet

Table 1. Summary of kinematic data for backstroke swimming

Subject 1

No.	SP (1) ART (1) CSP (1) AS (R)	Digital data (sec 10^{-2}) 7054 85 236	PP (R)	RP (R)	CPP (R)	CPP (L)	RP (L)	PP (L)	Digital data (sec 10^{-2})	AS (L)	No.
		206		38			36		271		
1	132	244	94		27	31		94	307	130	1
		338		38			37		401		
2	130	376	92		25	30		96	438	133	2
		468		38			38		534		
3	138	506	100		28	34		99	572	137	3
		606		35			36		671		
4	136	641	101		30	35		100	707	136	4
		742		36			34		807		
5	134	778	98		29	35		100	841	134	5
		876		36							

6	136	912	100		29					941	138	6
		1012		42		33	38			979		
										3700		
22	161	3618	119	41	40	38	41	119		3741	160	21
		3660								3860		
23	158	3779	117	42	40	37	40	116		3900	156	22
		3820								4016		
		3937			37	39	40	118		4056	158	23
24	158	3979	116	41						4174		
		4095			38	39	41	118		4215	159	24
25	159	4136	118	42						4333		
		4254			37	37	44	124		4377	168	25
26	160	4296	118	40						4491		
44		4414			37	40	43			4534	161	26
27	160	4454	120									

Subject 2

Left header block: SP (2) = 6983, ART (2) = 70, CSP (2) = 228

No.	AS (R)	Digital data (sec 10^{-2})	PP (R)	RP (R)	CPP (R)	CPP (L)	RP (L)	PP (L)	Digital data (sec 10^{-2})	AS (L)	No.
		236		42			37		300		
1	117	278							337	126	1
		353	75	37	22	16	41	79	416		
2	118	390							457	121	2
		471	81	41	26	14	43	80	537		
3	125	512							580	122	3
		596	84	38	25	16	43	79	659		
4	121	634							702	122	4
		717	83	39	25	15	45	79	781		
5	123	756							826	125	5
		841	84	36	25	15		80			

idx												
6										906		
	124	877	88		29	15	44			950	122	
		965										
		6045										
				41					6131			
37	173	6086	132		45	44	43		6174			170	36
		6218		40				127	6301			
38	167	6258	127		43	46	38		6339			163	37
		6385		40				125	6464			
39	168	6425	128		39	45	44		6508			168	38
		6553		40				124	6632			
40	169	6593	129		39	46	44		6676			175	39
		6722		41				131	6807			
41	169	6763	128		44	44	40		6847			165	40
		6891		44				125	6972			
		6935							7054			6

Table 2. Distribution of mean swimming speed ($\bar{V}$) for Subject 2

PEC no.	Distances (m)	Digital data (sec 10^{-2})	Time intervals (sec 10^{-2})	$\bar{V}$ (msec^{-1})
		0		
	5.00		214	2.34
1		214		
	10.00		650	1.54
2		864		
	10.00		658	1.52
3		1522		
	7.50		513	1.46
4		2035		
	7.50		519	1.44
5		2554		
	7.50		560	1.34
6		3114		
	5.00			
6				
	7.50			
5		3991		
	7.50		527	1.42
4		4518		
	7.50		555	1.35
3		5073		
	10.00		762	1.31
2		5835		
	10.00		763	1.31
1		6598		
	5.00		385	1.30
		6983		

(25.0 x 3.0 x 0.2 cm) pasted on both sides with reflection foil was affixed to the forehead side of the swimmer's water-polo cap ("helmet").

By passing the PEC, the produced signal voltages (dark-bright-contrast) were triggered in digital circuits for each PEC and then connected by on-gate circuits.

These methods were used to study two swimmers performing a 100-m backstroke race in a 50-m indoor swimming pool. The results are presented in Tables 1 and 2.

Description of Some Characteristic Parameters

It is possible that the phenomenon: ART (1) > ART (2) is caused by a change of the hand grip during the measured time, as can be seen from the analog recording strip. CSP (1) and CSP (2) show similar differences, as found in previous research by the author that was confined to starting only. Some selected groups of times of arm-stroke phases indicate, on the one hand, interindividual characteristics that are paralleled with a different swimming performance of both swimmers. On the other hand, intraindividual changes were found, as in Table 1. Of special interest is the increase of AS (1), whereas the quotient of PP/RP (1) does not change to the same degree on the first lap but only increases slightly on the second lap. The speed distance data (Table 2) show a ladder-shaped descent within the first lap. The swimmer (S 2) starts on the second lap with the mean speed of the first lap and finally reaches one level below that mean within the last 25 m. The swimmer (S 1) shows within the last 25 m a relative maximal speed, a result which was verified by preliminary studies.

RESEARCH DURING COMPETITION

The maintenance of the competitive situation during biomechanical research in swimming is dependent on the solution of two measurement problems: (a) a noncontact method that permits the swimmer freedom of movement, and (b) duplication of measurements. Although telemetry was not used, our initial attempt at solving the first problem met with satisfactory results. The influence of the measuring instruments on the swimming performance can be seen by comparing the experimental times with those recorded under competitive conditions. The test performances of both swimmers immediately after their swimming practice were only 1 sec slower than their last performance in practice and about 10 percent below their personal records swum at the German championships 2 wk later. Consequently, the test apparatus and procedures did not affect their performance appreciably.

The results indicated that one measurement system is adequate. This includes at least one starting signal, two starting instruments, four pairs of electrodes, two helmets, and six photoelectric cells. Plotting of the measured quantities is performed on a six-channel recorder.

ON-LINE RECORDING

In our preceding research, we were successful in printing in digital form the measured quantities by using an analog tape recorder running at slow speed. A universal counter is triggered by the starting signal. The pulses that follow cause the printer to print the available data from the BCD output of the counter. An interpretation of the data is possible by means of a simultaneously running six-channel analog recorder. Thus, a few minutes after the experiment, the data are ready for on-line information. The application of the system during the training process must await the availability of norms for the respective parameters.

REFERENCES

Maglischo, C. W., and E. Maglischo. 1968. Comparison of three racing starts used in competitive swimming. Res. Quart. 39: 604–609.

Miyashita, M. 1971. An analysis of fluctuations of swimming speed. *In:* First International Symposium on Biomechanics of Swimming. pp. 53–58. Université libre de Bruxelles, Brussels.

Rohrbach, C. 1967. Handbuch für elektrisches Messen mechanisher. Grössen, Düsseldorf.

Author's address: *Dr. Fredrich Krüger, Institut für Sport und Sportwissenschaften, Johann Wolfgang Goethe University, 6 Frankfurt-am-Main (West Germany)*

A kinematic analysis of world-class crawl stroke swimmers

J. H. Welch

New Mexico State University, Las Cruces

The purpose of this study was to describe and analyze, by kinematic techniques, selected movements of the trunk and upper extremity as displayed by
world-class crawl stroke swimmers engaged in competition. A principal task
was the ascertainment of instantaneous angular velocity and acceleration
characteristics of the underwater stroke. Joint action and stroke technique
were described relative to these kinematic parameters which are derivatives of
displacement data obtained through cinematographic techniques.

The specific problem was to determine, through objective measurement,
similarity or lack of similarity in the motion of the segments studied, as well
as to describe the technique used by the subjects.

The analysis took two primary approaches.

1. Analysis by segment. The analysis consisted of a comparison of each of
the four segments among all subjects. The segments, i.e., the trunk, brachium
(upper arm), antebrachium (forearm), and hand, were separated for examination according to breathing and nonbreathing sides.

2. Analysis by subject. This portion of the analysis involved a study of the
integrated actions of the limb segments within each subject, as displayed by
their velocity curves. These curves showed the sequence of action or the
pattern of motion used by the subjects individually.

METHODS

Subjects used in the study were 10 participants in the 1972 National Colle-
giate Athletic Association Swimming Championship Meet conducted at the
United States Military Academy, West Point, N.Y., March 23—25. All subjects
analyzed were world-class crawl-stroke swimmers. Seven of the subjects
(David Edgar, Steven Genter, Gary Hall, Jerry Heidenreich, John Kinsella,
John Murphy, and Mark Spitz) were members of the 1972 United States

Olympic Team. The remaining three (James McConica, Paul Tietze, and John Trembley) participated in Pan-American, World University, or other international competition. The subjects were photographed by the investigator during actual competition.

The motion pictures of the swimmers were taken for frame-by-frame analysis with a spring-driven, governor-controlled, 16-mm Kodak 100 camera with a 25-mm lens at a rate of 34 frames per sec. The photographic procedures met established criteria for films used in quantitative analysis (Dillman, 1970).

The photographs used in the kinematic analysis were taken from the profile view. Motion is "seen" by the camera only in a plane perpendicular to the line from the lens axis. Thus, the camera recorded subject motion as though the segments were projected onto a single plane—in this case, the propulsive plane. It was accepted that all upper-extremity motion that acts propulsively (i.e., contributed to propulsion) took place in this plane and was recorded on the film. It is in this plane that the predominant motions of crawl-stroke swimming occur.

The first step in measuring angular displacement involved making tracings of the limb segments. Specifically, the task, as outlined by Plagenhoef (1971), was to: (a) locate and record the positions of the joint centers of the segment involved (trunk, hip, and shoulder; brachium, shoulder, and elbow; antebrachium, elbow, and wrist: and hand, wrist, and long axis of the hand), (b) connect the joint centers by lines in order to form the component link system, and (c) measure the angle of the limb (segment) from the right horizontal axis to the nearest one-half degree by use of a drafting machine.

A displacement curve was obtained by plotting segmental angular change of position (in degrees) relative to the starting position against time. Computerized calculations of the instantaneous velocity and acceleration values were derived mathematically from the angular displacement curve. In order to calculate the angular velocity and acceleration values, it was necessary to smooth the raw displacement curve computed from drafting-machine measures of the joint motion plot. This was accomplished mathematically by use of the program that fitted a polynomial equation to the data by the least squares method of curve fitting. The object was to find the polynomial equation that would provide the "line of best fit." Polynomials ranging from 4 to 7° were applied to the raw angular displacement curves.

Angular displacement, velocity, and acceleration curves for all four segments on both the breathing and nonbreathing sides were analyzed for each subject independently. It was from this kinematic profile of each subject that the analysis was made.

Fundamental to the descriptive analysis of the data were the segment reference points around which the analysis focused. The point selected for each segment served as a common reference with which to compare the motions of the different subjects. The analysis of all segment motion began early in the stroke, when the hip as well as the upper extremity joints entered the field of view. Because this point of origin varied somewhat among subjects, it was desirable to find a means of aligning the stroke movements.

The reference marks for the brachium, antebrachium, and hand were points 45 and $90°$ from the horizontal that indicate the early and middle portions of the stroke. Angles recorded for use by the computer program were measured counter-clockwise from the right horizontal. The reference positions used in the discussion, however, simply were converted from the printout data to show where the segments had completed 45 and $90°$ of the underwater stroke as measured from the water surface.

The reference point for trunk analysis was chosen more arbitrarily. It was where the hip to shoulder angle was least, i.e., the point in the complete stroke cycle at which the shoulder was lowest in the water. The action occurred early in the underwater stroke and served as a convenient point to align the trunk motion of the subjects.

ANALYSIS–BY–SEGMENT FINDINGS

The Trunk

Seventeen of the 20 segments analyzed showed downward rotation of the trunk during the initial part of the underwater stroke as the hand and arm were directed forward and downward into the water. All subjects displayed maximal displacement as the succeeding segment, the brachium, completed the underwater stroke, i.e., once the trunk started to rotate upward, it continued in that direction for the remainder of the underwater stroke. Eighteen of 20 segments showed maximal angular velocity during the upward motion of the segment. Maximal angular velocities attained by the trunk were of a low magnitude, inasmuch as the highest trunk velocity at any point was only 3 rad per sec.

The Brachium

Of all segments studied, subjects showed the greatest similarity in the angular displacement patterns of this segment.

Seventeen of the 20 brachii reached maximal angular velocity at or near the $90°$ mark. Although subjects showed a consistent pattern in reaching maximal velocity at the $90°$ mark, they displayed divergent slopes prior to and following their respective peaks. All of the subjects reached their maximal angular acceleration prior to the $90°$ mark. Nineteen of the 20 segments analyzed showed a distinct period of deceleration at some point after the brachium passed that point. In the single exception (Heidenreich), the angular acceleration curve approached zero after the $90°$ point.

The Antebrachium

Of the 10 antebrachii analyzed on the breathing and nonbreathing sides, 18 of 20 showed a nearly linear angular displacement slope prior to the 45° reference mark.

Of the total of 20 segments analyzed, 18 showed a minor-major velocity peak sequence. The initial or minor peak occurred prior to the 45° mark, and the major angular velocity peak occurred after the same reference point. Logically, it was found that 18 of the 20 segments studied showed an acceleration-deceleration-acceleration sequence pattern comprising part or all of their respective angular acceleration curves. Eighteen of the 20 segments showed the deceleration phase to occur at the time when the segment passed the 45° mark.

The Hand

In observation of both sides (breathing and nonbreathing), 18 of the 20 segments analyzed showed a basically linear angular displacement pattern prior to and immediately following the 45° mark.

ANALYSIS–BY–SUBJECT FINDINGS

Brachium-Antebrachium Interaction

Relative to the antebrachium, the minor velocity peak evidenced by all subjects prior to the 45° mark showed that the antebrachium was moving at a higher velocity than the brachium, thereby positioning the limb to avoid a "pressing" action in the stroke. (This action is one in which force applied to water with the palm and anterior aspect of the antebrachium is directed downward, i.e., toward the bottom of the pool.) For some subjects, the velocity of the antebrachium was appreciably greater than that of the brachium; others showed a small velocity difference and for only a brief period. However, once the elbow was flexed, pressing action early in the stroke was avoided as long as the antebrachium moved at a rate equal to or faster than the brachium. This was the case for all subjects.

Antebrachium-Hand Interaction

Maximal angular velocity of the hand in 14 of 20 segments was reached prior to or immediately following the 45° reference position. This observation, along with that of the prevalent pattern of the antebrachium (18 of 20 segments) of showing a peak velocity during the same phase, suggested that the subjects, during the first 45°, moved the antebrachium and hand into a propulsive position, in which the palm and anterior antebrachium were directed backward.

Brachium-Antebrachium-Hand

The brachium in 19 of the 20 segments analyzed reached the 90° mark before the antebrachium or hand.

As discussed above, while the antebrachium and hand moved at a relatively high rate during the first 45°, these segments did not precede the brachium during the second 45° portion of the arc. When the brachium reached maximal velocity, the velocity of the antebrachium in 17 of 20 cases was approximately zero. This indicates that, once the brachium began to move, the musculature of the shoulder joint was the prime mover of the hand and arm.

Brachium-Antebrachium

In all subjects, when the brachium accelerated to maximal velocity, the antebrachium decelerated; as the brachium decelerated, the antebrachium accelerated. This phenomenon has been termed by the investigator the "mirror image" effect, referring to the velocity and acceleration curves of adjacent segments concurrently in opposite directions.

Antebrachium-Hand Interaction

The hand showed a pattern (19 of 20 segments) of early acceleration followed by deceleration. Once the hand reached low or zero velocity, it was a unit of the antebrachium. Upon reaching the vertical position, in 15 of 20 cases the hand was decelerating as the succeeding segment, the antebrachium, accelerated. Rather than attributing this to the lag associated with acceleration of the succeeding segment, it was interpreted that the hand had reached the advantageous propulsive position and that further wrist flexion was not consistent with sound mechanics of the stroke. In some cases, the wrist was extended to maintain the palm in the vertical or propulsive position.

SUMMARY

The following conclusions were made on the basis of the analysis.
1. Results based on analysis by segment. (a) Although displacement, velocity, and acceleration curves showed common patterns of motion, subjects produced individual differences relative to the magnitude and duration of the motion. (b) There were no major differences between patterns of motion on the breathing and nonbreathing sides. (c) Of all segments studied, subjects

showed the greatest similarity in the displacement pattern of the brachium. On the basis of the high performance level of these subjects, one may conclude that the action of the shoulder joint may be the one most related to effective propulsion, i.e., success in crawl-stroke swimming.

2. Results based on analysis by subject. (a) The sequence of joint action enabled the hand and forearm to reach their maximal linear velocity during the latter portion of the underwater stroke. (b) On the basis of the kinematic analysis, the musculature of the shoulder joint appeared to be the prime mover of the entire upper extremity during the propulsive phase of the underwater stroke. (c) Elbow and wrist action during the acceleration of the brachium serve to maintain the anterior antebrachium and palm in a propulsive position. (d) As the brachium decelerated after the $90°$ mark, the antebrachium accelerated to its maximal velocity. (e) The angular motion of the hand decelerated as it passed the $45°$ mark and approached the vertical position. When the antebrachium accelerated to its maximal angular velocity peak, the apparent purpose of wrist movement was to keep the hand in the vertical or propulsive position.

REFERENCES

Dillman, C. J. 1970. Photographic instrumentation. Paper read at the Eastern District Association of Health. Physical Education and Recreation Convention, New York, N.Y. March 27.

Plagenhoef, S. 1971. Patterns of Human Motion. Prentice-Hall, Englewood Cliffs, N.J.

Author's address: *Dr. John H. Welch, Department of Physical Education, New Mexico State University, Box 3M, Las Cruces, New Mexico 88003 (USA)*

A comparative analysis of the take-off employed in springboard dives from the forward and reverse groups

D. I. Miller
University of Washington, Seattle

The components of springboard dives from either the forward or reverse groups include the approach steps, hurdle, take-off, flight, and entry into the water. Because the initial conditions of the flight, namely, the angle of projection, velocity of the mass center, and angular momentum, are established during the take-off, this phase plays a major role in determining the success of the dive. For convenience of analysis, the springboard take-off, which was the focus of the present investigation, can be divided into a press and a lift.

The press occurs between the simultaneous two-foot contact and the maximal depression of the springboard. The lift, which follows, coincides with the recoil and continues until the instant that the diver's feet leave the board.

Although there are readily observable differences between the position of the diver at final board contact when he initiates a forward 2-1/2 as opposed to a reverse 2-1/2 somersault, the question of where in the take-off these variations begin to make their appearance has not been answered completely. One prominent coach (Armbruster, Allen, and Billingsley, 1973) contends that they can be traced back to the landing from the hurdle. Most writers, however, deal with the take-off in general terms and postpone any discussion of differences among dives until the completion of the lift. With such considerations in mind, the objectives of this study were threefold: to compare the springboard take-offs of selected dives from the forward and

This study was supported in part by a research grant from the University of Saskatchewan, Saskatoon, Canada.

reverse groups; to contribute descriptive data to the existing but rather limited body of research information on this skill (Bergmaier, Wettstein, and Wartenweiler, 1971; Darda, 1972); and to serve as one of the pilot projects preceding the filming of springboard diving at the 1976 Montreal Olympics in conjunction with the proposed Biomechanical Archives of Olympic Performance.

PROCEDURE

A 16-mm Locam camera operating at 62 frames per sec was employed to film the take-offs of four skilled divers, two male and two female (Table 1). The subjects performed forward and reverse dives and forward 2-1/2 somersaults from a 3-m Duraflex springboard. In addition, the women performed reverse 1-1/2 somersaults, while the men performed reverse 2-1/2 somersaults. Three trials of each dive were filmed and subsequently analyzed on a Vanguard motion analyzer.

RESULTS AND DISCUSSION

Examination of body segment angles as functions of time immediately preceding board contact and through the press and lift revealed a basic pattern common to all take-offs and certain differences characterizing not only the forward and reverse groups but also specific dives within these groups. As the diver descended from the hurdle, he prepared for impact with the board by dorsiflexing his ankles, flexing his knee and hip joints, and inclining his trunk slightly forward from the vertical. Leaving the stretched position above the head, the arms began to circle down behind the hips.

Table 1. Subject information

Subject no.	Age (yr)	Sex	Height (cm)	Weight (kg)	1972 Canadian Intercollegiate	1972 Olympics	Dives performed*
1	20	Female	168	57	First	Canadian team	101A, 105B 301A, 303A
2	20	Female	161	59	Second		101A, 105B 301B, 303B
3	23	Male	178	74	First	Canadian team	101A, 105B 301B, 305C
4	22	Male	175	69	Second		101A, 105B 301B, 305C

*These numbers correspond with the F.I.N.A. designation in which the first digits (1 and 3) refer to the forward and reverse groups, respectively, and the third digit indicates the number of half-somersaults performed. A, B, and C specify layout (straight), pike, and tuck positions.

Although a two-way analysis of variance revealed variations among the divers at board contact, only in the cases of a limited number of trunk and shoulder angles were there any statistically significant differences ($p \leqslant 0.05$) among the dives. Inasmuch as no consistent pattern was observed from one subject to another, these differences were attributed to a lack of proficiency in performance rather than to the requirements of specific dives.

Each diver initially contacted the springboard with the balls of his feet and with the line of gravity falling behind the metatarsophalangeal joint. Following the landing, the ankles continued to dorsiflex until the soles of the feet were in complete contact with the board. The knee and hip joint also continued to flex, but only for a period of 0.05–0.10 sec, which was sufficient to absorb the force of impact. They then began a period of rapid extension to assist in driving the springboard downward. By the time the board had reached its lowest point, the arms had been circled upward beyond shoulder level, the largest portion of the hip extension had been completed, and the line of gravity was in front of the metatarsals.

Analysis of the body segments at the point of maximal springboard depression (Table 2) indicated that the statistically significant differences among the dives were confined largely to the ankle and trunk angles. In all cases, the least dorsiflexion of the ankle occurred in the multiple reverse somersault dives and corresponded with the smallest deviation of the trunk from the vertical. As dorsiflexion increased through the reverse dive, forward dive, and forward 2-1/2 somersault, there was a related increase in the forward flexion of the trunk. While the differences between the hip angles of Subject 1 in the reverse 1-1/2 somersault layout and Subject 2 in the forward 2-1/2 somersault pike and in their other dives were statistically significant ($p \leqslant 0.05$), these variations were not characteristic of all the divers and were attributed to the fact that the two dives noted were the most difficult for these subjects to perform.

During the lift phase of the take-off, the variations among the dives became more pronounced. In forward-rotating dives, the knees continued to extend. After a brief period of extension initiated during the press of the forward 2-1/2 somersault, the hips began to flex. The amount of trunk forward flexion was related positively to the number of somersaults being attempted. Dives from the reverse group were characterized by continued extension of the hip joint. The trunk approached and, in all reverse multiple-somersault dives, passed the vertical. During the latter portion of the board recoil, the knees began to flex. In all dives, the ankles plantarflexed strongly during the final section of the lift.

As already stated, the line of gravity was in front of the metatarsophalangeal joint prior to maximal depression of the springboard. Inasmuch as this relative orientation was maintained for the remainder of the take-off in all dives, this indicated that the horizontal component of board reaction force during the lift always was directed toward the end of the board in a way that

Table 2. Selected body segment angles at critical points during the springboard take-off*

Angle	Dive	Subject 1			Subject 2			Subject 3			Subject 4		
		TD	MD	LC	TD	MD	LC	TD	MD	LC	TD	MD	LC
Knee	105B	122	147	187	122	140	186	124	141	193	123	143	178
	101A	133	142	188	126	141	186	121	146	195	125	151	180
	301	128	143	170	129	149	159	123	150	176	130	147	170
	303/305	127	150	121	127	152	128	127	153	136	134	152	123
Hip	105B	119	154	115	118	137	95	114	153	130	130	156	119
	101A	133	152	164	117	151	168	107	157	181	131	166	159
	301	127	154	201	115	155	187	116	156	185	139	162	197
	303/305	118	165	205	120	159	193	108	161	190	127	162	197
Trunk with vertical	105B	27	19	75	29	33	101	28	15	61	18	17	70
	101A	17	13	28	28	14	30	32	12	20	15	6	28
	301	17	8	-10	30	13	-5	24	12	3	8	6	-8
	303/305	23	3	-42	26	12	-24	31	8	-25	20	7	-34

*Average angles in degrees calculated over three trials at touchdown (TD), maximal springboard depression (MD), and last contact (LC).

Table 3. Duration of the springboard take-off (sec)

Dive	1	2	3	4
Forward 2-½ (105)	0.39	0.38	0.44	0.43
Forward dive (101)	0.43	0.41	0.44	0.45
Reverse dive (301)	0.42	0.42	0.46	0.46
Reverse 1-½/2-½ (303/5)	0.38	0.38	0.43	0.43

(Subject columns: 1, 2, 3, 4)

would cause reverse rotation. The vertical component on the other hand, acted upward behind the mass center in a direction that would produce forward rotation of the body. The ultimate direction of the rotation, therefore, depended on the relation of the resultant forward and upward springboard force to the position of the diver's mass center. Since the distance of the line of gravity in front of the base of support at last contact was related directly to the amount of forward rotation (2, 11, 20, and 31 cm in Subject 3 for dives 305C, 301B, 101A, and 105B, respectively), the resultant passed behind the mass center in dives for the forward group and in front in dives for the reverse group.

The duration of the take-off varied from 0.38 to 0.46 sec, with the board contact of the male divers being longer than that of the females (Table 3). There appeared to be a tendency for the divers to "rush" the take-offs of dives that were more difficult for them to perform. This was particularly evident in the forward 2-1/2 and reverse 1-1/2 somersaults of the women and in the reverse 2-1/2 somersaults of the men.

CONCLUSION

The analysis of position-time data of the springboard take-off indicated that variations in body orientation related to a particular dive had made their appearance by the time the springboard was maximally depressed. Because of the limited number of subjects, however, and their level of performance in the optional dives, this conclusion cannot be stated unequivocally for highly skilled divers in general without further supporting evidence. Examination of the films from the proposed Biomechanical Olympic Archives should shed further light on this question.

REFERENCES

Armbruster, D. A., R. H. Allen, and H. S. Billingsley. 1973. Swimming and Diving. 6th Ed. C. V. Mosby, St. Louis.

Bergmaier, G., A. Wettstein, and J. Wartenweiler. 1971. Diving measurement and analysis of the take-off. *In:* L. Lewillie and J. P. Clarys (eds.), First International Symposium on Biomechanics of Swimming. pp. 243–247. Université libre de Bruxelles, Brussels.

Darda, G. E. 1972. A method of determining the relative contributions of the diver and springboard to the vertical ascent of the forward three and one-half somersaults tuck. Unpublished Ph.D. dissertation, University of Wiscomsin, Madison, Wis.

Author's address: *Dr. Doris I. Miller, School of Physical and Health Education, University of Washington, Seattle, Washington 98105 (USA)*

Skiing
and
skating

A telemetry system for the measurement, transmission, and registration of biomechanical and physiological data, applied to skiing

P. A. Neukomm and B. Nigg
Swiss Federal Institute of Technology, Zürich

Telemetric equipment enables the biomechanical researcher to avoid using cables between the subject and registration devices. Very often, radio transmitting equipment with the following specifications is required: (a) small and lightweight devices on the subject, but with a strong transmitter with a long battery life; (b) many data channels for the transmitting of quick signals, but with high accuracy (amplitude, phase, linearity) of the experimental procedure. Even if there are no financial limitations, all of these requirements cannot be satisfied; therefore, it is the duty of the engineer to find a good compromise.

TELEMETRIC SYSTEM

Our basic idea was to develop a telemetric system that could be handled easily by any researcher without specialized knowledge of electronics and to which every type of transducer could be connected.

Our telemetry system consists of two independent, wireless transmitter-receiver subsystems, the Data-Telemetric-System and the Radio-Control-System (see Figure 1).

The Data-Telemetric-System (see Figure 2) consists of: (a) seven different or equal transducers (1) (accelerometer, goniometer, positionmeter, electrocardiogram, electroencephalogram, or electromyogram pick-up); (b) as many as seven different or equal preamplifiers for the detection, amplification, and calibration of the transducer signals, (2) (plug-in models); (c) seven-channel transmitter (3) with the following specifications: transmitting power, 0.5 W (equal radius of action over 2 km); transmitting frequency, 231.5 MHz (equal

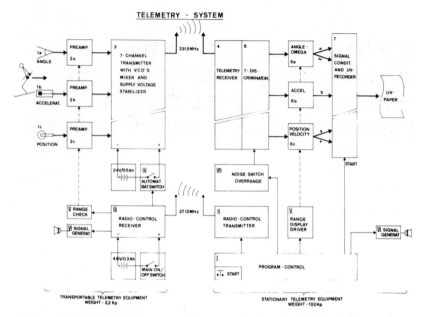

Figure 1. Schematic telemetry system (the numbers in the diagram are explained in the text). *PREAMP.*, preamplifier; *AUTOMAT. BAT.*, automatic battery; *GENERAT.*, generator; *CONDIT.*, conditioner; *ACCEL.*, acceleration display

length of antenna only, 32 cm); data channels IRIG 11 through IRIG 17 (FM-FM system with channel band width 0–110 Hz up to 0–790 Hz with a linearity better than 1 percent). It is made by CONTRAVES, Switzerland; (d) FM receiver (*4*) (Defense Electronics GPR-20); (e) seven discriminators (*5*) (Defense Electronics SCD-11); (f) final data processors (*6*), such as the Angle-Omega device, acceleration display, position, and velocity meter, and other equipment, which also can be used without telemetry by direct transducer connections; (g) signal conditioner (*7*) with ultraviolet recorder.

The Radio-Control-System consists of: (a) Program-control (*I*). This is the heart of the whole system. It drives the starter, checks the data transmissions, and controls the calibration and recording of the measurement signals. (b) Radio-control transmitter (*II*). This transmitter has a power of 0.5 W, works on 27.12 MHz, and has a radius of action over 1 km. (c) Radio-control receiver (*III*). This type of receiver is very small and normally is used for RC-model airplanes. (d) Automatic battery switch (*IV*). The data transmission runs only in the moment of data recording. During 40 hr, the data transmission can be switched on by the program-control, so the total transmitting time of about 1 hr can be used to optimal advantage. (e) Range-check (*V*). Each preamplifier has six ranges; and, by a PDM transmitting over the normal data channels, the selected ranges will be indicated on the panels of the data processors. (f) Signal-generator (*VI*). The start of a data transmission run is indicated by a whistling signal. (g) Noise switch and overrange display

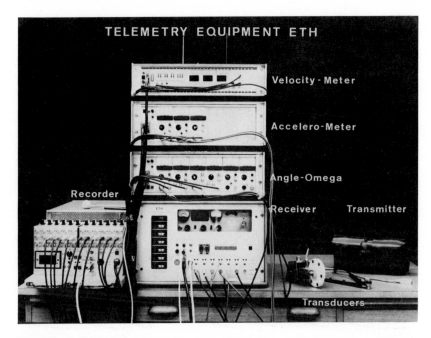

Figure 2. A photograph of the telemetry equipment in a ready-for use state. *Right,* the transmitter and the transducers. The position-transducer (*white wheel with spikes*) will be mounted on the ski and rolls over the snow surface. This transducer gives us, by on-line calculation, the momentary position and velocity of the skier racing downhill with an accuracy of about 2 percent.

(*VII*). Because the noise of the receiver (which occurs when no transmitter is on the chosen band) may damage the sensitive galvanometers of the recorder, it is necessary to interrupt the line to the data processors. In addition, each overrange in the data channel will be indicated in order to avoid incorrect measurements.

This system is constructed in such a way that even rough conditions (temperature, dust, vibrations, etc.) cannot disturb a good and fully automatic registration of physiological data. Because of the exchangeable preamplifier, any kind of transducer may be connected without recalibration. The weight of all of the transportable equipment, which is carried by the examination subject, amounts to 2.2 kg without transducers. Because the equipment is carried on the hip, its effect on the subject is negligible for many types of sport activities.

APPLICATION

This telemetric system was used in the first measurements concerning influences of force on the human body in skiing. The external influences were measured with accelerometers affixed to the head (vertex), the hip (ilium),

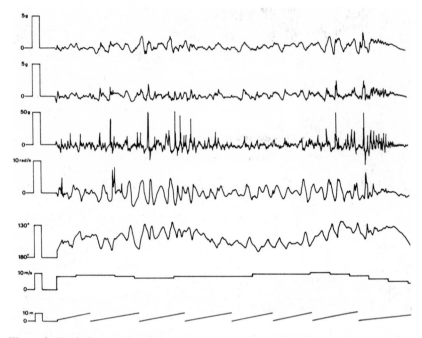

Figure 3. Typical example of the measurement for a skier skiing downhill. *Top to bottom rows:* acceleration, head; acceleration, hip; acceleration, shinbone; angular velocity, knee; knee angle; ski velocity; ski position.

and the shinbone (tibia). The knee angle and the ski position were measured at the same time. These measured data were transmitted by the described telemetry system (see Figures 3 and 4).

The first measurements were made on a slope with a very small incline (skiing slope for beginners) on a test ski run.

The acceleration at the shinbone increases quadratically with the increase of the ski velocity, whereas the acceleration at the hip and the head increases linearly. This process seems to be a natural protective function of the human body. The damping increases with the increase of the external influences.

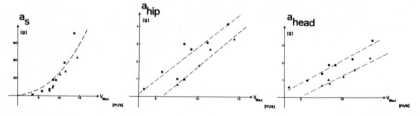

Figure 4. Relationship between the ski velocity and the acceleration at the shinbone, the hip, and the head. *m/s,* m per sec.

Table 1. Order of magnitude of the different accelerations in skiing

	Velocity (km/h)	Order of magnitude of the acceleration		
		shinbone (g)	hip (g)	head (g)
Deep snow	30	4–6	0.5–1.5	0.5–1
Prepared ski run, powder snow	30	30–50	1–3	0.5–2
Prepared ski run, hard snow	30	30–60	1–3	0.5–2

These results, given in Table 1, clearly show the difference between skiing on a prepared ski run and in deep snow. On a prepared ski run, the external influences are about 10 times greater than in the deep snow, even with the same skiing velocity.

A comparison between these results and results obtained by similar measurements in (normal) running shows that the measured accelerations at the shinbone are about seven times larger in skiing than in running in a gymnasium, but the results at the hip are approximately equal in both cases. These results allow the interpretation that running in a gymnasium and skiing downhill have nearly the same influence on the upper part of the body and the spine. The greater external forces in skiing, therefore, must be absorbed by the joints, the tendons, and the muscles of the legs. This means that the legs suffer the largest stress in skiing downhill on a prepared ski run.

REFERENCES

Neukomm, P. A. 1973. Some modern methods and equipment for the measurement of angles and position in biomechanics research. Paper presented at the International Symposium: The Problems of Motion-Biology in Track and Field, Budapest.

Nigg, B. and P. A. Neukomm. 1973. Erschütterungen beim Skifahren, Tagung der Schweizerischen Arbeitsgemeinschaft für Biomed. Technik, St. Gallen.

Author's address: Mr. Peter Neukomm, Laboratory of Biomechanics, Swiss Federal Institute of Technology, Zurich (Switzerland)

Biomechanics of the human leg in alpine skiing

E. Asang

Technische Universität, Munich

Ten years ago, the continually increasing number of injured skiers inspired us at Munich to begin to consider a special program for the prevention of injuries in alpine skiing. All commonly used measures for prevention of injuries resulting from a dangerous fall by the skier were apparently insufficient. Also, release bindings provided only limited protection for the leg, because in 63 percent of the cases of injured skiers, the binding did not release. No definite information was available concerning how to adjust the release bindings for maximal safety. In addition, the mechanical and dynamic characteristics for safe release bindings were not known. These two problems are also currently of great interest in the United States, which has an increasing number of alpine skiers. A subcommittee on ski safety from The American Society for Testing and Materials began extensive work on this subject last year.

In the traumatology of alpine skiing, the tibia is the most frequent site for the most severe of the typical skiing injuries; it seems to be the weakest link in the lower extremity. Therefore, biomechanical research on the injury threshold of this body segment was begun. Initially, we had to determine the relationships of individual dimensions, sex, and age to the biomechanical characteristics of this bone.

PROCEDURES

After 20 years (1950–1970) of extensive statistical research on skiing injuries by the author, the first experimental studies of the research team consisted of

This research was supported by the Deutsche Forschungsgemeinschaft (German Science Foundation) under the program "Eigenschaften des menschlichen Beins" (properties of the human leg). In addition, financial support was provided by Internationaler Arbeits-

mechanical measurements of the individual limits of loading in bending (Posch, 1970) and twisting (Engelbrecht, 1970) with nondenatured shinbones. Dynamic impact resistance as well as hardness were determined on specimens of the same bones. Spectroscopical analyses also were included in these studies (Ter Welp, 1970). In the meantime, these results were confirmed by an electronic measuring system using more than 500 shinbones. With the assistance of the E.D.P. system, knowledge of the individual characteristics of the human tibia was expanded. Measurements of static and dynamic muscle forces provided important results to be compared with information on the loading limits of bones.

FINDINGS

Basically, there was no difference between the properties and behavior of the human tibia on the basis of sex. However, there were differences between men and women in the frequency of some types of skiing injuries.

Strength properties of the bone at advanced ages (over 70) were approximately half those at the age of 20. During youth (younger than age 17), strength limits also were nearly half those of adults. Obviously, this difference is due to the smaller dimensions of the bones of children. From ages 17 to 18, the values increased to those of adults (Lange, 1973). Between ages 20 and 60, strength properties of the tibia remain nearly constant. An increase of the shape-related moment of inertia compensates for the decreasing material strength of the aging bone substance. In advanced age, an immense dispersion of the injury limits is a result of the differences between well-trained and untrained people as a consequence of the different degrees of participation in activity (Guttenberger, 1973).

The analysis of a diagram of typical slow twisting (Figure 1) demonstrates the elastic, plastic, and fracture behavior of the human tibia. The elastic behavior was found to be surprisingly constant; elastic angles of twist were approximately 12°. This generalization is most important because the elastic limit of the tibia is the lower limit for injury of the leg. During the period of plastic deformation, the structure of the bone is being altered. As evidence of this fact, the fracture limit was reduced considerably by fatigue after repeated loadings higher than the elastic limit.

The ratio between twisting and bending strength is about 1 to 3. The forward (ventral) bending load test diagrams (Figure 2) show the dependence of the elastic threshold and the ultimate loading limit at the fracture level on the cross-section of the tibial shaft (Watzinger, 1973). The calculation of the

kreis Sicherheit beim Skilauf—IAS (International Association for Safety in Skiing) and by Bayerisches Staatsministerium fur Arbeit und Sozialordnung (Bavarian State Ministry for Labor and Social Affairs).

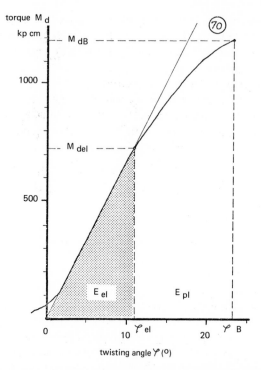

Figure 1. Analysis of a diagram of typical twisting of the human tibia. M_{dB} = fracture torque; M_{del} = elasticity torque; E_{el} = elastic deformation energy; E_{pl} = plastic deformation energy; $E_{el} + E_{pl}$ = total fracture energy; kp, kilopond.

moments of inertia and the section moduli serve as evidence of this fact. These calculations were made possible with the utilization of E.D.P. (Wittmann, 1973). It is impossible, however, to measure the cross-section of the tibial shaft on living persons. Therefore, the frontal diameter at the tibial head was chosen as a measure, in vivo. The comparison of the diameter versus elastic and fracture limits and the moments of inertia versus the same loading limits verifies similar statistical relationships.

In backward (dorsal) bending motion, the fracture limit was 15 percent less and the elastic threshold 10 percent less than in forward bending. This result is caused by the external cross-section of the shaft.

In the skier's fall, the most typical and most dangerous situation is a combined bending and twisting of the leg. Therefore, we imitated this type of accident by experimental twisting together with high bending loads (Höpp, 1973). The combined loading diagrams do not differ from those of purely torsional loading. However, the ultimate fracture limits and elastic thresholds decreased about 12–15 percent as the result of the additional bending loads. Consequently, we chose the elastic threshold of combined loading as the injury threshold in twisting of the leg (Figure 3).

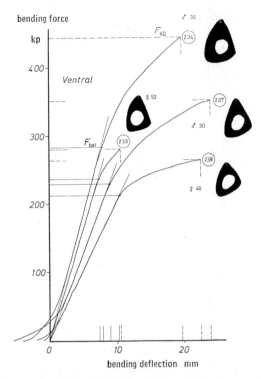

Figure 2. Typical forward (ventral) loading diagrams of the human tibia with corresponding cross-sections at the fracture levels. F_{bB} = bending fracture force; F_{bel} = bending elasticity force.

Bending and twisting limits shown in Figure 3 are related statistically to 80 percent of the total. The in vivo measurement of the diameter at the tibia head results in a greater dimension than that of the bare bone because of the cutaneous and subcutaneous tissue. It differs in men by about 0.5–1 cm, and in women by 1.0–1.5 cm. By means of this measurement at the tibial head, an individual prognosis of the personal injury limits of the leg is provided.

The protective function of the release bindings is most important to prevent typical skiing injuries of the leg (upper limit of release adjustment). However, skiing is possible only if the binding consistently holds the skier's boot in all normal situations, uphill and downhill. This holding force is the lower limit of the release adjustment. An inadvertent release of this coupling precipitates the danger of an atypical ski injury. Such injuries eventually may be more serious (i.e., fractures of the skull or of the vertebral column) than most typical ski injuries (leg injuries). Therefore, because of the design requirements of release bindings, we had to research active static and dynamic muscle forces in steering the ski and muscle reactions to disturbing impacts

torsional and bending moments

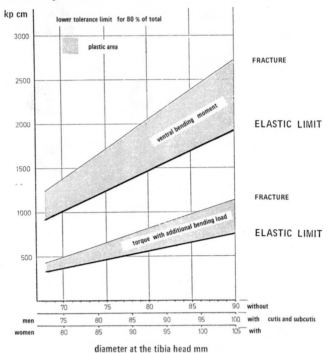

diameter at the tibia head mm

Figure 3. Elastic threshold limits of the shinbone are injury thresholds of the human leg in bending and twisting (combined loading); individual prognosis by measurement at the tibia head in vivo.

produced by the ground in downhill skiing, as well as to measure these impacts (Grimm, 1973; Krexa, 1973).

Biomechanical examinations of muscle forces of the legs of children, adolescents, and adults were highly interesting (Scherm, 1973; Schipek, 1973; Schwesinger, 1970). Simultaneously recorded electromyograms were used as a measurement of muscle actions and reactions. Static muscle forces in general corresponded to the statistically lower tolerance limit (80 percent of total) of the elastic threshold of one's tibia calculated by the measured diameter at the tibial head. After these results are known, this relationship becomes more feasible if one considers the human leg as a single functional unit. It is possible to make an indefinite number of these measurements on live persons. They provide a verification of the results obtained from the measurements made on the tibias of over 500 cadavers.

Surprisingly, the dynamic muscle forces (at the heel) were found to be above the calculated injury thresholds of the bones (in slow bending load). Forces ranging from 0.07 to 0.1 sec were up to 64 percent higher than the static muscle forces. Muscle forces exerted for less than 0.05 sec usually exceeded the calculated slow loading limit of the bone. The shortest dynamic muscle forces measured were 0.01 sec in duration.

The behavior of the entire shinbone also was examined on impacts (Kuhlicke, 1973). Impacts of 0.1-sec duration generally revealed an increase of about 5 percent in the fracture limits and in the elastic thresholds, in comparison with the slow loading. With impacts of 0.01-sec duration, the fracture limits were almost 20 percent higher than in slow bending.

In addition to the impact behavior of the bare bone, the shock absorption capacity of the human leg as a functional unit finally was tested and electronically recorded. These tests were conducted under conditions at different impact loadings, at various speeds, with tensed or relaxed muscles, and with different joint positions. With correctly adjusted release bindings, these experiments were not dangerous to a man who was being used as a "human dummy."

These experiments in the laboratory gave us the experience necessary for field tests and measurements in alpine skiing (Wittmann, 1973). All of this biomechanical work has been conducted with the main purpose of obtaining better protection for the millions of skiers throughout the world by means of a useful injury prevention program.

REFERENCES

Asang, E. 1970. 20 Jahre Skitraumatologie und Grundlagenforschung zum Verletzungsschutz im alpinen Skisport. Habilitationsschrift, Technische Universität, München.

Asang, E. 1972. Verletzungsschutz beim Skisport. Sport. Sportmed. 8: 209–219.

Asang, E. 1973. Experimentelle und praktische Biomechanik des menschlichen Beins. Med. Sport 8: 245–255.

Engelbrecht, R. 1970. Experimentelle Untersuchungen zur Torsionsfestigkeit des menschlichen Schienbeins. Inaugural dissertation, Technische Universität, München.

Grimm, C. M. 1973. Ergebnisse telemetrischer Messungen von mechanischen und elektromyographischen Signalen beim alpinen Skilauf und ihre Auswertung. Inaugural dissertation, Technische Universität, München.

Guttenberger, R. 1973. Veränderungen der Elastizität und Bruchfestigkeit des menschlichen Schienbeins im fortgeschrittenen Lebensalter. Inaugural dissertation, Technische Universität, München.

Höpp, H. 1973. Elastizität und Bruchfestigkeit der menschlichen Tibia bei kombinierten Biege- und Drehbelastungen". Inaugural dissertation, Technische Universität, München.

Krexa, H. 1973. Grundlagen zur telemetrischen Messung von elektromyographischen und mechanischen Signalen beim alpinen Skilauf. Inaugural dissertation, Technische Universität, München.

Kuhlicke, V. 1973. Elastizität und Bruchfestigkeit der menschlichen Tibia bei dynamischen Biege belastungen. Inaugural dissertation, Technische Universität, München.

Lange, J. 1973. Elastizität und Bruchfestigkeit der Tibia von Kindern und Jugendlichen. Inaugural dissertation, Technische Universität, München.

Posch, P. 1970. Experimentelle Untersuchungen zur Biegebruchfestigkeit des menschlichen Schienbeins. Inaugural dissertation, Technische Universität, München.

Scherm, G. 1973. Vergleichende Untersuchungen statischer sowie dynamischer Drehkräfte des menschlichen Beins im Mechanogramm und Elektromyogramm. Inaugural dissertation, Technische Universität, München.

Schipek, E. 1973. Vergleichende Untersuchungen statischer sowie dynamischer Zugkräfte des menschlichen Beins im Mechanogramm und Elektromyogramm. Inaugural dissertation, Technische Universität, München.

Schwesinger, G. 1970. Untersuchungen zur Zug- und Drehmuskelkraft des menschlichen Beins. Inaugural dissertation, Technische Universität, München.

Ter Welp, P. 1970. Experimentelle Untersuchungen zur Schlagfestigkeit und Härte des menschlichen Schienbeins. Inaugural dissertation, Technische Universität, München.

Watzinger, P. 1973. Elastizität und Bruchfestigkeit der menschlichen Tibia bei Biegebelastungen in ventraler und dorsaler Richtung. Inaugural dissertation, Technische Universität, München.

Wittmann, G. 1973. Biomechanische Untersuchungen zum Verletzungsschutz im alpinen Skisport. Inaugural dissertation, Technische Universität, München.

Author's address: *Dr. Ernst Asang, Belgradstrasse 5, 8 Munchen 40 (West Germany)*

Biomechanical research on release bindings in alpine skiing

G. Wittmann

Technischer Überwachungsverein Bavaria, Munich

There is little doubt that active participation in sport is of great importance for the maintenance of human health and recreation. However, activity in sport often includes a high risk of injury. In any kind of sport, there are typical injuries, especially in alpine skiing, in which an abundance of injuries occur (Asang, 1970).

When the skier loses his ability to control the active and passive forces between ski and ground, he falls. Typical injuries occur when the long lever of the ski generates external loads on the leg in such a way that specific biological tolerances are exceeded. The frequency of dangerous falls may be reduced essentially by physical training and improvement in skiing performance. A basic requirement for prevention of injuries, however, must be the interruption of the force transmission between the ski and the leg before it reaches a dangerous level. This is only one of the functions of a release binding.

The second basic function of the release binding is the transmission of all forces necessary for steering and all disturbing forces not dangerous for the leg.

The release binding is efficient in injury prevention only if it presents certain technical properties for a reliable release and if it is adjusted to the correct release value for the individual skier. To determine this individual release value, it is necessary to know the tolerance of the leg to withstand forces—especially of the tibia—and the forces generated during skiing.

Only extended biomechanical research can answer the questions of human tolerances and the forces imposed during ski runs. Data from such research will provide the basis for recommendations in order to establish: (a) correct settings of the binding for individual skiers and (b) proper construction of bindings. Furthermore, the results of such research can be applied in the establishment of standards and a testing program for release bindings. The aim of the investigations described was to determine the resultant forces generated during practical alpine skiing.

METHOD

The external loads in the functional chain, leg-boot-binding-ski, transmitted in both directions, create reaction forces on the binding. One must distinguish between the active steering forces generated by the muscle actions of the leg and the passive disturbing forces caused by slope conditions. Both typical forces are superimposed and represent the resultant force, which depends on the personal characteristics of the skier and on different external conditions.

The reaction forces on the binding that arise during practical skiing under various conditions have been determined and statistically evaluated by means of electrical transducers, telemetric measuring equipment, and the E.D.P.

Strain-gauge force transducers were placed between ski and normal re-lease-binding elements. They were capable of measuring lateral forces (right and left) on the toe and on the heel, vertical forces (up and down) on the toe and on the heel, and the vertical force on the bale (down). The signals were transmitted by wire to radio-transmitter equipment. The weight of the complete system carried by the skier was about two kg and caused no disturbing effect and no danger for the test skier.

The receiving equipment located in the center of the test field consisted of an FM receiver, an ultraviolet recorder, and a tape recorder. All signals were recorded continously and simultaneously (Figure 1). To obtain appropriate statistical results, the tape-recorded signals were classified according to their amplitudes (cumulative frequency analysis, Figure 3).

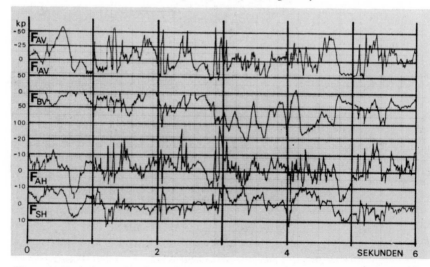

Figure 1. Typical recorder chart of the reaction forces at the binding during skiing on a rough and hard slope at a high speed.

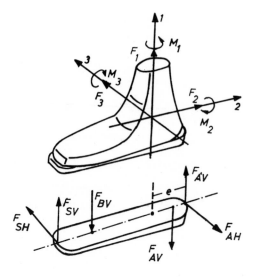

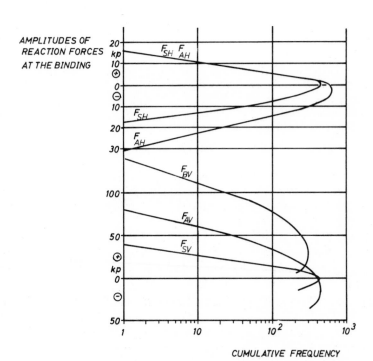

Figures 2 and 3. Characteristic frequency distribution curves of the different reaction forces at the binding. The maximum of the amplitudes represents the minimum of the retention level under certain conditions. F_{AH}, lateral force on the heel; F_{SH}, lateral force on the toe; F_{AV}, vertical force on the heel; F_{BV}, vertical force on the bale; F_{SV}, vertical force on the toe; kp, kilopond.

The tests were performed by five skiers with different physical character-istics and skiing abilities. Two ski instructors, two "high-speed" skiers, and one beginner skied on a difficult terrain under various snow and slope conditions. In addition, electromyographic measurements of the active muscle potentials of the leg were recorded simultaneously (Grimm, 1973; Krexa, 1973). In this manner, it was possible to separate active steering forces from passive disturbing forces. During the total test program, each test subject had to complete a run consisting of a vertical drop of 2000 m.

FINDINGS

The principal results of the described field measurements (Wittmann, 1973) were as follows:
1. The steering forces on the binding reached their maximum under condi-tions of high snow, which removed resistance (deep or severe snow), on a difficult terrain, and at high skiing speed. The shortest time of acting was about 0.10 sec.
2. The impact-disturbing forces reached their maximum on a rough and hard slope, on a difficult terrain, and at a high skiing speed. Under unfavorable conditions, the disturbing forces exceed the level of the steering forces. Their shortest duration is about 0.01 sec.
3. The cumulative frequency distributions showed a typical configuration for each of the different reaction forces at the binding. This characteristic makes it possible to describe the influence of the various conditions such as skiing style, speed, snow, etc. The cumulative frequency distribution curve enables us to define the resultant levels under certain boundary conditions.

The results of the biomechanical research relative to the load capacity of the human tibia and to the resultant level, as evaluated by the field measure-ments, make it possible to define the individual release value for the adjust-ment of the ski release binding.

The individual requirements of the skier concerning the necessary resul-tant forces must be satisfied within a certain range. This range takes into consideration the important factors such as weight, speed, and skiing perfor-mance, as well as the protective muscle function (i.e., the training condition and the age). With a suitable evaluation system, the static release value may be chosen so that the risks of injuries (the amount above the tolerance level) as well as the chances for inadvertent release (amount below the resultant level) are minimized.

The essential condition for efficient injury prevention is that the ski release binding releases reliably under all circumstances. This involves numer-ous technical properties, especially the dynamic properties of the binding, which ensure that all external loads acting on the leg for a duration capable of producing an injury result in a release of the binding. For impact loads, however, which are not dangerous to the leg because of their short duration, a binding release should not occur.

Inadvertent releases must be expected if, in a binding system, the statically adjusted release value is set at a level within the dynamic range. This system must be adjusted statically to a higher value to avoid such inadvertent releases. Thus, the release value may dangerously approach or even exceed the injury level.

The behavior of the release for the force values within the dynamic range for a given static adjustment is one of the essential qualities and safety features of a ski binding.

It is therefore necessary to develop a test method by which it is possible to determine clearly the behavior of the release binding at specific static settings within the dynamic range of forces. Preliminary tests show that, for this purpose, it is necessary to consider the "total system" of the skier. Inasmuch as real dynamic tests, for instance, those from impact loading, are not dangerous, it seems advisable to use the best dummy possible, namely, the human skier (Wittmann, 1973).

The results of the biomechanical research relating to the injury and retention levels were applied to practical injury prevention by the International Association for Safety in Skiing (founded in 1967 in Munich).

The IAS-Adjustment Table for Ski Bindings (Figure 4) provides a high degree of protection against typical injuries if appropriate release bindings are set on the correct value. This has been proven statistically for several years on thousands of skiers.

The IAS-Specification No. 100, relating to the technical characteristics, includes the standards and the test program for a useful release binding. After a successful test in the laboratories of the Technischer Überwachungsverein Bavaria, Munich (an official testing and control organization for industrial and product safety appointed by the Government of the Federal Republic of West Germany), bindings that meet this standard are certified with the IAS label.

All biomechanical research of this type will contribute to the common aim of reducing to a significant degree the injury risk in alpine skiing everywhere.

REFERENCES

Asang, E. 1970. 20 Jahre Skitraumatologie und Grundlagenforschung zum Verletzungsschutz im alpinen Skisport. Habilitationsschrift, Technische Universität, München.

Asang, E. 1973. Funktelemetrische Messungen beim Skifahren. Med. Sport 13: 1.

Grimm, C. M. 1973. Ergebnisse telemetrischer Messungen von mechanischen

International Association for Safety in Skiing, D-86 Bamberg, Hainstr. 18

IAS-Adjustment Table for Ski Bindings

frontal diameter at the tibia head		torsional release; horizontal force at the boot tip		forward release; vertical force upwards at the heel		LIPE
cm	in	kp	lbs	kp	lbs	
		Children and juniors up to 18 years				
6	2³/₈	4...5	10...15	20...25	50...65	2
6,5	2¹/₂	5...7	15...20	25...30	65...80	2-3
7	2³/₄	7...9	20...25	30...35	80...95	3
7,5	3	9...11	25...30	35...40	95...110	4
8	3¹/₈	11...13	30...35	40...50	110...125	5
8,5	3³/₈	13...15	35...40	50...60	125...145	5-6

				Adults up to approx. 50 years						
		women		men		women		men		w	m
cm	in	kp	lbs	kp	lbs	kp	lbs	kp	lbs		
7,5	3	11...12	16...20	12...13	20...24	45...50	90...100	55...60	115...125	4	5
8	3¹/₈	13...14	22...26	14...15	26...30	55...60	105...115	65...70	130...140	5	6
8,5	3³/₈	15...16	28...32	16...17	32...36	65...70	120...130	75...80	145...155	6	6-7
9	3¹/₂	17...18	34...38	18...19	38...42	70...75	135...145	80...85	160...170	6-7	7
9,5	3³/₄	19...20	40...44	20...21	44...48	75...80	150...160	85...90	175...185	7	8
10	3⁷/₈	21...22	46...50	22...23	50...54	80...85	165...175	90...95	190...200	8	9
10,5	4¹/₈	23...24	52...56	24...25	56...60	85...90	180...190	95...100	205...215	9	10
11	4³/₈	25...26	58...62	26...27	62...66	90...95	195...205	100...105	220...230	10	11
11,5	4¹/₂	27...28	64...68	28...29	68...72	95...100	210...220	105...110	235...245	10-11	12

Explanation of the table

1. The diameter of the leg is measured with a slide rule as follows: Palpate the head of the fibula at the outer side below the knee. Measure the transverse diameter of the tibia head directly above this point. The measurement should be taken while the skier is sitting with his knee bent.

2. The values recommended for the adjustment are then read on the table corresponding to the measured diameter.
Example: Adult, female up to 50 years.

    Diameter measured: 3³/₈ in
    Lateral force: 28 ... 32 lbs
    Heel force: 120 ... 130 lbs
    Set value: 30 lbs; 125 lbs

3. Persons over 50 should set the values shown one or two lines above the measured diameter
Example: Adult, male over 50 years.

    Diameter measured: 3³/₄ in
    Effective diameter: 3¹/₂ or 3³/₈ in
    Lateral force: 38 ... 42 lbs
        or 32 ... 36 lbs
    Heel force: 160 ... 170 lbs
        or 145 ... 155 lbs

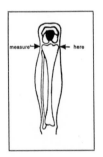

measure →   ← here

4. Well trained adults under 50 can set values obtained by adding ¹/₄ in to the measured diameter.

5. In case of a turntable binding the measured diameter can be increased by ¹/₄ in for determining the lateral release force.

6. It is recommended that adults over 60 should use short skis.

7. Only reliable testing equipment should be used for the adjustment. It should be checked from time to time for accuracy. Since friction between the boot and binding elements often affects the binding system, adjustment should always be made with moist or greased contact surfaces.

8. An antifriction device must be used under the ball of the foot.

9. Values essentially higher than those given must be avoided. They are dangerous. If they are necessary, however, this generally indicates lack of suitability or faulty fitting of the binding.

10. The adjustment of the binding should be checked from time to time. If the skis, boots or bindings are changed the adjustment must always be re-checked.

11. It is essential to protect the binding against dirt and corrosion.

Druck: Meisenbach KG, Bamberg

Figure 4. The IAS-Adjustment Table for Ski Bindings.

und elektromyographischen Signalen beim alpinen Skilauf und ihre Auswertung. Inaugural dissertation, Technische Universität, München.

Krexa, H. 1973. Grundlagen zur telemetrischen Messung von elektromyographischen und mechanischen Signalen beim alpinen Skilauf. Inaugural dissertation, Technische Universität, München.

Wittmann, G. 1973. Biomechanische Untersuchungen zum Verletzungsschutz im alpinen Skisport. Inaugural dissertation, München.

Author's address: *Dr. Gerhard Wittmann, Technischer Überwachungsverein Bayern e.V., Buro, Laboratorien und Versandanschrift, 8 München 21 (West Germany)*

Computerized method for determination of ice-skating velocity

K. Nicol

Johann Wolfgang Goethe Universität, Frankfurt

Partial times of skaters were measured at two events of speed skating in Inzell, Germany, in the winter of 1972–1973. The treatment of these data, as with many biomechanical analyses, is rather time consuming at present and therefore does not provide immediate results. Therefore, it was desirable to develop a computer system that would produce information within 2 min after a contest.

APPARATUS

The apparatus used was similar to that used by Kuhlow (1974) 2 yr ago. Unfortunately, it was not possible to measure time intervals for each track separately because of the increased possibility that the skaters would collide, especially on the curves. It was not possible to install photoelectric cells (PEC's) between the lanes, so that the light beam necessarily went across both lanes (see Figure 1). In the analysis, we had to coordinate the signals with the lanes or the corresponding skaters. Only the PEC at the finish provided data for each skater after he had passed the 100- or 500-m mark. Altogether, 20 PEC's were installed and connected by cable with the recorder. Their signals could be differentiated on the basis of the amplitudes of the impulses.

The hardware consisted of an analog-to-digital converter with one channel for signal input, a real-time clock, a teletype for listing of results, and a computer capable of servicing the equipment.

DESCRIPTION OF THE PROGRAM

Figure 2 shows a block diagram and the successive steps of the analysis. This program is written in Basic FORTRAN and contains 250 statements. Its main

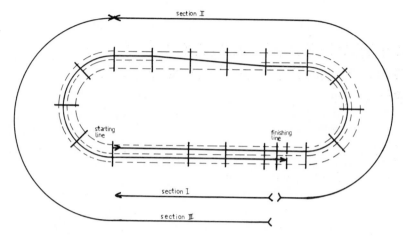

Figure 1. Situation of PECs, track of competitor starting on the inner lane and sections of evaluation.

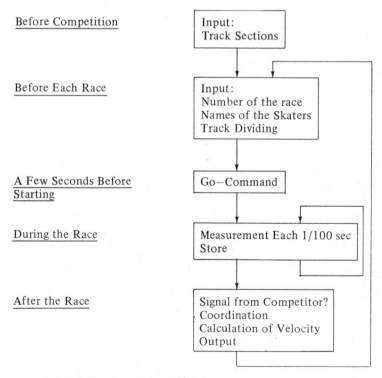

Figure 2. Block-diagram and time-table.

functions were: (a) analytical registration of the amplitude and the moment of the appearance of the signal, (b) separation of interfering signals, and (c) coordination of signal and corresponding competitor. These three problems are discussed in more detail in the following sections.

Measurement of Time

During analysis of the time of signals, many problems had to be solved. Although a time resolution of 0.01 sec was required, the clock of the computer was limited to a resolution of 0.1 sec. Moreover, the analog-to-digital converter could not process a new signal until 0.1 sec had elapsed after the inquisition of the timer.

During the competition, the timer could not be sampled after each signal because many signals would not have been detected. Thus, after each analog-to-digital converter inquisition, the detected signal was stored. Following the storage, a delaying loop was passed that required only 0.01 sec for the total procedure. With the use of this timer, which was controlled by the real-time clock of the computer, the partial times were calculated.

Separation of Interfering Signals

The second function of the program is the separation of signals that were caused not by competitors but by journalists, coaches, and those skaters who were not participating. The separation of those signals could be accomplished on the basis of the length of the signals because the covering time of the PEC by the skaters is considerably different from that by pedestrians. Interfering signals that could not be detected by that method were separated when the above-mentioned coordination was carried out.

Coordination

The coordination is by far the most time-consuming part of the program. Its fundamental principles are shown in Figure 3, in which the abscissa represents the time and the ordinate represents the running distance. It is important to note that coordination always is measured along the inner lane. If the velocity of both competitors is constant and if the one who starts on the inner lane (IS) is the faster skater, the resultant diagram will look like that in Figure 3,a. The parts of the diagram with the flat gradient indicate that the skater is on the outside curve. It is one of the functions of the program to coordinate the distance-speed measurements of the PEC's and the corresponding diagram. It is observed that this can be managed quite easily because the curves, apart from the starting distance, are far away from each other.

The situation is different if the competitor in the outside lane (OS) is the faster skater. In this case, the starter on the inner lane overtakes along the first curve and the starter on the outside lane overtakes on the back straight

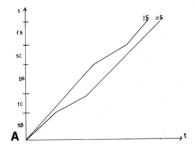 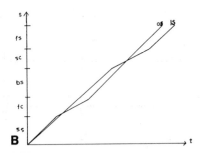

Figure 3. Distance (measured at the inner lane) versus time. Track speed is constant.
    Symbols: ss = starting stretch, fc = first curve, bs = back straight, sc = second curve, fs = final stretch
    Fig 3a: IS is the faster one;
    Fig 3b: OS is the faster one.

or in the second curve. This means that, together with the starting stretch, there are three critical intervals.

Until now it has been assumed that the competitors will keep constant track speeds; in reality, this is not the case. Results reported by Kuhlow (1973, 1974) indicated that: (a) characteristic speed variations for starters on the inner and outer lanes varied by ± 10 percent (apart from the first 50 m), and (b) there were individual variations of ± 10 percent.

The characteristic variations can be considered when the predicted values are computed; the individual ones, however, can lead to miscoordinations. However, this happens only under the worst circumstances. The purpose of the procedure, which now is described more exactly, is to eliminate this source of error.

The steps of coordination are: (a) the start begins from those cells the coordination of which is empirically known; (b) the skating distance is in various sections. The analysis of the sections starts at that point where the distances between the competitors is assumed to be the maximum; and (c) larger sections without known coordination are analyzed, starting from both ends.

From the application of these principles to the analysis of the 500-m distance, the following sections resulted (see Figure 1): Section I was from the 100-m mark back to the start, Section II from the 100-m mark to the end of the back straight, and Section III from the finish back to the end of the back straight. Thus, the PEC at the end of the back straight is analyzed twice. If the two coordinations do not coincide, a commentary is listed by the computer.

For each PEC, here indexed as PEC_n (See Figure 4), the following coordination procedure was carried out. If the analysis was done according to the direction of skating (Section II), then the predicted partial times $T_{exp,n,in}$ and $T_{exp,n,out}$ were calculated from the partial times at PEC_{n-1}

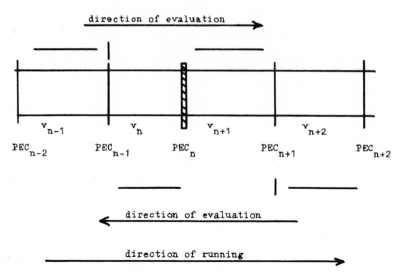

Figure 4. Symbols of calculation of predicted values.

and the means of the velocities v_{n-1} and v_{n+1}. If one of these values was not known, then the predicted values were used (1).

All signals that were caused by the competitors (this can be found by the length of signal) were arranged according to PEC's. For each PEC, they were labeled separately. From these signals, a specific combination of T_{sig,n,i_0} and T_{sig,n,j_0} was determined for which the deviation from the predicted values

$$(T_{sig,n,i} - T_{exp,n,in})^2 + (T_{sig,n,j} - T_{exp,n,out})^2$$

was a minimum. By this equation, T_{sig,n,i_0} could be coordinated with IS and T_{sig,n,j_0} with OS as the partial time at the PEC_n. When the section was analyzed against the direction of skating, the procedure was analogous.

What are the reasons for miscoordinations? First, both signals can be permutated. This can happen when both distance/speed curves (a) are very close together and (b) have a concave profile (see Figure 5).

Then it is possible that the predicted values are nearer to the wrong curve and thus are being coordinated incorrectly. This can happen also to the following signals within a section. In this case, however, the predicted values with high probability are placed between the two curves (see Figure 5). If this happens three times in succession, the program assumes a permutation and goes back three PEC's. The permutation is cancelled and the program starts to compute again on the corrected basis.

Second, it is possible that an interfering signal is coordinated. However, one can expect that only those of the interfering signals (they are distributed homogeneously over the total plane of distance and speed) that are in the very neighborhood of the curve are miscoordinated. According to the speed variation, the probability for this occurrence is beyond 1 percent.

PRETEST

In order to test the program, partial times were computed on the basis of velocities that varied statistically from interval to interval. The simulated runs were analyzed by the program and tested for error coordinations. In line with our expectations, the permutations of signals resulted exclusively when both were very similar. A transference of the error to the following PEC's was stopped by the previously mentioned control. Thus, a maximal miscoordination of only two was possible. The coordination of an interfering signal was observed very rarely. The error was smaller than the error of permutation. The total error term

$$T_{coordinated} - T_{true} : T_{interval\ of\ PEC}$$

(the mean over all lanes and intervals) was at speed variations of ± 25 percent under 1 percent.

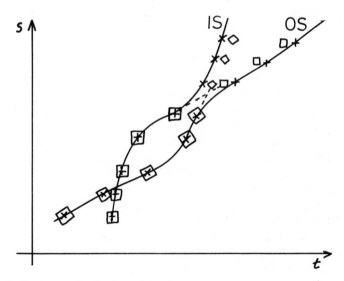

Figure 5. Wrong coordination by permutation.
    Symbols: □ time predicted for IS   + signal of IS   ◇ time predicted for OS   X signal of OS

ANALYSIS OF COMPETITIONS

After the pretest, we analyzed 500-m competitions of the previously mentioned events, which had been stored on tapes. Table 1 shows a partial
listing produced 2 min after the completion of the event. In addition to the
partial times (Column 3) and the velocities of each of the 24 intervals
(measured in m per sec, Column 4), Table 1 contains the velocities of the first
and the second part of each straight and curve (Column 6) as well as the
ratios of the velocities to the mean (Columns 5–7). Column 6 is important
because its values characterize each run individually. Above that, the spatial
lead of the leading skater was computed at every PEC (Column 9). Usually,
it cannot be observed very exactly because of the different lengths of curves.

FUTURE DEVELOPMENTS AND SUMMARY

We plan to develop two aspects of this method. First it will be adapted to
greater distances. Then, surely, miscoordination will be reduced because of
the reduced number of overtakings. In addition, the system will be adapted to
accommodate more than two lanes and consequently will be used in running
events as well.

A computer program was developed to analyze short-time registration of
the 500-m skating track. At every lane of the track, photoelectric cells were
installed at the finish; at 18 other points of the track, the beam of light of the
photoelectric cells crossed both lanes. The coordination of signal and lane
and the separation of interfering signals are the main problems of the
program. This coordination is obtained by comparing the measured time with
the time predicted by interpolation. In discrimination of the signals of the

Table 1. Partial listing of performance data for one event

Section	Skater	Time (sec)	Velocity (m/sec)	Ratio V/X̄	Velocity (m/sec)	Ratio V/X̄	Velocity (m/sec)	Distance between skaters (cm)
8	1	15.23	12.36	1.05	12.63	1.08	11.74	−176.0
8	2	15.60	12.26	1.03	12.77	1.07	11.89	173.0
9	1	16.82	12.59	1.07	0.00	0.00	11.74	−243.0
9	2	17.35	13.24	1.11	0.00	0.00	11.89	278.0
10	1	18.41	12.59	1.07	12.59	1.07	11.74	−307.0
10	2	19.15	12.87	1.08	13.05	1.10	11.89	328.0
11	1	20.21	12.66	1.08	0.00	0.00	11.74	−410.0
11	2	20.84	13.49	1.13	0.00	0.00	11.89	462.0
12	1	22.04	12.45	1.06	0.00	0.00	11.74	−526.0
12	2	22.61	12.88	1.08	0.00	0.00	11.89	550.0

two skaters, their different angular velocities in the curve were very helpful. Subsequently, the interval velocities and velocities in certain sections as well as the effective spatial lead of the leading skater were computed. The results are available within 2 min after the end of each competition. The system shows great promise for application in the biomechanical analysis of a variety of sports.

REFERENCES

Kuhlow, A. 1973. Geschwindigkeitsverteilung beim 500 m und 1000 m Eisschnellauf von Maennern und Frauen. *In:* Informationsheft zum Training. No. 14. Deutscher Sportbund.

Kuhlow, A. 1974. Analysis of competitors in world's speed skating championship. *In:* R. C. Nelson and C. A. Morehouse (eds.), Biomechanics IV. University Park Press, Baltimore, and S. Karger, Basel.

Author's address: *Dr. Klaus Nicol, Institut für Sport und Sportwissenschaften, Johann Wolfgang Goethe Universität, 6 Frankfurt-am-Main, Ginnheimer Landstrasse 39 (West Germany)*

Analysis of competitors in world speed-skating championship

A. Kuhlow

Johann Wolfgang Goethe Universität, Frankfurt

Although the biomechanics of running and the applicability of the results of such investigations have attracted the attention of researchers for many decades, the author was unable to locate any studies concerned with skating. Consequently, this investigation represents an initial attempt to study some characteristics in speed skating.

The author has been interested in recording the changes in speed during skating. A second purpose of this study was to provide a mathematical-statistical model for predicting the influence of mean speed within respective track sections (curves and straights) on the complete performance (time for the skating event).

METHOD

At the 2nd World Speed-Skating Championship held in Inzell, Germany, in 1971, our research team had the chance to analyze male and female skaters in 500-m and 1000-m sprint events. This article is confined to analysis of male skaters performing in the 500-m event.

Ten photoelectric cells (PEC's) were placed around the speed-skating track as indicated in Figure 1. When the ankle of the sprinter passed the PEC, the time elapsed from the starting line to the position of the respective PEC was recorded. Because it was of interest to investigate the change in mean speed within the starting stretch as well as in the final stretch, PEC's were placed at 28.50, 57.00, 85.50, and 114 m along the straight sections. The electrical impulses from the starting gun and the finish line were transmitted from the official chronograph to our recorder. Altogether, there were 14 track sections (S 1–S 14) of different lengths. Even over the shortest distance

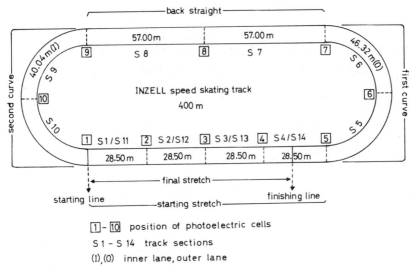

Figure 1. Diagram of speed skating track with positions of the photoelectric cells designated by numbers.

of 28.50 m, measurement error was limited to 0.05 m per sec, which represents an acceptable level of precision.

Forty highly skilled male skaters ranging in performance times from 38.89 to 43.60 sec for the 500-m event served as subjects. This group of sprinters was classified, according to level of athletic achievement (time for the 500-m event), into two groups—"top level" and "low level" sprinters—in order to determine whether there were any differences between their mean speeds within the 14 track sections. A further criterion for classifying them in groups was their starting position on either the inside or outside lane. This classification system resulted in four groups, as shown in Table 1.

RESULTS

The data derived from the analysis are summarized in Table 2. The following characteristics can be noted:

1. An increase in mean speed ($\bar{v}$) within the initial stages of the sprint to the end of the starting stretch (S 1–S 14), with maximal speed in S 4.
2. A decrease of $\bar{v}$ within the first half (S 5) of the first curve.
3. An increase of $\bar{v}$ within the second half (S 6) of the first curve.
4. No apparent trend for $\bar{v}$ within the back straight (S 7, S 8).
5. An increase of $\bar{v}$ within the first half (S 9) of the second curve when in the

Table 1. Classification system for subjects

	Groups	N	Means (sec)	Range of performance (sec)
G 1	Inner lane; top-level sprinters	10	39.70	38.98–40.42
G 2	Inner lane; low-level sprinters	10	41.65	40.49–43.19
G 3	Outside lane; top-level sprinters	10	39.87	38.89–40.32
G 4	Outside lane; low-level sprinters	10	41.50	40.53–43.60

outside lane, but a decrease of v̄ within the first half (S 9) of the second curve when in the inner lane.

6. A decrease of v̄ from S 10 along the final stretch to the finish line.

These characteristics of the speed-distance curves were observed in all groups studied.

In order to get additional information about the speed-distance curve of the competitors, these data were submitted to analyses of variance to evaluate the differences in v̄ within curves and straights among representatives of top- (G 1, G 3) and low-level (G 2, G 4) sprinters. Analysis of variance revealed the following differences for skaters starting in the inner lane (G 1 versus G 2).

1. The 0.01 level of significance between groups within the starting stretch indicates a higher increase of v̄ and a higher speed maximum in favor of G 1.

2. Contrary to the situation in starting stretch, there was no statistical difference between the groups in their performance around the first curve. Whereas G 1 decreased its maximal speed considerably within the first half of the curve, G 2 did so only slightly. Even a higher v̄ of G 1 on the second half of the inner lane could not compensate adequately for the rapid decrease of v̄ within S 5.

3. Similar to the situation in starting stretch, higher v̄ of G 1 is observed within the back straight, the outside lane, and the final stretch ($p < 0.01$).

Results for skaters starting in the outside lane (G 3 versus G 4) were as follows:

1. Although there were significant differences among both groups for v̄ within the starting stretch and a comparable decrease of v̄ within the first half

Table 2. Mean speed ($\bar{v}$) within the track sections (S 1–S 14)

Track sections		Inner lane				Outside lane			
		G 1		G 2		G 3		G 4	
S i*	Straights curves	M	SD	M	SD	M	SD	M	SD
1		7.2	0.2	7.0	0.2	7.1	0.2	6.9	0.3
2	Starting	11.2	0.1	10.4	0.6	11.1	0.2	10.7	0.3
3	stretch	12.3	0.2	12.0	0.7	12.1	0.2	11.8	0.5
4		14.3	0.6	13.4	0.6	14.1	0.7	13.5	0.8
5	First	12.6 (I)	0.3	12.4 (I)	0.5	12.9 (O)	0.3	12.4 (O)	0.4
6	curve	13.7 (I)	0.2	13.2 (I)	0.4	13.9 (O)	0.3	13.4 (O)	0.4
7	Back	13.8	1.0	13.1	0.3	13.9	0.2	13.4	0.3
8	straight	13.4	0.2	13.2	0.4	14.0	0.2	13.4	0.3
9	Second	13.9 (O)	0.3	13.3 (O)	0.3	13.7 (I)	0.2	13.2 (I)	0.3
10	curve	13.4 (O)	0.3	13.2 (O)	0.4	13.3 (I)	0.1	13.2 (I)	0.4
11		13.4	0.2	13.0	0.3	13.4	0.2	12.9	0.3
12	Final	13.2	0.3	12.7	0.3	13.3	0.2	12.9	0.4
13	stretch	13.3	0.2	12.6	0.3	13.2	0.5	12.6	0.4
14		12.9	0.7	12.6	0.4	13.4	0.4	12.6	0.4

*S i, number of track section; M, means of speed (m per sec); SD, standard deviations; (I), inner lane; (O), outside lane.

of the first curve (outside lane), this did not cause a significant difference of $\bar{v}$ within the whole inner lane. It may be concluded from this result that maximal speed within S 4 does not differ among the groups. Taking into account this latter finding, it is attractive to hypothesize an optimal speed for this outside lane that is anticipated by the top-level sprinters at the end of the starting stretch and that therefore exerts a negative influence on maximal speed in order to reduce to a minimum the difference between maximal and curve speeds.

2. A detailed analysis of $\bar{v}$ within the first and second half of the second curve (inner lane) reveals a successive approximation of $\bar{v}$ among groups from S 8 to S 9 to S 10. Thus, no significant differences were observed within the second curve.

Treatment of the data was continued by subjecting them to factorial and regression analyses for predicting the quantitative influence of the time needed for the respective track sections on the complete performance (time for the 500-m event). Those track sections that by factor analysis were found to be relatively independent of one another were selected as regressors. Within the limitations of our mathematical-statistical model, the following findings seem warranted for all groups. The improvement of the skaters' time for the 500-m event is determined to a high degree by the time needed to cover S 7, followed in descending order by S 6, S 13 + S 14, and S 3 + S 4. Although no generalization can be made beyond the present findings, in summary it is emphasized that the data from this study suggest that the best training methods are those that work efficiently in developing $\bar{v}$ within the back straight and the curves, and secondarily $\bar{v}$ within starting and final stretch. It is difficult to interpret adequately these results because of the limited nature of this study. Coaches as well as researchers are therefore in general agreement that in further investigations measurement should be made of some of the key factors (e.g., leg strength and racing technique, especially turning technique, length of gliding phase, etc.) involved in developing the phenomenon "speed" in order to optimize the performance in speed skating.

SUMMARY

At the 2nd World Speed-Skating Championship, mean speed ($\bar{v}$) was recorded within 14 track sections of different lengths by means of 10 photoelectric cells placed around the speed-skating track. In order to determine whether there were any differences in speed-distance curves of the 40 competitors analyzed, the participants were classified into four groups according to level of athletic achievement (time for the 500-m event) and their starting lane. Furthermore, those track sections were identified within which $\bar{v}$ of our groups differed significantly. The data were submitted to a mathematical-statistical model for predicting the quantitative influence of v̄ within respective track sections on the time for the 500-m event. From our results, recommendations were made for optimizing the training process.

REFERENCES

Kuhlow, A. 1973. Geschwindigkeitsverteilung beim 500 m und 1000 m Eisschnellauf von Männern und Frauen. *In:* Informationsheft zum Training (1973) no. 14. Deutscher Sportbund.

Author's address: *Dr. Angela Kuhlow, Institut für Sport und Sportwissenschaften, Johann Wolfgang Goethe Universitat, 6 Frankfurt-am-Main 90 (West Germany)*

Relative impact-attenuating properties of face masks of ice hockey goaltenders

R. W. Norman, J. C. Thompson, Y. Sze, and D. Hayes
University of Waterloo, Waterloo

One of the most dangerous positions in ice hockey is goaltending, a function which exposes the player to the possibility of being struck on the face by a frozen, sharp-edged, hard rubber puck that weighs 4 oz (114.6 g) and travels at speeds as high as 110 mph (177 km per hr). It has been only in the past 10 years that most goaltenders have worn protective face masks. Because of the lack of objectively founded design recommendations, a wide array of shapes, sizes, and materials have been used in these masks. A sampling of masks appears in Figure 1; a brief description of the features of those tested is given

Figure 1. Samples of masks tested, a shim, and the modified puck used as for striking.
Figure 2. The strain-gauge instrumented head model.

Financial support for this program was provided by a University of Waterloo Research Grant and by the Department of Kinesiology.

Table 1. Mask characteristics

Mask	Characteristics
A	High-density butyrate plastic, leather-covered foam shims
B1	High-density polyethylene, rubatex shims, shape same as B2
B2	High-density ABS, rubatex shims, shape same as B1
P; PS	Polyethylene, no shims; same mask with rubatex shims
F; FS	Fiberglass, no shims; same mask with rubber shims
W	Wire cage, thick rubber, leather, plastic pad on forehead, chin

in Table 1. For improvement of comfort, fit, and protection, some masks have thin "shims," also seen in Figure 1, made of closed cell foam, rubber, felt, leather, or combinations of these materials. These shims may be attached to the inside of the mask by the manufacturer or they may be supplied separately, to be attached by the player. Many goaltenders do not even use the shims but wear one of the popular, face-molded masks directly against whatever support points it happens to contact.

The objectives of the masks are to protect against lacerations and to deflect, distribute, and reduce impacts from pucks and sticks before the force reaches the face. Unfortunately, neither governmental nor industry-wide procedures to evaluate the degree to which masks meet these objectives have been established as yet.

METHODS

The purpose of this study was to investigate the relative impact-attenuating qualities of some of the commercially available face masks. To the knowledge of the authors, no other published study of this nature exists.

Each mask was tested on the instrumented head model (Figure 2) designed by Sze et al. (1973). Validation experiments established that the transient force histories on the model and on a human head agreed well and that data extrapolation to realistic puck momentum levels was permissible.

Dynamic loads were induced on the masks by striking them with a 33.3-oz (942-g) puck (Figure 1) impacting at about 4 ft per sec. Constancy of impact location, magnitude, and direction was achieved by swinging the puck on guy wires. Eight masks were struck in three different locations directly over contact points: (a) the center of the forehead, (b) the right side of the forehead approaching the side of the eye and temple, and (c) the right cheek. Striking contact points modeled the extreme condition experienced during play. The probes were arranged on the masks to coincide with contact points observed when the mask was worn by one of the authors. The tests provided puck deceleration histories plus such injury-related information as impact times, load dispersion, load on "bottomed" points, peak forces, and transient force rise times. For brevity, only peak force data are presented here.

RESULTS AND DISCUSSION

The histograms on the left in Figure 3 illustrate the magnitudes of the highest loads transmitted through the masks to the face probes for blows to the three locations indicated. The loads are expressed in terms of lbs of peak force per lb sec of impact momentum and correspond roughly to an impact expected from a puck travelling at about 85 mph (137 km per hr).

In every case, the point of highest loading on the face was, as might be expected, directly beneath the site of impact. It should be noted from the histograms that the magnitude of the load transmitted depends on the site of impact and the mask observed. Generally, a blow to the cheek or center forehead produced a higher peak force than a blow to the right forehead. The radius of curvature of most masks was smaller at the latter location, hence, a tangential component of the force was developed that reduced the magnitude of the impact passing through it to the face and often caused the mask to shift (not unrealistically). This was the case for Masks P and F particularly, which could be worn without shims. For these masks, the right forehead peak forces were lower when the masks were worn without shims than when protective shims were inserted. No shifting occurred when the masks were struck directly on the cheek and center forehead. Here the data show that shims attenuated the blow considerably, at least at the low-impact momenta generated during the tests.

None of the face-molded masks could be considered to provide good protection against all of the blows. The closest to this ideal was Mask PS, although its center forehead load approached 250 lbs. Hodgson and Naka-mura (1968) have shown that pulses with durations longer than 6 msec and of 150–250 lbs magnitude produced zygoma fracture.

Data for the wire mask were not easily comparable in histogram form because it was supported only on the center forehead and chin. The histo-gram shows peak force for a blow to the center forehead to be second lowest of the masks studied. For a direct blow to a wire on the chin, the peak force transmitted was 144 lbs per lb sec, again very low.

The histograms on the right in Figure 3 illustrate whether a mask was effec-tive in distributing a blow instead of concentrating it on the site of impact. To avoid getting a false impression, these histograms must be evaluated in con-junction with the magnitudes of the forces transmitted and presented in their corresponding peak-force histogram. For example, all of the load transmitted through the wire mask goes to the center forehead. There is no distribution of the load. However, the magnitude of the force transmitted is low. On the other hand, from 95 to 100 percent of the load transmitted through Masks B1 and B2 from a blow to the cheek goes straight to the cheek, and the magnitudes of the loads are high, about 375 lbs per lb sec.

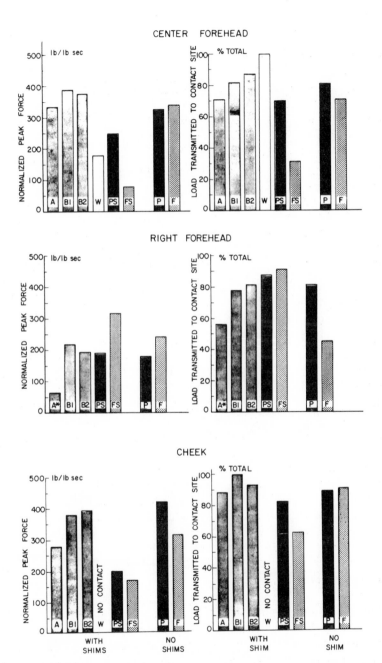

Figure 3. Four blows received by the contact point directly beneath the point of impact specified; largest peak force values observed, expressed as lbs per lb sec of impact momentum (*left*) and percentage of total load transmitted (*right*). *, glancing blow. See Table 1 for explanation of abbreviations.

The generalization that can be drawn is that more than 75 percent of the total load transmitted through a mask from a direct blow to a contact point arrives on the contact point. In most cases, the magnitude of the load is high as well. The masks do not distribute these types of loads well. In a few cases, they do dissipate the load before it gets to the face, as in the wire mask.

Table 2 illustrates the percentage of the total peak force transmitted that was received by various contact points. The data presented are from conditions in which more than 20 percent (arbitrary) of the transmitted load was spread to sites other than the immediate contact site. Loading the masks between contact sites usually resulted in a distribution of the load over the two closest sites, as expected. However, the contact sites of the face-molded masks were in close proximity to one another and often on relatively weak places on the skull, the side of the forehead close to the sphenoid bone, or on the zygoma (Hodgson and Nakamura, 1968). The wire mask was supported on the center of the forehead and chin, both relatively strong bones. A nose-level blow induced about 120 lbs per lb sec to the chin and 75 lbs per lb sec to the forehead area. Both blow distribution and reduction were effective.

The thin shims (usually less than 0.25 in) appear to provide protection in masks for blows to the center forehead and cheek, as seen in Figure 3 for Masks PS and FS, compared with Masks P and F for normalized peak force. This is so probably only for blows under the low-momentum conditions of the test. Data presented by Sze et al. (1973) for masks with shims have shown impact times that regress to those of the same mask with no shims as the impact momentum increases. Several shim materials have been tried, but all seem ineffective at puck speeds above 20 mph (32.3 km per hr). Once bottoming occurs, all masks cease to distribute loads. Moreover, flexible regions of masks (e.g., regions of shallow curvature) require only minor loads to "bottom out." Thus, apparent gaps between the mask and the face under

Table 2. Percentage of total transmitted peak force distributed to sites indicated

Mask*	Impact site	Forehead			Cheek		Chin
		Left	Center	Right	Left	Right	
FS	Cheek			25		63	12
F	RF†		33	45		22	
FS	CF‡	31	26	31	6	6	
F	CF	15	70	15			
PS	CF	6	70	6	9	9	
A	CF	6	71	6			17
B1	RF		22	78			

*See Table 1 for explanation of abbreviations.
†Right forehead.
‡Center forehead.

no-load conditions may not provide the protection expected. For example, Mask PS bottomed at the center forehead position under a static load of only 55 lbs (25 Kg).

CONCLUSIONS

The wire mask achieved the objective of both force dispersion and reduction and at the same time prohibited loads from reaching the potentially vulnerable right forehead and cheek. The face-molded masks differed in their abilities to attenuate impacting forces, but most permitted loads that were quite high to penetrate to the contact point beneath the area struck. More than 75 percent of the load transmitted was received by the contact point directly under the site of impact. Very little disperson of the load occurred. It was found that shims appeared to attenuate the peak force reaching the face for some blows at low-impact momentum; however, they all bottom at low loads and probably have little effect at normal puck momentum. Finally, it was shown that the shape of the mask, particularly the reinforcing effects of ridges, can play a much more important role in distributing loads than does the type of mask material.

ACKNOWLEDGMENTS

The technical assistance of Mr. Wolfgang Michenfelder and preliminary work by Mr. Derek Humphreys are greatly appreciated.

REFERENCES

Hodgson, V. R., and G. S. Nakamura. 1968. Mechanical impedance and impact responses of the human cadaver zygoma. J. Biomech. 1: 73–78.
Sze, Y., J. C. Thompson, R. W. Norman, and D. Hayes, 1973. Procedures for the evaluation of hockey goaltender's face masks. Paper presented at the ASME Symposium, November.

Author's address: *Mr. R. W. Norman, Department of Kinesiology, Faculty of Human Kinetics and Leisure Studies, University of Waterloo, Waterloo, Ontario, N2L3G1 (Canada)*

Miscellaneous

Studies of motion and motor abilities of sportsmen

V. M. Zatziorsky
Central Institute of Physical Culture, Moscow

Two directions of research have developed within contemporary bio-mechanics of sport. The first is based on the use of the techniques of classical mechanics, with the utilization of the ideas of the theory of automatic control and of some other branches of knowledge. The second direction emphasizes statistical analysis of the relationships among the biomechanical parameters.

DYNAMICS METHODS

In the first case, an analysis is made of the cause-and-effect relationships that accompany performance of motions in precise ways. Two major groups of tasks are of special significance. These are: (a) determination of forces that operate during a movement utilizing the application of dynamics. This re-quires calculation of the changes in force and acceleration; and (b) the task of optimization.

The most simple solutions are found for tasks that can be reduced to the movement of a material point. Such cases occur as an exception, however. For instance, solutions can be found for the tasks to optimize the descent trajectories in alpine skiing by considering an athlete as a material point moving down a mountain slope.

A much more promising direction is that of mechanical analysis when the human body is viewed as a system of interlinked segments. In such a case, the task is to determine the forces and the moments of forces that operated during a movement by recording the position of these segments in space and by calculating their accelerations. The founder of biomechanics in the Soviet Union, Prof. N. A. Bernstein, worked extensively on this task. Excellent progress has been made in this area in the past few years. However, consider-

ably more research is needed before acceptable levels of accuracy are achieved. If such a development occurs in biomechanics, it must be regarded as a significant increase in quality of its development. However, for this to happen, it is necessary to cope with three problems:
1. The biomechanical behavior of the human body is complex and has not been studied extensively. One must keep in mind the possible dislocation of the internal organs, the resilient and dissipative forces, the shifting of the instantaneous axes of rotation in the joints, etc.
2. There are no experimentally developed methods available that make it possible to register movements of any complexity with acceptable accuracy.
3. For motions with a great number of human body segments involved, the mathematical task of calculating the forces by means of accelerations remains unsolved.

STOCHASTIC METHODS

Such difficulties compel us to look for other ways besides the detailed analysis of the mechanics of motion. The more natural one seems to be to use stochastic methods, for example, multidimensional statistics (the so-called "statistical biomechanics"). An example of such an application is presented. In the case of broad jumpers who had performances of 680–818 cm, a number of biomechanical characteristics of the jump were registered and matched with the achievements of these athletes in this event. The correlation matrix is given in Table 1.

Table 1. Correlation coefficients between certain motor characteristics of movements in top-level long jumpers (N = 43)

Characteristics	1	2	3	4	5	6
1. Performance						
2. Velocity of approach	0.843					
3. Length of last stride	0.557	0.696				
4. Take-off time	−0.833	−0.641	−0.029			
5. Maximal knee angle	0.420	−0.520	−0.422	−0.202		
6. Stroke scope	0.164	0.124	0.029	−0.240	0.049	
7. Body angle during amortization	0.079	0.009	−0.010	−0.261	0.128	−0.088

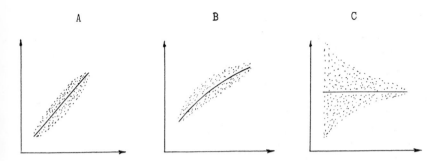

Figure 1. The most frequently encountered types of relationships between biomechanical characteristics (*ordinate*) and performances (*abscissa*).

Of course, the use of multidimensional statistics to analyze biomechanical data will require special research to fix the boundaries and the possibilities of these methods in such specific situations. However, the point is never to reduce the situation to applying the standard formulas to new sets of data or to intercorrelate large numbers of derived biomechanical parameters. On the other hand, inasmuch as it is impossible to give a detailed analysis of a vast majority of complex movements, we have no tool to establish relationships among biomechanical characteristics other than the stochastic methods.

The greatest difficulty in using stochastic methods lies in the diversity of the forms of links between the biomechanical characteristics of movements (Figure 1). For the distribution in Figure 1, *A*, there are no difficulties in using the traditional methods of multidimensional statistics, whereas in Figure 1, *B*, complicated mathematical methods are required. As for Figure 1, *C*, there are at present virtually no satisfactory methods of multidimensional statistics to analyze such situations. According to our observations, in speed and force events such as jumps, throws, and so on, the links of the type shown in Figure 1, *A* are the most typical of the time characteristics of motion, those of the type shown in Figure 1, *B* are typical of force characteristics, and those of the type shown in Figure 1, *C* are typical of angular characteristics.

The objects of study in statistical biomechanics of sport usually are the patterns of sports events as well as of the relationships among (a) various characteristics of motion, (b) performances and characteristics of motion, (c) performance in various tasks, and (d) performances and characteristics of motion, on the one hand, and special features of a performer or external conditions, on the other.

It is concluded that both mechanical and statistical approaches are necessary for the complete scientific development of biomechanics of sport.

Author's address: *Dr. Vladimir M. Zatziorsky, Central Institute of Physical Culture, Moscow (Russia)*

Control of biodynamic structures in sports

D. Donskoi, V. Nazarov, and H. Gross

Central Institute for Physical Culture, Moscow

This investigation has been aimed at developing general theoretical propositions concerning the pattern of movements in sports, working out methods for mathematical simulation of their manifestation, and verifying the efficiency of the methods that control the biodynamic structures in sports exercises.

The modern theory of biomechanics is based on the methodological principles of the theory of the construction of movements (Bernstein, 1967). A further development of this theory in sports is aimed at defining the principles of the structural nature of the movements, inquiring into the properties of biomechanical systems, developing the foundations for the rearrangement of movement systems (Donskoi 1968, 1973), and searching for the most effective criteria (Gross, 1972) for improving the technical performance in sports. Bernstein's school continues its work on theoretical substantiation of movement control by using methods of both mathematical (formalized) and physical (content) simulation. Complex methods of recording both kinematic and dynamic characteristics have been used.

The program of movements in sports exercises with preset kinematics (gymnastics) has been studied (Korenev, 1972; Nazarov, 1971). For exercises with an optimal dynamic structure (skiing, swimming), the phase structure was investigated to evaluate the rationality of dynamic structure variants.

The position of the body is conditioned by: (a) a location of its center of gravity, (b) an orientation of the body in space, and (c) a pose for the body.

The program of movement consists of: (a) a general program of position and orientation (the natural movement without any pose alteration), and (b) a program of pose alteration (controlling movements).

The location is described by generalized coordinates, while the orientation by Euler's right angle coordinates constantly is fixed in relation to the body and converted to those permanently oriented.

The controlling movements to alter a pose produce additional forces and moments when interacting with external bodies (or without them). The standard program (always obligatory) is realized by the principal controlling movements, while the variations or adjustment to specific dynamic situations are realized by correcting controlling movements (a compensation for deviations exceeding the allowable).

The analytical description of the gymnast's movements is presented as Lagrange's system of equations with indefinite coefficients. In addition, it is necessary to have the equations of connection, which should define the restrictions being imposed on the body links. These are connections that will depend on the construction of the motion apparatus, the external bodies, and the preset program. The principal controlling function, the necessary controlling forces and moments of forces, and the principal and correcting movements are defined in such a way. With the data obtained, it is possible to build an optimal sequence of training.

In order to determine the rationale for an optimal program for attaining an extreme objective, the step-by-step modelling also has been defined. First, a kinetic model of the movement system must be built in the form of synchronized graphs of characteristics.

The criterion of efficiency, which is an indispensable and sufficient value to show that the aim is approached (the velocity of operational points), is selected in terms of the task of action. The most complicated part of this is to single out from the system (subsystem) elements in the time range, i.e., phases. This is performed according to the following rules. Each phase: (a) is essentially different from the adjacent ones; (b) has strictly defined boundaries; (c) comprises all the movements of the human body links for a given time; (d) is characterized by its particular task in the solution of the whole system of movements; and (e) has concrete requirements for the task to be solved.

After the phase composition has been established (the system analysis), the study of the efficiency of structural interconnections follows.

The correlation models showing the closeness of connections between phase characteristics, their influence on the efficiency criterion, and the degree of significance in the system of movements (the system synthesis) must finally be built.

Also, the system analysis of spatial elements (subsystems) as elementary actions of the body links should be carried out. In doing this, a study should be made of the characteristic task of each of the spatial elements (by general phases and with respect to the whole system of movements) as well as requirements which ensue from making such a study.

Typical of both methods is a step-by-step modeling which substantiates the whole complex of requirements in order to control the concrete biodynamic structures by means of applying the forces determined with regard to magnitudes, directions, times, and their sources.

REFERENCES

Bernstein, N. A. 1967. Co-ordination and Regulation of Movements. Pergamon Press, Oxford.

Donskoi, D. D. 1968. Bewegungsprinzipien der Biomechanik im Sport. *In:* J. Wartenweiler, E. Jokl, and M. Hebbelinck (eds.), Biomechanics I. pp. 150–154. S. Karger, Basel, and University Park Press, Baltimore.

Donskoi, D.D. 1973. Control exercised over the reconstruction of the movements' system. *In:* S. Cerquiglini, A. Venerando, and J. Wartenweiler (eds.), Biomechanics III, pp. 124–128. University Park Press, Baltimore, and S. Karger, Basel.

Gross, H. H. 1972. Objective data showing the accomplishments of the sports technics with the stabilization of dynamic structure. Summaries of the reports of the XII All-Union Conference on Physiology, Morphology, Biomechanics, and Biochemistry. pp. 230–232.

Korenev, G. V. 1972. On the human movements reaching a pre-set purpose. Automat. Telemech. 6.

Nazarov, V. T. 1971. The controlling movements of a sportsman in joints to produce rotation in the non-supporting condition. Theory Pract. Phys. Cult. 3.

Author's address: *Prof. Dr. D. Donskoi, Central Institute for Physical Culture, Moscow, 64 (USSR)*

Analysis of twisting and turning movements

B. Nigg
Swiss Federal Institute of Technology, Zürich

When a parallel-piped block of a homogeneous material with the length of side a, b, and c (a > b > c) is thrown into the air rotating around one of the three principal axes 1, 2, or 3, there is no amount of force for the whole system during the flight (see Figure 1). This means that the angular momentum A remains constant. Therefore, it is expected that the rotation beginning around one of these three axes should continue in the same axis. Figure 2 shows the result of the experiment.

The illustrations show that the expectations are realized only by the rotation around Axes 1 and 3. These two are the axes with the maximal (3) and minimal (1) values of the moment of inertia. The rotating system is in an unstable state of equilibrium by the rotation around the axis with the median value of the moment of inertia.

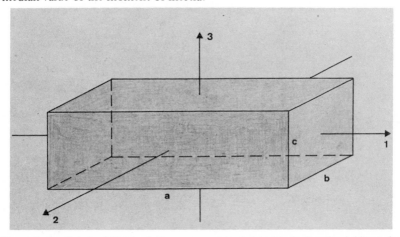

Figure 1. Parallel-piped block of a homogeneous material with the length of side a > b > c and the three axes 1, 2, and 3.

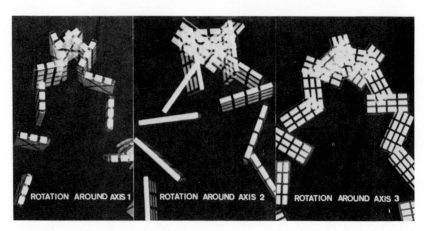

Figure 2. Rotation around the three principal axes.

PHYSICAL CONSIDERATIONS

Symbols

In the following equations, I_{ii} is the moment of inertia for a principal axis; E_{kr} is kinetic rotation energy; A is angular momentum; and ω is angular velocity.

Suppositions

The axes 1, 2, and 3 are principal axes $I_{11} < I_{22} < I_{33}$. For a principal-axis-system, the kinetic rotation energy is (Ziegler, 1956):

$$E_{kr} = \tfrac{1}{2}\, I_{11} \cdot \omega_1^2 + \tfrac{1}{2}\, I_{22} \cdot \omega_2^2 + \tfrac{1}{2}\, I_{33} \cdot \omega_3^2. \qquad 1$$

The relationship between the angular momentum A and the angular velocity ω is:

$$A_i = I_{ii} \cdot \omega_i. \qquad 2$$

Equations 2 and 1 yield:

$$1 = A_1^2/[I_{11} \cdot 2E_{kr}] + A_2^2/[I_{22} \cdot 2E_{kr}] + A_3^2/[I_{33} \cdot 2E_{kr}]. \qquad 3$$

Equation 3 represents an ellipsoid surface in the angular momentum area where $\sqrt{I_{ii} \cdot 2E_{kr}}$ are the particular semiaxes. On the other hand, the equations of any axis of Rotation "a" are:

$$A_a = I_a \cdot \omega_a$$
$$E_{kr} = \tfrac{1}{2} I_a \cdot \omega_a^2$$

$$E_{kr} = \tfrac{1}{2} \cdot \frac{A_a^2}{I_a} \cdot \qquad\qquad\qquad\qquad 4$$

This angular momentum with respect to the axis "a" can be reduced to:

$$A_a^2 = A_1^2 + A_2^2 + A_3^2 \cdot \qquad\qquad\qquad\qquad 5$$

Equations 5 and 4 yield:

$$1 = A_1^2/[I_a \cdot 2 \cdot E_{kr}] + A_2^2/[I_a \cdot 2 \cdot E_{kr}] + A_3^2/[I_a \cdot 2 \cdot E_{kr}]. \qquad 6$$

Equation 6 represents the surface of a sphere in the angular momentum area where $\sqrt{I_a \cdot 2 \cdot E_{kr}}$ is the radius.

Equations 3 and 6 must satisfy the considered rotation. This means that the mathematical resolution of our problem is the intersecting curve of the surface of the sphere and the surface of the ellipsoid.

Special Case 1

The momentary rotation axis coincides with the principal axis 1.

$$I_{11} = I_a \text{ and } \omega_1 = \omega_a \rightarrow \sqrt{I_{11} \cdot 2E_{kr}} = \sqrt{I_a \cdot 2E_{kr}} \cdot$$

The surface of the sphere and the surface of the ellipsoid have two common points of contact. All other points of the ellipsoid surface fall outside of the sphere surface because $I_{11} < I_{22} < I_{33}$.

A small variation of the momentary angular velocity direction signifies a small variation in the stable state of equilibrium. This means that the rotation around the axis with the minimal value of the momentum of inertia is a stable rotation.

Special Case 2

The momentary rotation axis coincides with the principal axis 3.

$$I_{33} = I_a \text{ and } \omega_3 = \omega_a \rightarrow \sqrt{I_{33} \cdot 2E_{kr}} = \sqrt{I_a \cdot 2E_{kr}} \cdot$$

The surface of the sphere and the surface of the ellipsoid also have two common points of contact. All other points of the sphere surface fall outside of the ellipsoid surface. In this second special case, the rotation around the axis with the maximal value of the momentum of inertia is also a stable rotation (Figure 3).

Special Case 3

The momentary rotation axis coincides with the principal axis 2.

$$I_{22} = I_a \text{ and } \omega_2 = \omega_a \to \sqrt{I_{22} \cdot 2E_{kr}} = \sqrt{I_a \cdot 2E_{kr}} \ .$$

The sphere surface and the ellipsoid surface have two intersecting curves. All points of these two intersecting curves are possible solutions.

A small variation of the momentary angular velocity direction signifies a small variation in the unstable state of equilibrium. The axis 2 moves away from the angular momentum direction. This means that a rotation around the axis with the median value of the momentum of inertia is in an unstable state of equilibrium (Figure 3).

APPLICATION TO HUMAN MOVEMENT

Hochmuth (1967) presented the moments of inertia for principal inertia axes of the human body across the center of gravity. In the standing position, the axis with the median value of the moment of inertia is the transverse axis of the human body. The rotation around the transverse axis is that of the layout somersault. If, in this movement, the angular momentum vector does not coincide with the transverse axis direction, rotations around the other two axes will begin and the athlete is forced to correct his movement. This undesired rotation around the two other axes increases with the increase in angular velocity.

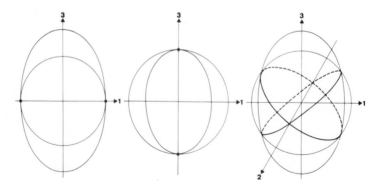

Figure 3. The mathematical solution in the angular momentum area.

It appears also that the layout somersault is not as simple as it seems. The difference in the difficulty between a layout and a twist somersault appears to be less than generally is acknowledged (i.e., as a basis for competition evaluation). The influence of this unstable equilibrium which disturbs the movement must be minimized by the athlete (gymnast, springboard diver, or trampolinist). One possibility consists of extending the arms laterally. Thus, the moment of inertia around the sagittal axis will be almost doubled and the rotation around this axis will be slower.

The moment of inertia around the transverse axis in the normal somersault is larger than the moment of inertia of the other two. Thus, the normal somersault involves a stable rotation.

The influence of the unstable equilibrium is amplified in the twist somersault by the arm movement. The arm is brought down across the body to begin the twist movement. Therefore, the moment of inertia around the lateral axis will be asymmetrical, and the momentary rotation axis of the twist somersault "moves away faster" from the angular momentum vector.

It is expected that the unstable equilibrium by rotations around the lateral axis will have a selective influence, inasmuch as it is probable that only athletes with good motor ability can compensate for these variations.

REFERENCES

Ziegler, H. 1956. Mechanik II. Birkhäuser, Basel.
Hochmuth, G. 1967. Biomechanik sportlicher Bewegungen. Berlin.
Nigg, B. 1973. AK Biomechanik. Vorlesungsmanuskript, ETH, Zürich.

Author's address: Mr. Benno Nigg, Biomechanics Laboratory, Swiss Federal Institute of Technology, Plattenstrasse 26, 8032 Zurich (Switzerland)

Experimental determination of forces exerted in tennis play

T. R. Kane, W. C. Hayes, and J. D. Priest
Stanford University, Stanford

In a recent clinical study of 84 of the world's best tennis players, elbow injuries in significant numbers, marked increases in muscle and bone dimensions in the dominant upper extremity, and pronounced postural assymetry of the shoulders were found. These findings are striking when one considers that in the United States alone approximately 11.5 million people play tennis and that pain in the region of the lateral humeral epicondyle, or "tennis elbow," poses an unsolved problem.

The hypothesis that the phenomena mentioned are related to the exertion of forces by the player comes to mind readily; but it is difficult to determine the exact nature of this relationship. However, it seems clear that two kinds of information must play central roles in the endeavor to do so: knowledge of the forces exerted by the player and precise descriptions of the medical phenomena of interest. This study is concerned with means for obtaining information about the forces.

THEORY

The forces giving rise to the motion of a tennis racket can be divided into gravitational forces, forces associated with the collision between the ball and the racket, and forces exerted on the racket by the player's hand. In Figure 1, these three groups of forces are represented respectively as follows: a force of magnitude mg, applied at point P, the mass center of the racket, where m is the mass of the racket and g is the acceleration of gravity; a force C, applied at point R, the point at which the ball strikes the racket; and a force $\underline{F}$, applied at a point Q, together with a torque $\underline{T}$. In accordance with Newton's Second Law and the Angular Momentum Principle, the quantities m, g, $\underline{C}$, $\underline{F}$, and $\underline{T}$ are related to each other in such a way that

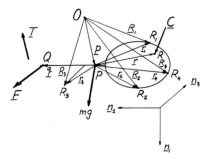

Figure 1. Schematic representation of racket. See text for explanation.

$$mg + \underline{C} + \underline{F} = ma \qquad (1)$$

and

$$\underline{r} \times \underline{C} + \underline{q} \times \underline{F} + \underline{T} = I \cdot \alpha + \omega \times I \cdot \omega \qquad (2)$$

where a is the acceleration of P, $\underline{r}$ is the position vector of R relative to P, $\underline{q}$ is the position vector of Q relative to P, I is the central inertia dyadic of the racket, and α and ω are the angular acceleration and the angular velocity of the racket, respectively. To solve Equations 1 and 2 for $\underline{F}$ and $\underline{T}$, the two quantities that characterize the forces exerted by the player, one must know all other terms in these equations. In particular, one must be able to evaluate the terms involving a, α, and ω. It now will be shown how one can do this by using four triaxial accelerometers.

In Figure 1, $R_1 \ldots , R_4$ designate points fixed on the racket, and $\underline{r}_1, \ldots , \underline{r}_4$ are the position vectors of these points relative to P. If R_4 is not coplanar with $R_1, \ldots R_3$, then $\underline{r}_4$ can be expressed as

$$\underline{r}_4 = \lambda_1 \underline{r}_1 + \lambda_2 \underline{r}_2 + \lambda_3 \underline{r}_3 \qquad (3)$$

where $\lambda_1 \ldots , \lambda_3$ are constants given by

$$\lambda_1 = \frac{\underline{r}_2 \times \underline{r}_3}{\underline{r}_1 \times \underline{r}_2 \cdot \underline{r}_3} \cdot \underline{r}_4, \lambda_2 = \frac{\underline{r}_3 \times \underline{r}_1}{\underline{r}_1 \times \underline{r}_2 \cdot \underline{r}_3} \cdot \underline{r}_4, \lambda_3 = \frac{\underline{r}_1 \times \underline{r}_2}{\underline{r}_1 \times \underline{r}_2 \cdot \underline{r}_3} \cdot \underline{r}_4 \qquad (4)$$

Moreover, if $\underline{R}_1 \ldots , \underline{R}_4$ and $\underline{P}$ are the position vectors of $R_1, \ldots , R_4$ and P relative to a fixed point 0, then

$$\underline{r}_i = \underline{R}_i - \underline{P} \ (i = 1, \ldots , 4) \qquad (5)$$

so that, from Equation 3,

$$\underline{R}_4 - \underline{P} = \lambda_1 \, (\underline{R}_1 - \underline{P}) + \lambda_2 \, (\underline{R}_2 - \underline{P}) + \lambda_3 \, (\underline{R}_3 - \underline{P}) \qquad (6)$$

or

$$\underline{P} = \frac{\lambda_1 \underline{R}_1 + \lambda_2 \underline{R}_2 + \lambda_3 \underline{R}_3 - \underline{R}_4}{\lambda_1 + \lambda_2 + \lambda_3 - 1} \qquad (7)$$

Now, a is simply the second time-derivative of $\underline{P}$. Hence, from Equation 7,

$$a = \frac{\lambda_1 a_1 + \lambda_2 a_2 + \lambda_3 a_3 - a_4}{\lambda_1 + \lambda_2 + \lambda_3 - 1} \qquad (8)$$

where $\underline{a}_1, \ldots, \underline{a}_4$ are the accelerations of $R_1, \ldots, R_4$. Once $\underline{a}_1, \ldots, \underline{a}_4$ have been found from accelerometer readings, one can evaluate a by using Equations 4 and 8.

Turning to the terms in Equation 2 involving α and ω, one can take advantage of the following fact: If $\underline{n}_1, \underline{n}_2, \underline{n}_3$ form a dextral set of orthogonal unit vectors, each parallel to a central principal axis of inertia of the racket, and if I_j, α_j, and ω_j are defined as

$$I_j \overset{\Delta}{=} \underline{n}_j \cdot I \cdot \underline{n}_j \, , \alpha_j \overset{\Delta}{=} \alpha_j \cdot \underline{n}_j \, , \omega_j \overset{\Delta}{=} \omega \cdot \underline{n}_j \quad (j = 1, 2, 3) \quad (9)$$

then

$$I \cdot \alpha + \omega \times I \cdot \omega = [I_1 \alpha_1 - (I_2 - I_3)\omega_2 \omega_3] \, \underline{n}_1 + \ldots \quad (10)$$

where the *three dots* represent terms obtained by cyclic permutation of subscripts. Equation 10 shows that one needs a scheme for determining $\alpha_1, \ldots, \alpha_3$ and the products $\omega_2 \omega_3, \omega_3 \omega_1, \omega_1 \omega_2$. Now, the acceleration of point R_i can be expressed as

$$a_i = a + \alpha \times \underline{r}_i + \omega \times (\omega \times \underline{r}_i) \, (i = 1, \ldots, 4) \qquad (11)$$

and, if b_{ij}, r_{ij}, β_j, and γ_j are defined as

$$b_{ij} \overset{\Delta}{=} (a_i - a) \cdot \underline{n}_j \, , r_{ij} \overset{\Delta}{=} \underline{r}_i \cdot \underline{n}_j \quad (i = 1, \ldots, 4; j = 1, 2, 3) \qquad (12)$$

$$\beta_1 \overset{\Delta}{=} \omega_2 \omega_3 \, , \beta_2 \overset{\Delta}{=} \omega_3 \omega_1 \, , \beta_3 \overset{\Delta}{=} \omega_1 \omega_2 \qquad (13)$$

$$\gamma_1 \overset{\Delta}{=} \omega_2{}^2 + \omega_3{}^2 \, , \gamma_2 \overset{\Delta}{=} \omega_3{}^2 + \omega_1{}^2 \, , \gamma_3 \overset{\Delta}{=} \omega_1{}^2 + \omega_2{}^2 \qquad (14)$$

then Equations 11 are equivalent to the twelve equations

$$\alpha_2 r_{i3} - \alpha_3 r_{i2} + \beta_2 r_{i3} + \beta_3 r_{i2} - \gamma_1 r_{i1} = b_{i1} \quad (i = 1, \ldots, 4) \qquad (15)$$

$$\alpha_3 r_{i1} - \alpha_1 r_{i3} + \beta_3 r_{i1} + \beta_1 r_{i3} - \gamma_2 r_{i2} = b_{i2} \quad (i = 1, \ldots, 4) \qquad (16)$$

$$\alpha_1 r_{i2} - \alpha_2 r_{i1} + \beta_1 r_{i2} + \beta_2 r_{i1} - \gamma_3 r_{i3} = b_{i3} \quad (i = 1, \ldots 4) \qquad (17)$$

Consequently, once nine of the Equations 15–17, e.g., those corresponding to $i = 1, 2, 3$, have been solved for α_j, β_j, and γ_j ($j = 1, 2, 3$), one can express the righthand member of Equation 2 entirely in terms of known quantities by using Equations 10 and 13.

EXPERIMENTS

The theory presented in the preceding section was used to evaluate the righthand members of Equations 1 and 2 for two test subjects, one an expert player, the other a novice, each performing several forehand and backhand ground strokes. To this end, four triaxial accelerometers were attached to a racket at noncoplanar points, and these were connected with long leads to charge amplifiers which, in turn, were connected to a recorder, as shown in Figure 2. The accelerometers were very small and the post carrying one of the accelerometers (in order to remove this accelerometer from the plane determined by the remaining three) was made of aluminum. Thus the inertial properties of the racket, previously determined by weighing the racket and performing oscillation tests, were not altered substantially by the presence of the accelerometers and the post. The recorder produced 12 traces during the

Figure 2. Test equipment.

performance of a stroke, each trace corresponding to an acceleration component a_{ij} defined as

$$a_{ij} \triangleq a_i \cdot \underline{n}_j \quad (i = 1, \ldots, 4; j = 1, 2, 3). \tag{18}$$

DATA REDUCTION

The 12 values of a_{ij} associated with a given instant during a test run were used to evaluate dimensionless measures of the righthand members of Equations 1 and 2 defined as follows:

$$\tilde{F}_i \triangleq |\, m a \cdot \underline{n}_i \,| \,/\, |\, mg \,| \quad (i = 1, 2, 3) \tag{19}$$

$$\tilde{T}_i \triangleq |\, (I \cdot \alpha + \omega \times I \cdot \omega) \cdot \underline{n}_i \,| \,/\, |\, mgL \,| \quad (i = 1, 2, 3) \tag{20}$$

where L is the distance between P and Q (Figure 1). This was done for a sufficient number of instants during a given run to permit the construction of smooth plots of $\tilde{F}_i$ (*normalized force components*) and $\tilde{T}_i$ (*normalized torque components*) versus time t, as shown in Figure 3 and 4. In addition, $\tilde{F}$ (*normalized force resultant*) and $\tilde{T}$ (*normalized torque resultant*), defined as

$$\tilde{F} \triangleq (\tilde{F}_1{}^2 + \tilde{F}_2{}^2 + \tilde{F}_3{}^2)^{1/2} \tag{21}$$

and

$$\tilde{T} \triangleq (\tilde{T}_1{}^2 + \tilde{T}_2{}^2 + \tilde{T}_3{}^2)^{1/2} \tag{22}$$

were calculated for each set of values of $\tilde{F}_i$ and $\tilde{T}_i$ ($i = 1, 2, 3$). These quantities furnish direct measures of the magnitudes of the righthand members of Equations 1 and 2. Figure 5 shows a sample plot of $\tilde{F}$ and $\tilde{T}$ versus t.

CONCLUSIONS

Analysis of the data obtained justified the following qualitative and quantitative conclusions.

For a particular test subject, curves such as those in Figures 3–5 are readily reproducible, both for forehand and backhand ground strokes. Curves corresponding to the same stroke performed by two different players have the same general shapes, but differ from each other in detail; that is, they display different maxima and minima, different slopes, etc. Hence, the curves can be used both to analyze a particular stroke of one player and to compare the ways in which different players execute the same stroke.

Table 1 shows extreme values of normalized force components, torque components, force resultants, and torque resultants obtained with the expert

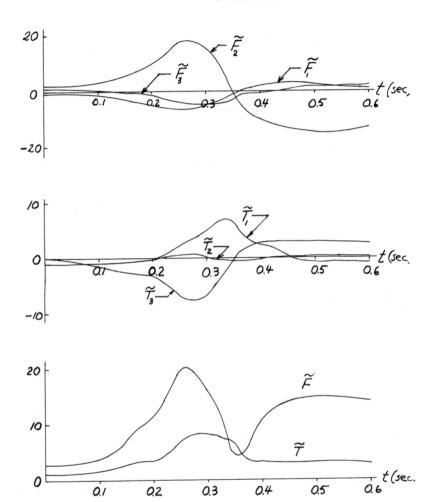

Figure 3. Normalized force components.
Figure 4. Normalized torque components.
Figure 5. Normalized force and torque resultants.

Table 1.

	$\bar{F}_1$	$\bar{F}_2$	$\bar{F}_3$	$\bar{T}_1$	$\bar{T}_2$	$\bar{T}_3$	$\bar{F}$	$\bar{T}$
Forehand	−7	18	−5	7	1	−8	21	9
Backhand	5	25	−4	5	1	8	26	9

player. The racket used in this study had a mass m ≈ 0.4 kg, and the distance between P and Q had the value L ≈ 0.3 m. Hence, the normalized forces and torques in Table 1 can be converted to actual forces and torques, expressed respectively in Newtons (N) and Newtonmeters (Nm) by multiplication with 0.4 × 9.8 = 3.9 and 0.4 × 9.8 × 0.3 = 1.2. For example, the maximal magnitudes of the righthand members of Equations 1 and 2 thus are found to have had the values 101.4 N and 10.8 Nm, respectively. The first of these can be compared with the "maximal implusive force F" of Hatze (H. Hatze, personal communication, 1973), expressed in Newtons by the formula

$$F \approx 1.5 \, V^{1.2}$$

where V is the speed of the ball, in m per sec (ms^{-1}), at the instant of impact. With F = 101.4 N, this relationship yields V = 33.5 ms^{-1}, a rather high speed. Thus, it appears that forces required merely to swing a tennis racket can have magnitudes comparable to or even exceeding those arising from impact between ball and racket.

Author's address: *Dr. T. R. Kane, Department of Applied Mechanics, School of Engineering, Stanford University, Stanford, California 94305 (USA)*

Biomechanics of bowling

Y. Murase
Nagoya-Gakuin University, Aichi

M. Miyashita and H. Matsui
University of Nagoya, Nagoya

S. Mizutani and H. Wakita
Mie University, Nagoya

Although many books and papers describing bowling techniques have been published, there is little information dealing with the mechanics of bowling. The present study was undertaken to investigate bowling performances by analyzing with cinematography the motion of the bowling ball during the approach and delivery.

PROCEDURES

Eight righthanded subjects (skilled and unskilled) were asked to deliver the ball 10 times at all 10 pins. Their physical characteristics and average scores are shown in Table 1. Each performance was filmed from the side with a 16-mm D. B. Millken-55 motion camera. The right side of the subject was toward the camera in all instances. According to the timing marks (0.10 sec) placed on the film by a signal generator, the speed of the camera was 43.3 frames per sec. The correction factor for film image size to real-life size was obtained by photographing yardsticks arranged every 500 mm along the approach lane. The 16-mm film was analyzed with the help of an NAC motion analyzer, which enlarged the image 15 times and projected it on an X-Y coordinate screen. Measurements were made of the horizontal and vertical displacement of the ball and of the bowler's right shoulder. The weight of the ball used throughout the experiments was 15–16 lbs.

RESULTS

Trajectories of the Ball

The center of gravity of the ball was transposed in every frame from 16-mm film onto the same paper for 10 deliveries. The trajectories of the ball for eight subjects are shown in order of their average scores, as shown in Figure 1. Those figures show that generally the ball was held at waist or hip height, was

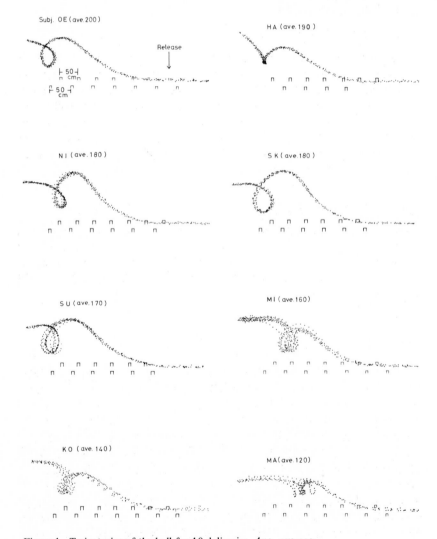

Figure 1. Trajectories of the ball for 10 deliveries. *Ave.*, average.

pushed away, dropped to the bottom of the swing, elevated to the top of the backswing arc, and swung down and forward to the point of release. The small deviations in the ball trajectories indicate that the high-average subject was able to deliver the ball consistently, while the large deviations indicate that the low-average subject delivered the ball in a somewhat different way during each trial. These results indicate that the higher the average scores, the more often the ball takes the same course during the approach and delivery. In other words, improvement in average score depends on the degree of consistency in approach and release.

Height of the Backswing and Ball Speed

The mean values of the height of the backswing were calculated (Table 1). The values for all subjects ranged from 0.7 to 1.34 m. There was a positive relationship between the height of the backswing and the average score, except with Subject H. A. The ball speed immediately after the ball release was calculated from the distance that the ball covered during three frames (69 m per sec). The values ranged from 6.31 to 7.76 m per sec. Judging from the results in the present investigation, there is no distinct relationship between the ball speed and the average score.

Speed of the Right Shoulder

Since it is difficult to measure the speed of the entire body during approach, the speed of the right shoulder was measured. The speed varied with respect to the step. The fluctuation was almost 0.9–1.9 m per sec. The mean speeds during approach were calculated, as shown in Table 1. They ranged from 2.34 to 3.28 m per sec.

DISCUSSION

Broer (1966) points out that the problem in bowling is that of applying force, which gives the ball sufficient velocity in an effective direction. The present study was undertaken to analyze this problem from the point of view of the ball motion during approach and delivery. It is important to control the direction because a deviation of one degree in the path of the ball results in a change in location of the ball by 1.05 ft at the head pin (Broer, 1966). Therefore, the ball should be delivered accurately to strike the pins in the same place each time. The coordination of the bowling arm and the feet during approach and the delivery form at the foul line affect the direction in which the ball is thrown. The ball trajectories obtained for 10 deliveries indicate that the higher the average score of the subject, the less the deviations in the trajectories of delivery (Figure 1). These results imply that the

Table 1. The average scores and stature of the subjects, height of the backswing, and speed of the ball and shoulder

Subject	Average score	Stature (cm)	Height of the backswing (m)	Calculated speed; $V = \sqrt{2gh}$ (m/sec)	Ball speed after release (m/sec)	Shoulder speed (m/sec)		
						Minimum	Maximum	Mean
O. E.	200	166	1.10	4.63	6.76	1.84	3.25	2.70
H. A.	190	170	0.87	4.14	7.76	2.22	3.85	2.97
N. I.	180	168	1.27	4.99	7.72	1.94	2.91	2.34
S. K.	180	163	1.34	5.12	7.30	2.10	3.03	2.58
S. U.	170	168	1.21	4.86	6.31	2.27	3.30	2.69
M. I.	160	165	0.92	4.24	7.15	1.51	3.45	2.67
K. O.	140	179	0.96	4.33	7.38	2.21	3.77	3.28
M. A.	120	163	0.70	3.69	7.70	1.97	3.20	2.76

coordination of the bowling arm and the feet during approach and the form of delivery were quite consistent each time for the high-average subjects. On the other hand, timing in pushaway, backswing, and forward and downward swing during approach were not consistent each time in the case of low-average subjects. The force to give the ball sufficient speed consists of (a) gravity, which depends on the height of the backswing, (b) the momentum of the body gained during the approach, and (c) muscle contractions, which increase the speed of the downswing. These different kinds of force applied to the ball cannot be measured directly. Therefore, the amount of force applied to the ball during approach and delivery was estimated. The results show that there was no distinct difference in the final ball speed between the skilled and unskilled subjects (Table 1). That is, the ball speed after the release was almost constant, regardless of the average scores. The velocities obtained would yield the generally accepted time of from 2-1/8 to 2-3/4 sec (Bellisimo and Neal, 1971) during which the ball travels from the foul line to the head pin. However, the height of the backswing differed greatly among the subjects. Assuming that there is no resistance in a straight pendulum swing during delivery, the speed was calculated by the equation: $V=\sqrt{2gh}$ (where V is the ball speed, g is the acceleration caused by gravity, and h is the height of the backswing). The values ranged from 3.69 to 5.12 m per sec (Table 1). The ratio of this speed to the actual ball speed after release ranged from 47.9 to 77.0 percent. The high-average subjects, exept for Subject H. A., showed the higher percentages, from 64.6 to 77.0, whereas the low-average subjects showed the lower percentages, from 47.9 to 59.3. In other words, the low-average subjects and Subject H. A. generated the larger momentum of the body through either the approach or the greater force of the arm movement. Indeed, the shoulder of Subject H. A. moved faster than those of the other high-average subjects. The ball speed, A, minus the speed of the shoulder, B, and the estimated speed from the acceleration of gravity, C, was calculated in relation to time for all subjects, respectively. The results are shown in Figure 2. In the case of the highly skilled Subject O. E., the speed B increased parallel with the speed C almost until the moment of the ball release. Assuming that a certain resistance existed in joints and muscles, it was observed that gravity and inertia brought the ball down and forward to the point of release without muscular force in Subject O. E. On the other hand, the speed B curve crossed the speed C curve approximately 0.1 sec before the ball release in the case of Subject M. I. This means that Subject M. I. forced the downswing in the earlier part of delivery. Also, in the case of unskilled Subject M. A., the speed B decreased to almost zero at first and then increased rapidly. This means that Subject M. A. carried the ball in the first half of the downswing and that thereafter he forced the downswing positively in order to give the ball sufficient speed.

   Broer (1966) pointed out problems in bowling, such as speed and direction. As for speed, there was no difference among the subjects employed in

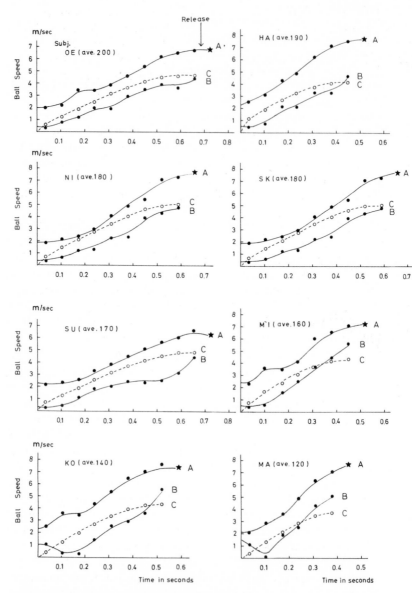

Figure 2. Ball speed of the forward and downswing. A, actual ball speed; B, actual ball speed minus speed of the shoulder;* C, estimated ball speed from the equation, $V = 2gh$.

$$*|\vec{V}_B| = \sqrt{(V_Bx - Vsx)^2 + (V_By - Vsy)^2}$$
$$\vec{V}_B = (V_Bx, V_By): \text{Ball velocity}$$
$$V_S = (Vsx, Vsy): \text{Velocity of the shoulder.}$$

this study, although their average scores were not the same. On the other hand, the results indicated that the factor of accuracy is more important.

Armbruster, Irwin, and Musker (1967) advised the bowler not to force a delivery, but to let the weight of the ball work. The arm merely serves as a pendulum. The results obtained by the analysis of the ball motion during approach and delivery were in agreement with that description of the bowling technique.

REFERENCES

Armbruster, D. A., L. W. Irwin, and F. F. Musker. 1967. Basic Skills in Sports for Men and Women. 4th Ed. C. V. Mosby, St. Louis.
Bellisimo, L., and L. L. Neal. 1971. Bowling. Prentice-Hall, Englewood Cliffs, N. J.
Broer, M. R. 1966. Efficiency of Human Movement. 2nd Ed. W. B. Saunders, Philadelphia.

Author's address: Mr. Y. Murase, Nagoya Gakuin University, Kamishinano-cho, Seto City, Aichi-Ken (Japan)

Kinematic and kinetic analysis of the golf swing

J. M. Cooper, B. T. Bates, J. Bedi, and J. Scheuchenzuber
Indiana University, Bloomington

The action of swinging a club in golf is a two-arm underhand striking pattern. It is believed that this classification is defensible because the left arm (for a righthanded golfer) moves from a horizontal position parallel with the trunk axis at impact (Cooper and Glassow, 1972). The sequence of joint movements consists of forward action of the left knee, hip rotation, spinal rotation, shoulder action, and, finally, wrist and hand action.

However, it is believed that questions concerning body weight shift, head action, forces generated by the feet, the action of the arms during backswing and downswing, and many other factors have not been answered satisfactorily by the results of previous studies.

PURPOSE

The purpose of this study was to analyze, by means of cinematographic and force-recording techniques, the swing of highly skilled golfers using various clubs.

PROCEDURES

Five members of the Indiana University's Big Ten Conference championship golf team served as subjects and performed in a laboratory setting using the seven-iron, three-iron, and driver as clubs. Each subject was filmed at 200 frames per sec with a LOCAM camera. Forces during the swing were recorded for each foot on separate force platforms which had the capacity to detect horizontal and vertical forces for each plate up to 300 lbs. Both of these force platforms were mounted on a larger single force platform capable of responding to vertical forces up to 1500 lbs. Also, the large platform could detect horizontal and lateral forces up to 500 ft-lbs and a maximal torque of 250 ft lbs. All recording equipment was synchronized by means of an electronic signaling device so that comparisons among the various data could be made. Data were collected from the film with a digital coordinatograph and were processed by computer with the use of a special program (FILMDAT).

ANALYSIS OF DATA

The data were analyzed under the following headings: (a) the relationship between the position of the center of gravity as computed by the segmental method and the forces generated, (b) the shifting of body weight during the action, (c) the horizontal and vertical forces generated by each foot and the total body, and the relationships among various force components, (d) the angular positions and velocities of selected body parts (especially the arms) and the club during backswing and downswing, (e) the ball and club velocities immediately prior to and after impact, and (f) the angle of trajectory of the ball. Comparisons were made among the mean values of all golfers using the three clubs.

FINDINGS

Results related to the mean center of gravity shift, as recorded cinemato-graphically, and to body weight shift and total body forces, as recorded by both the individual foot force plates and the single force platform, are generalized as follows (see Figures 1 and 2):
1. For all clubs, the force was distributed approximately equally between both feet; also, the line of gravity fell about midway between the two feet at the beginning of the downswing.
2. There was a great similarity in the shift of forces between the two feet and the location of the line of gravity until about at the point of impact for all clubs. There was a force shift of about 25 percent toward the front foot as impact was reached. The line of gravity shifted to a point three-fourths of the distance toward the front foot. The weight distribution was about 75 percent on the front foot and 25 percent on the rear foot at impact.
3. After the impact point was reached, the line of gravity continued to shift toward the front foot for all clubs evaluated. This shift appeared greatest in the seven-iron and least in the driver.
4. After the impact point was reached, the force distribution between the feet continued at nearly the same value for the seven-iron (75 percent, front; 25 percent, rear) while reverting to a more balanced position (50 percent, front; 50 percent, rear) in the case of the driver. This would indicate that a

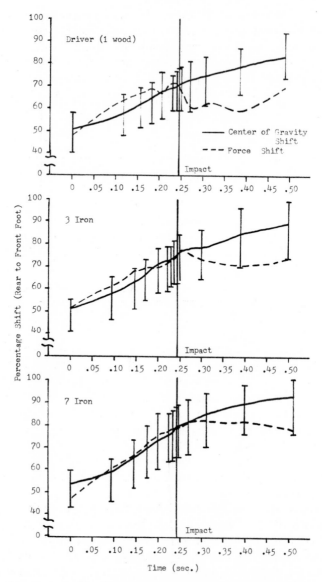

Figure 1. Percentage of shift of forces and line of gravity during golf swing.

more stable position was assumed by the golfer when using the less lofted club.

5. At the end of the follow-through, the force distribution was 70–80 percent on the front foot, depending on the club selected, with the value for the driver nearer 70 percent and that of the seven-iron nearer 80 percent.

6. During the use of the three- and seven-irons, the maximal force shift to the

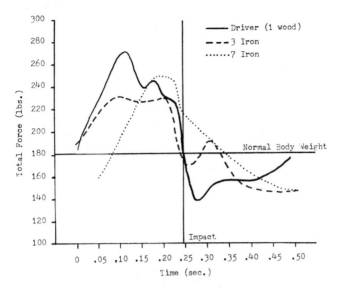

Figure 2. Total vertical force during golf swing (foot force plates).

front foot occurred after impact. However, when the driver was used, the maximal force shift occurred prior to the club's reaching the point of impact.
7. The magnitude of the total force exerted in the vertical direction, as recorded by the foot force plates, reached a maximal value between the start of the downswing and a horizontal club position prior to impact, depending on the club used. The magnitude of this force ranged from 150 percent of the total body weight when the driver was used to 133 percent of the total body weight when the three-iron was used.
8. In all cases examined, the total force produced in the vertical direction decreased rapidly during the impact phase of the swing (a club position of 45° prior to and after impact).
9. The minimal total force exerted in the vertical direction against the force plates occurred after the impact point and was approximately 80 percent of the total body weight, thus indicating that the centrifugal force of the club had pulled the body upward. The minimal total force was reached at the end of the follow-through for the two irons; but for the driver, this minimum was reached at a club position of 45° after impact.

Results related to rotational forces as recorded by means of the single force platform are as follows (see Figure 3):
1. The rotational forces changed from clockwise to counter-clockwise prior to impact.
2. The maximal clockwise force was achieved early in the downswing phase, whereas the maximal counter-clockwise force occurred early in the follow-through.

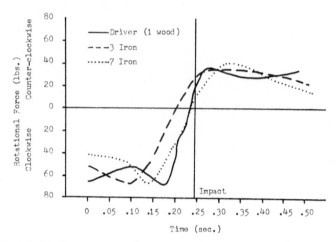

Figure 3. Rotational components of force during golf swing.

3. The occurrence of the torque changes from clockwise to counter-clock-wise indicated that a turning of the body occurred prior to impact so that a more advantageous striking position could be assumed.

Results related to head position, club head velocity, and angle of projection are as follows:

1. Although the minimal vertical head positions of the golfers differed, that for each golfer was consistent for all clubs used. This low head position occurred for three golfers prior to impact, for one at impact, and for one immediately after impact.

2. The minimal vertical position of the total body center of gravity occurred considerably prior to impact and before the club reached a position parallel to the ground.

3. Linear club head velocities indicated that all golfers appeared to accelerate the driver and the three-iron up to the point of impact, with a corresponding deceleration occurring immediately after impact. However, in the case of the seven-iron, deceleration started well before impact occurred. The mean velocities for the three clubs from a position of 45° before impact to 45° after impact are presented in Table 1.

4. The linear velocity of the club head in the impact zone was highest for one golfer (rated best by his coach) for all swings analyzed (this golfer had the highest ball velocity for the three clubs also). Another golfer (rated second best) had the lowest club head velocity in the impact zone and, correspondingly, the slowest ball velocity.

5. The mean angles of projection of the ball for the three clubs were 11.25, 11.45, and 19° for the driver, three-iron, and seven-iron, respectively.

6. The mean ball velocities for the three clubs were as follows: for the driver,

Table 1. Mean club head velocity

Time (sec)	Driver velocity (ft/sec)	(m/sec)	Time (sec)	Three-iron velocity (ft/sec)	(m/sec)	Time (sec)	Seven-iron velocity (ft/sec)	(m/sec)
-0.015	134.28	40.93	-0.015	124.83	38.05	-0.015	122.75	37.41
-0.005	149.35	45.52	-0.005	131.24	40.00	-0.005	114.32	35.16
0.000*	120.25	36.65	0.000*	109.07	33.24	0.000*	96.50	29.41
+0.005	95.72	29.18	+0.005	87.35	26.63	+0.005	80.48	24.53
+0.025			+0.030			+0.015		

*Point of impact.

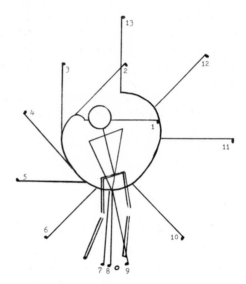

Figure 4. Golf swing analysis. Frames selected for analysis at 45° intervals (club angle). Prominent frames: *1*, top of back swing; *7*, one frame prior to impact; *8*, impact; *9*, one frame after impact; *13*, end of the follow-through.

205.36 ft per sec (62.59 m per sec); for the three-iron, 192.02 ft per sec (58.53 m per sec); and for the seven-iron, 167.95 ft per sec (51.19 m per sec).

CONCLUSIONS

The following conclusions were drawn from the results of the analysis of the data gathered in this investigation. First, there is some evidence to support the concept of "hitting with the left side," in that there was a shift of body weight to the front foot during the swing. However, the weight shift was not complete; therefore, the concept is probably not as conclusive as once was thought by golf teachers and performers. Second, the different behavior of the golfers while using the seven-iron was unusual, in contrast with their behavior with the other clubs. This difference could have been caused by the golfers' altering their motions in an attempt to accentuate the action of this particular club so as to hit down into the ball, or it could have been that the golfers were trying to gain greater control of the club, emphasizing accuracy rather than force.

RECOMMENDATIONS

The following recommendations are made for consideration in future investigations. First, because, the golf swing does not occur only in a vertical plane, biaxial cinematographic techniques should be employed to increase the accuracy of the film data. Second, a similar study should be conducted using world-class golfers as subjects to ascertain whether the same patterns are characteristic of their movements, as was found in this study. Third, the study should be replicated using a greater number of subjects to increase the validity of the findings. Finally, in future studies, the filming speed should be increased (possibly to as high as 1,000 frames per sec), because it was difficult to determine the point of impact with a filming speed of 200 frames per sec.

REFERENCES

Broer, M. R., and S. J. Honty. 1967. Patterns of Muscular Activity in Selected Sports Skills. Charles C Thomas, Springfield, Ill.

Cochran, A. and J. Stobbs. 1968. The Search for the Perfect Swing. J. B. Lippincott, Philadelphia.

Cooper, J. M., and R. B. Glassow. 1972. Kinesiology. C. V. Mosby, St. Louis.

Jorgensen, T., Jr. 1969. On the dynamics of the swing of the golf club. Amer. J. Phys. 38.

McIntyre, N. and R. Snyder. 1962. The truth about weight shift. Golf Dig. 13.

Rehling, C. H. 1955. Analysis of techniques of the golf drive. Res. Quart. 26: 80.

Slater-Hammel, A. T. 1948. Action current study of contraction movement relationships in golf stroke. Res. Quart. 19: 164.

Williams, D. 1967. The dynamics of the golf swing. Quart. J. Mech. Appl. Math. 20.

Author's address: Dr. John M. Cooper, Biomechanics Laboratory, School of Health, Physical Education and Recreation, HPER Building, Indiana University, Bloomington, Indiana 47401 (USA)

Application of biomechanics to the fencing lunge

C. Simonian and E. L. Fox

Ohio State University, Columbus

For at least 100 years, the manner of execution of the fencing lunge has remained essentially unchanged. Other than empirical refinements made to suit individual styles, very little research has been done on fencing techniques in general or on the lunge in particular. Yet, the lunge is one of the most important movements in fencing and is given great emphasis by fencing teachers.

Close examination of the traditional lunge suggested some ways in which biomechanical principles could be applied to lengthen the lunge distance. It was hypothesized that: (a) if the fencer's body were in motion just before the rear leg began its force application, the force would be more effective; (b) if the fencer's center of gravity were lowered as the lunge began, the horizontal component of applied force would be increased; and (c) if the leading lower leg underwent a vigorous pendulum-like motion, it would tend to reach out to a more distant base of support. It was the purpose of this study to devise a lunge technique which in theory would utilize these principles to produce a longer lunge.

METHODS

Traditional Lunge

Representative fencing texts, both past (Corbesier, 1873; Hutton, 1891; Prevost and Jollivet, 1891) and present (Anderson, 1970; Castello and Castello, 1962; Crosnier, 1951), were found to be in essential agreement as to both lunging form and the source of force for the lunge. The lunge typically is described as being a sequence of movements which begins with an extension of the sword-arm, as required in the rules to establish the right to attack. This is followed by a forward movement of the leading foot and of the body

that results from a vigorous extension of the rear leg. Some authors specifically advocate that the leading leg be extended at the knee just after the
sword-arm is extended.

Experimental Lunge

As with the traditional lunge, the experimental lunge begins with an extension of the sword-arm. Simultaneously with this extension, there is a short,
rapid flexion of the leading leg at the knee. This flexion removes the forward
base of support of the fencer in the on-guard position and causes the center
of gravity to fall in an arc, the axis of which is the rear foot. With the body's
inertia thus overcome and the center of gravity lowered, the rear leg then is
extended forcefully. During the rear-leg extension, the front lower leg swings
forward from its flexed position to one of full extension at the knee. Finally,
the front lower leg once again is flexed at the knee so that the heel can strike
the floor to establish a new base of support as the lunge is completed.

PROCEDURES

The subjects were 20 male undergraduate students enrolled in an advanced
fencing course in the physical education program at The Ohio State University. Each subject already had completed a beginning fencing course in which
the traditional lunge had been taught. In the first week of the advanced
course, the subjects were tested for distance on three maximal traditional
lunges; the mean distance was recorded as the test score. All lunges were
executed from an on-guard stance in which the subject's feet were placed on
either side of lines 45.7 cm apart. Lunge distance was measured as the
distance moved forward by the leading heel. The rear foot was not permitted
to slide forward, inasmuch as this would increase lunging distance, and no
lunge was counted if the subject lost his balance before a measurement could
be taken. The subjects were divided into two groups on the basis of the lunge
distances.

In the next class period, the experimental lunge was introduced to one
group, while the control group continued to practice the traditional lunge but
with an emphasis on stretching out the lunge as far as possible. From that
point on, each group did 25 lunges at the beginning of each class period;
additionally, each group used its own method of lunging during the regular
class period.

By the ninth class meeting, the experimental group was judged to have
learned the new lunge form sufficiently, and the subjects were given posttests
on lunge distance to determine gains made by each group. Each subject made
three maximal lunges of the type he had been practicing, and the distances
were recorded. All subjects were observed carefully to assure that the correct

Table 1. Pretest, posttest and gains in lunge distance (cm) for the control and experimental groups

	Control group			Experimental group		
Subject	Pretest	Posttest	Gain	Pretest	Posttest	Gain
1	91.4	97.8	6.4	94.0	112.3	18.3
2	81.3	83.1	1.8	78.7	93.5	14.8
3	78.7	85.6	6.9	76.2	88.9	12.7
4	76.2	81.8	5.6	74.9	79.2	4.3
5	76.2	91.9	15.5	73.7	81.8	8.1
6	72.4	83.1	10.7	71.1	89.7	18.6
7	68.6	80.8	12.2	66.0	89.7	23.7
8	66.0	71.1	5.1	63.5	73.9	10.4
9	64.8	75.7	10.9	55.9	80.0	24.1
10	64.8	72.9	8.1			
11	52.1	65.5	13.4			
$\bar{X}$	72.0	80.8	8.8	72.7	87.7	15.0
S.D.	10.4	9.3	4.1	10.7	11.2	6.8
			$p < 0.001$	$p < 0.05$		$p < 0.001$

form was used. Any false starts, improper executions, and other unacceptable lunges were not recorded and had to be repeated.

RESULTS AND DISCUSSION

The individual and group results of the experiments are shown in Table 1. As indicated by the independent t test, there was no significant difference between groups in pretest lunge distances. Each group gained significantly ($p < 0.001$, paired t test) in their lunge distances, with the experimental group showing a significantly greater gain ($p < 0.05$) than the control group.

With no other factors identifiable to account for this significant difference between groups, it can be concluded reasonably that the modified lunge, as described here, can be used by a fencer who desires to increase the length of his lunge. However, it also should be noted that not all fencers will wish to change their present lunge techniques simply to gain additional distance. For example, there is no evidence known to us to indicate that a long lunge is even related to success in fencing. It was our main purpose here to illustrate one application of the principles of biomechanics to a sports skill so as to effect a positive change.

REFERENCES

Anderson, B. 1970. All About Fencing. Stanley Paul and Co. London.
Castello, H., and J. Castello. 1962. Fencing. The Ronald Press, New York.

Corbesier, A. 1873. Theory of Fencing. Government Printing Office, Washington, D.C.

Crosnier, R. 1951. Fencing with the Foil. A. S. Barnes Co., New York.

Hutton, A. 1891. The Swordsman. H. Grevel and Co., London.

Prevost, C., and G. Jollivet. 1891. L'Escrime et le Duel. Libraire Hachette, Paris.

Author's address: *Dr. Charles Simonian, School of Health, Physical Education and Recreation, The Ohio State University, Columbus, Ohio 43210 (USA)*

Electromyo-
graphy

The anatomical aspects of current electromyographic observations

T. G. Simard
Université de Montréal, Montreal

The specialist in gross anatomy who is interested in the structural and functional muscles of specific postures and movements should be impressed by the recent observations obtained with the aid of electromyography. Since the valuable work of G. B. Duchenne of Boulogne, modern methods of kinesiologic evaluations have been developed that have led to new conclusions on the normal and abnormal behavior of motor unit activity. The isolation and maintenance of motor unit activity were demonstrated by normal and handicapped persons, and the possible ability to command the final common pathway, the motor unit, served to create both theories and applications. The utilization of willed discharges of action potentials in motorizing prosthetic and orthetic systems is of particular importance in orthopedics, neurology, and rehabilitation.

This study concerns basic morphological aspects of concepts derived from electromyography. The concept and characteristics of single motor unit action potentials, recent studies on ability to command their fine motor performance, and special observations on the functional anatomy of limb articulations will be treated. An original solution for studying a postural achievement from combined electromyographic, electrogoniometric, and osteokinematic techniques is summarized.

THE MOTOR UNIT TERRITORY AND THE CHARACTERISTICS OF MOTOR UNIT ACTION POTENTIALS

The "motor unit," a term used first by Sherrington in 1925 (Eccles and Sherrington, 1930), is an entity composed of a motor nerve cell and the group

Financial support for this research has been provided by the Medical Research Council of Canada.

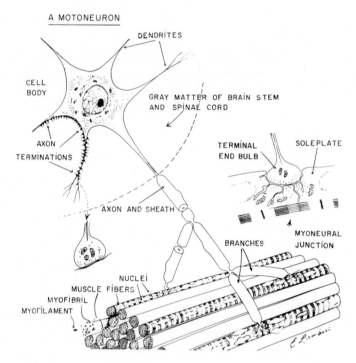

Figure 1. The motor unit.

of muscle fibers that it innervates (Figure 1). Motor nerve cells, or motoneurons, are found in the brain stem (motor nuclei for cranial nerves) and in the spinal cord (ventral horn cells for spinal nerves). A motoneuron is made up of a cell body, a number of dendrites, and one axon with its sheath. The surfaces of the dendrites and cell body are covered with as many as 1,000 axon terminations of other nerve cells from spinal and higher center projection fibers (Fulton, 1951). The axon, commonly called the nerve fiber, subdivides into a number of branches. Each branch terminates at an end plate by a terminal end bulb on the surface of the muscle fiber. In the region of innervation, the surface of the muscle fiber contains a specialized zone of modified sarcoplasm called "the soleplate." A very thin gap exists between the end bulb of an axon and the soleplate of a muscle fiber. The muscle fiber, a multinucleated cell, is specialized to conduct impulses and subsequently to contract.

The number of muscle fibers which are supplied by the terminal branches of a nerve fiber—within a motor unit—is variable from muscle to muscle and within the same muscle; it has been related to the quality of control exhibited by the muscle. For example, the innervation ratio of the small delicate laryngeal muscles is described as being 3:1 (Ruëdi, 1959) or that of the human lateral rectus muscle as 9:1 (Feinstein et al., 1955). These may be contrasted with the large innervation ratio of the medial head of the gastro-cnemius muscle of 2000:1 or with that of the tibialis anterior muscle of 562:1 (Feinstein et al., 1955).

The area within which the constituent muscle fibers of a motor unit are found is variable and depends on the number and the diameter of muscle fibers contained within the motor unit, as well as on the structure of a muscle. The territory of a motor unit can vary over a circular area 4–6 mm in diameter (Buchthal, Guld, and Rosenfalck, 1957). The muscle fibers of a motor unit can be found intermixed with other muscle fibers from different motor units and at different depths (Coërs and Woolf, 1959).

When an impulse travels along the nerve membrane and reaches the myoneural junction, release of acetylcholine through the presynaptic membrane and into the intercellular subneural space brings about changes in permeability and subsequent modifications in the difference of potential across the postsynaptic membrane of a muscle fiber (Porter and Bonneville, 1963). As a consequence of these changes, the membrane depolarizes and the muscle fibers of the same motor unit contract when an impulse arrives. The summation of electrical changes in the muscle-fiber membranes which occurs during each contraction and each recovery period produces a characteristic electrical wave or spike which is called a "motor unit action potential" (Figure 2).

By adequate technique of amplification and display, the characteristic properties of each motor unit action potential can be described by their shape, amplitude, duration, and frequency of discharges.

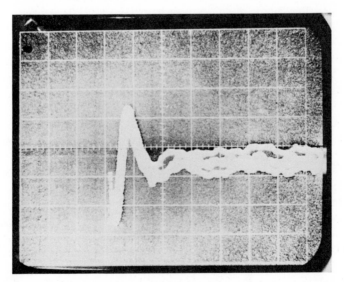

Figure 2. A motor unit action potential maintained in regular activity. Activity is superimposed by the trigger system of a monitor oscilloscope.

The shape of an action potential depends on the number of muscle fibers within a motor unit, on their approximately simultaneous contraction, on the magnitude of the electrical event, on the kind of electrode used, on the distance of the motor unit territory from the recording electrode, and on the response-band width utilized in the amplification system. The action potential may be diphasic or polyphasic.

The amplitude of a motor unit action potential is the peak-to-peak deflection of the highest component spike or wave which is produced. In general, it ranges from approximately 100 to 300 microvolts. The duration of an action potential is quite stable for each normal unit. It can be 4–14 msec in normal adults, but is significantly shorter in infants (Buchthal and Rosenfalck, 1962). The frequency of firing a normal, active, isolated motor unit is usually less than 16 per sec (Basmajian, Baeza, and Fabrigar, 1965), but it may reach 50 per sec during a strong contraction, in which the isolation of a motor unit no longer is seen in normal subjects (Adrian and Bronk, 1929). The interval between single repetitive action potentials is slightly irregular during a steady repetitive activity of a motor unit. In a voluntary contraction of a single motor unit the interval may be for as long as the individual wishes (Harrison and Mortensen, 1962).

ABILITY TO COMMAND FINE MOTOR RESPONSES

With the aid of electromyography and under the influence of the audiovisual feedback information on the activity of motor units, an excellent ability to command fine motor responses may be obtained by adequate training. The ability to command the frequency of discharges of single units, including specific rhythms, as the production of gallops, drum roll, and drum beat rhythms in the most skillful subjects (Basmajian, 1963), is significant, not only for physiological but also for therapeutic purposes.

Effects of movements (Basmajian and Simard, 1967) were studied, and methods of training individual muscle segments were developed by superimposing some factors known to influence motor performance. The method was developed further for training muscular activity during the treatment programs for amputees and for malformed and paralyzed patients (Simard, 1969; Simard and Ladd, 1969, 1971), in which an emphasis was given to the influence of the most important daily living movements over acquired, combined, segmental, neuromuscular, fine control of the motor responses. The method was adapted to evaluate the ability of the quadriplegic patient to command the motor responses. The ability to activate three muscle segments of a single muscle or several muscles in different combinations of activation and inhibition of one, two, and three segments simultaneously can be obtained with a step-by-step training method (Simard, Ladd, and Hamonet, 1973). The application of this method to motorization of an orthetic hand system is now under intensive investigation at the Service de Réeducation Fonctionnelle du Centre Hospitalier Universitaire de Créteil under the directive of C. Hamonet.

OBSERVATIONS ON THE FUNCTIONAL ANATOMY OF LIMB ARTICULATIONS

Angular joint displacements require a constant adjustment from chief synergists, stabilizers, and link actions of muscles. Certain segmental postures held at rest are governed by the forces of gravity and the nature of supportive surfaces.

Within the upper limb, the mechanism of shoulder, elbow, wrist, and hand activities was undertaken by several investigators; the most basic information was treated earlier (Basmajian, 1967). One should read the original work of Inman, Saunders, and Abbott (1944), in which the study of the mechhanism of the shoulder joint was discussed and degrees of arm flexion and abduction were related to the level of electromyographic activity. In descriptive anatomy, it is accepted that the chief muscles that act on the glenohumeral joint are the deltoid, pectoralis major, serratus anterior, and periscapular muscles, especially the rotator cuff. However, the activity of the deltoid and supraspinatus muscles in arm abduction has been surrounded by controversy. Today, we know that the complete paralysis of the supraspinatus muscle reduces the strength and endurance of arm abduction (Von Linge and Mulder, 1963), and, in a few cases of paralysis of the deltoid muscle, arm abduction is also possible (C. Hamonet, personal communication, 1973). Inman et al. (1944) found initial and secondary peak activity in the anterior deltoid and in the clavicular head of the pectoralis major muscles during arm flexion. The biceps brachii, especially the long head, also acts during this arm movement (Basmajian and Latif, 1957). The active depressors of the humerus demand further investigations because the exact results from the pectoralis minor, short head of biceps, and coracobrachialis muscles acting on the coracoid process and indirectly on the arm depression have not been established yet with electromyography. The rotator cuff muscles were shown by Inman et al. (1944) to act as a force couple during arm flexion and abduction. The adduction movement can be made with the activity of the pectoralis major and the latissimus dorsi muscles, but the teres major muscle seems to take an active part in holding a resisted backward arm adduction during a combined arm internal rotation (E. Paquette, D. Gravel, and T. Simard, unpublished observations, 1973). The teres major muscle shows individual behavior in our present investigation of the activity of link shoulder muscles. The synergy between the brachialis, biceps brachii, and brachioradialis muscles in elbow flexion is variable, and there is rarely any unanimity of action between them (Basmajian and Latif, 1957).

Observations of the effect of the speed, stability, position, individual morphology variables, and a way to apply resistance on the pattern of spatial and temporal activity must be taken into consideration in any research. The effect of speed on the activity of chief pronators of the forearm was considered by two authors who indicated firmly that the pronator quadratus muscle is the main pronator whatever the speed employed (Basmajian, 1967; Hollingshead, 1958). The ways to apply resistance and varying amounts of resistance have been investigated. For example, all authors agree that the anconeus muscle is active during extension of the elbow but that this muscle also is active in resisted pronation and supination of the forearm (Travill, 1962). These cocontractions in opposing positions for certain muscles under specific effects of resistance are of great importance in applying methods of treatment for physically ill patients.

Hand function has been the favorite subject of several investigators. Positions of firm and normal grips and pinch were studied and mean electromyographic activities were compared. Forrest and Basmajian (1965) demonstrated that the thenar muscles are slightly more active during the holding of a cup than during the holding of a glass. In fact, the lateral pinch utilized to hold the handle of a cup or to use a key is an important daily dynamic activity, and this may explain the greater demand of activity in holding this posture. In a present study of the upper extremity link system, we found that several hand and forearm movements affect the shoulder muscles (Paquette, Gravel, and Simard, 1973). These link actions may help to clarify the functional use of the upper extremity in daily living and may lead to modern methods of training the coordination, ability, and strength of the upper extremity.

EVALUATION OF MUSCLE, BONE, AND JOINT FUNCTIONS IN UNILATERAL BELOW-KNEE AMPUTEES

It is important to observe the behavior of several structures simultaneously in order to understand the functional interrelationship among them. A summary is presented of a method developed by A. Balakrishnan (biomedical engineer at the Artificial Limb Centre, Madras, India) and this author to evaluate the functional anatomy of the knee in below-knee amputees in order to give a general view of the principle.

Five muscles of the sound limb of seven patients were chosen for their role in knee flexion and positioning of the leg during squatting with feet flat on the ground. The muscles studied were the popliteus, tibialis anterior, rectus femoris, semimembranosus, and biceps femoris. Their electromyographic activities were detected at a predetermined anatomical region situated as close as possible to the neuromuscular hili by use of precise bony landmarks on the skin. A fine-wire bipolar electrode was inserted into the popliteus muscle with a hypodermic needle 1.5 in long. For the other muscles, surface miniature Beckman electrodes were utilized; these electrodes were connected to the electromyographic low-level preamplifiers, and the information was displayed on a monitor scope and recorded on linographic paper with a sensitive ultraviolet system (Honeywell, Type 1508).

To obtain simulataneous recordings of angular displacements of knee and ankle joints during the complete squatting movement and at the completion of the posture, goniometers were installed. A first potentiometer was applied at the functional knee center, suggested by Gardner and Chippinger (1969), on the lateral or medial epicondyle of the femur of the sound leg. The ankle goniometer involved the placement of a potentiometer over the center of the lateral or medial malleolus. The test for the stability of potentiometers was made following five squatting movements involving flexion and extension of the knee and dorsiplantar flexion of the foot. The displacements of the joints were transduced and amplified linearly through signal conditioning units, and the information was sent to the recording system, in which the electromyographic traces were related in time to the electrogoniographic traces.

Subjects were required to squat five times and to remain in the squat position for 5 sec at the completion of each movement. The electromyographic and electrogoniometric data during the five complete cycles of squatting were recorded.

To study bone displacement, osteokinematic observations also were made simultaneously. For example, the loci of the levels of the head, greater trochanter, lateral epicondyle of femur, and lateral malleolus during the complete cycle of squatting of subjects were studied. A wooden frame was built to study the space relationship of the body displacement with a stable structure (Figure 3). A single photograph of the complete movement was taken using a stroboscopic technique of multiple exposures on the same negative. Loci of movements were traced and body height in the squatting posture was derived from the analyses of the photographs.

The data from the utilization of this method currently being evaluated will point out the role of muscles in relation to body displacement.

The role of muscles in handicapped limbs in regard to dynamic posture can be analyzed utilizing similar techniques in basic research of this type. In conclusion, both the usefulness and the practicability of this approach are emphasized in combining electromyography with biomechanics. Together they offer many dividends for treating handicapped people.

ACKNOWLEDGMENTS

I wish to thank Dr. J. V. Basmajian, Director of Research, Emory University Regional Rehabilitation Research and Training Centers, Atlanta, Georgia, and Dr. P. Jean, Director of le Département d'Anatomie, Université de Montréal, Montréal, for their helpful comments in reading this paper.

320   Simard

Figure 3. Development of a wood frame for the dynamical postural study of squatting. Subject in full squatting position. Bony landmarks are installed for the osteokinematical study over specific region. *T.R.*, temporal region; *K.C.*, knee center; *G.T.*, greater trochanter; *L.M.*, lateral malleolus; *G.T.L.*, the greater trochanter level, in standing position.

REFERENCES

Adrian, E. D., and D. W. Bronk. 1929. The discharge of impulses in motor nerve fibres. J. Physiol. 67: 119–151.

Basmajian, J. V., and A. Latif. 1957. Integrated actions and functions of the chief flexors of the elbow: a detailed electromyographic analysis. J. Bone Joint Surg. 39-A: 1106-1118.

Basmajian, J. V. 1963. Control and training of individual motor units. Science 141: 440–441.

Basmajian, J. V. 1967. Muscles Alive: Their Functions Revealed by Electromyography. 2nd Ed. Williams & Wilkins, Baltimore.

Basmajian, J. V., M. Baeza, and C. Fabrigar. 1965. Conscious control and training of individual spinal motor neurons in normal human subjects. J. New Drugs 5: 78–85.

Buchthal, F., C. Guld, and P. Rosenfalck. 1957. Multielectrode study of the territory of a motor unit. Acta Physiol. Scand. 39: 83–103.

Buchthal, F., and P. Rosenfalck. 1962. Motor unit potentials at different ages. Arch. Neurol. 6: 336–373.

Coërs, C., and A. L. Woolf. 1959. The Innervation of Muscle, A Biopsy Study. Charles C. Thomas, Springfield, Ill.

Eccles, J. C., and C. J. Sherrington. 1930. Numbers and contraction-values of individual motor units examined in some muscles of the limb. Proc. Roy. Soc. 106B: 327–357.

Feinstein, B., B. Lindegård, E. Nyman, and G. Wohlfart. 1955. Morphologic studies of motor units in normal human muscles. Acta Anat. 23: 127–142.

Forrest, W. J., and J. V. Basmajian. 1965. Function of human thenar and hypothenar muscles: an electromyographic study of twenty-five hands. J. Bone Joint Surg. 47-A: 1585–1594.

Fulton, J. F. 1951. Physiology of the Nervous System. 3rd Ed. Oxford University Press, New York.

Harrison, J. F., and O. A. Mortensen. 1962. Identification and voluntary control of single motor unit activity in the tibialis anterior muscle. Anat. Rec. 144: 109–116.

Inman, V. T., J. B. de C. M. Saunders, and L. C. Abbott. 1944. Observations on the function of the shoulder joint. J. Bone Joint Surg. 26: 1–30.

Porter, K. R., and M. A. Bonneville. 1963. An Introduction to the Fine Structure of Cells and Tissues. Lea & Febiger, Philadelphia.

Ruëdi, L. 1959. Some observations on the histology and function of the larynx. J. Laryngol. Otol. 73: 1–20.

Simard, T. G. 1969. Méthode d'entraînement pré-prothétique chez les amputés du membre supérieur. Étude électromyographique. Laval Med. 40: 366–370.

Simard, T., and H. Ladd. 1969. Conscious control of motor unit with thalidomide children. An electromyographic study. J. Development. Med. Child Neurol. 11: 743–748.

Simard, T., and H. Ladd. 1971. Differential control of muscle segments by quadriplegic patients. An electromyographic procedural investigation. Arch. Phys. Med. Rehabil. 52: 447–454.

Simard, T., H. W. Ladd, and C. Hamonet. 1973. Electromyography in biomedical studies. Myoelectric control of ortheses. In: J. E. Desmedt (ed.), New Developments in Electromyography and Clinical Neurophysiology. Vol. 1, pp. 663–691. S. Karger, Basel.

Travill, A. A. 1962. Electromyographic study of the extensor apparatus of the forearm. Anat. Rec. 144: 373–376.

Von Linge, B., and J. D. Mulder. 1963. Function of the supraspinatus muscle and its relation to the supraspinatus syndrome. J. Bone Joint Surg. 45 B: 740–754.

Author's address: *Dr. Thérèse G. Simard, Département d'Anatomie de la faculté de médecine de la Université de Montréal, Montréal, P.Q. (Canada)*

Improved EMG quantification through suppression of skin impedance influences

G. Rau

Research Institute for Human Engineering, Meckenheim

Surface electromyography (EMG) can be used in studying electrical activities of large skeletal muscles which are situated almost directly under the skin. Surface EMG has been quantified by many investigators in various ways. However, independent of the manner of EMG processing, the electrical properties of the skin and their changes must be taken into account whenever surface electrodes are going to be applied. In the present investigation, the influence of electrical skin impedance on the EMG signal is discussed, and suggestions are given to decrease this influence which results in an improved repeatability of quantified EMG activity.

CAUSES OF VARIATIONS IN SKIN RESISTANCE

There are at least four basic factors which cause variations in skin resistance. First, the electrical resistance of the skin is complex, i.e., its value depends on the frequency of the alternating current passing through it. In a previous study (Rau and Vredenbregt, 1971) with a typical EMG electrode arrangement (electrode area, 50 mm^2; distance between the electrodes, 40 mm), the skin resistance was investigated in the range of frequencies between 5 Hz and 2 kHz. In lower frequencies, especially below 20 Hz, the skin impedance increased up to the order of magnitude of 10^6 ohms. Second, the skin resistance also has been observed to vary as a function of time (Andringa, 1971). Especially within the first 30 min after attaching the electrodes, the skin impedance decreased in several cases to about 50 percent of its initial value. However, for durations of a day or a week, variations of skin impedance also occurred with the same order of magnitude. Third, the range of variations was obtained at different skin sites on the arm and the leg. Values between 500 kilohms and 2.5 megohms were observed on the same subject at a measuring frequency of 10 Hz. Fourth, skin resistance at the same skin site of different subjects ranged from 200 kilohms to 2 megohms.

DECREASING THE INFLUENCE OF SKIN RESISTANCE

The complex skin resistance should be considered as an electrical filter which is in series with the input resistance of the EMG amplifier (Rau, 1973). Because the surface EMG signal spectrum is extended from about 1 Hz to 1 KHz (Chaffin, 1969) with its maximal amplitude at about 50 Hz, and still 20–30 percent of the maximal amplitude is at 10 Hz, the EMG signal is changed by the complex skin resistance. This influence can be expected to be reduced if a sufficiently high input impedance of the EMG amplifier is used.

To decrease the influence of skin resistance and its variations as described above, the input impedance ought to be at least 10 times the skin impedance. This requirement is fulfilled by a differential input stage for impedance transformation using two FET's and two very high gate resistors (220 megohms), as shown in Figure 1,a. Resistors of such high values are not easily available. Therefore, an alternative principle is given in Figure 1,b, in which the resistance is generated by a bootstrap (Bergey and Squires, 1971) and any value up to 1,000 megohms can be obtained using smaller values of the resistors. In this latter case, it was more difficult to get a satisfying high common mode rejection ratio.

The influence of the input impedance of the amplifier in relation to the skin impedance was investigated by using an arrangement of 4 electrodes, as indicated in Figure 2. A constant alternating current signal was applied to skin electrodes A and A', with the frequency changed in steps between 0.5

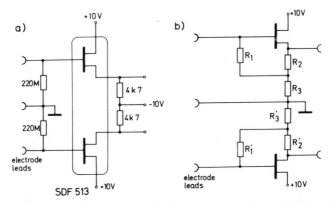

Figure 1. First stage of the high input impedance amplifier. $a)$ Well-known source follower; $b)$ bootstrap version, where the input resistance is $R_i = R_1 \cdot (R_2 + R_3)/R_2$. (Example: R_1 = 20 megohms, R_2 = 5 kilohms, R_3 = 100 kilohms; then $R_i \approx 400$ megohms).

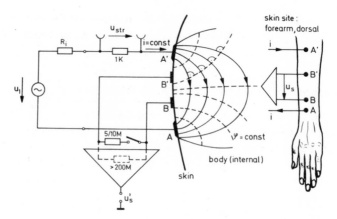

Figure 2. Measuring arrangement for considering the influence of the skin impedance and the input impedance by the amplifier on the EMG signal.

and 20 Hz. At EMG surface electrodes B and B', a signal was picked up by use of an amplifier with an input stage, as shown in Figure 1,a. The signal voltage amplitude of less than 10 mV corresponded to EMG amplitudes.

To demonstrate the advantage of the high input impedance of > 200 megohms, the results were compared with those obtained when the impedance was reduced to about 5 and 10 megohms by placing resistors parallel to the input leads of the amplifier. Decreases in amplitude, measured for different subjects by decreasing the input impedance, are shown in Figure 3,a for 5 megohms and in Figure 3,b for 10 megohms. The results demonstrate clearly that an input impedance of 5 or even 10 megohms is not sufficiently high, if the lower part of the EMG signal spectrum must be observed and therefore is unaffected by the skin impedance. The influence of changes in skin impedance dependent on time, skin site, or subjects, as mentioned above, also will be suppressed if the input impedance is high in comparison with the skin impedance and its variations.

METHODS

Sample measurements were made concerning the reproducibility of the relationship between the static extension torque of the wrist and the EMG activity of the extensor carpi radialis longus and brevis muscles while a preamplifier (Figure 1,a) with an input impedance of 440 megohms was used differentially. The torque was derived from the force measured at the metacarpal joints, and the EMG signal processing consisted of rectification and filtering by means of an averaging filter (linear third-order Paynter filter with t = 440 msec), as suggested by Gottlieb and Agarwal (1970) and by Kreifeldt (1971). Both signals were recorded on-line simultaneously on an x-y-recorder.

During the experiment, the arm-hand system was positioned and anchored very carefully in a specially designed apparatus which could be

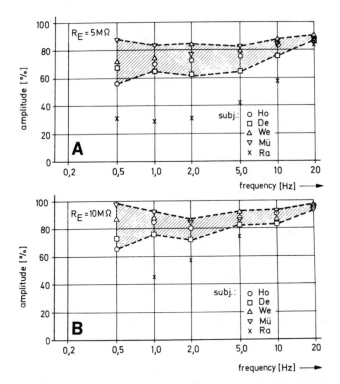

Figure 3. Variations of the signal amplitudes as a function of frequency, observed using an amplifier input impedance of *a,* 5 megohms and *b,* 10 megohms. *Subj.,* subject.

adapted to the individual anatomical properties of each subject. The subjects were instructed to increase the force exertion continuously within 3 sec up to a force level as high as possible, and then to release again abruptly. After three trials each, separated by a 2-min break, the electrodes were removed and the skin sites were marked in order to find the same electrode position again on the succeeding days of the experiment.

RESULTS AND DISCUSSION

The results from one subject are shown in Figure 4. In three successive single trials during one experimental session, the EMG-torque relationship was linear with only small deviations from each other. The mean curves of such triple sets, recorded every other day, are shown in Figure 4,*b.* Here, the deviations of the mean curves did not exceed those of the single measurement curves. Differences existed in the characteristics of the curves of the various subjects, but for each subject, the deviations were all about the same.

Even without normalizing the EMG values to a reference value, usually obtained by a reference measurement before each experiment (Vredenbregt and Rau, 1973), the reproducibility of the relationship was within 10–15

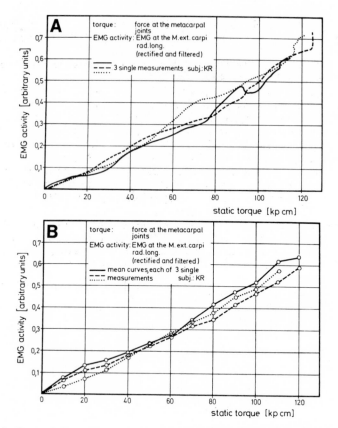

Figure 4. EMG-torque relationship at the wrist during static extensions: *a*, single trials; *b*, mean curves. *M. ext. carpi rad. long.*, extensor carpi radialis longus muscle.

percent or less during a week-long period. Besides the well-defined positioning of the arm-hand system, the rather high test-retest reliability obtained can be assumed to be caused by the suppression of variations in skin impedance, especially short-term variations which cannot be reduced easily by any means other than a comparatively high input impedance. The increased quantitative reproducibility of the relationship between the absolute strength of muscular contraction and the EMG activity might also turn out to be an improved measure for clinical applications which can be used to evaluate muscle function.

REFERENCES

Andringa, J. 1971. Onderzoek van de huidimpedantie en haar variatiegebied, gemeten met een bestaand elektrodensystem. Rept. no. 222, Inst. f. Perception Res., Eindhoven.

Bergey, G. E., and R. D. Squires. 1971. Improved buffer amplifier for incorporation within a biopotential electrode. IEEE Trans. Biomed. Eng. 430–431.

Chaffin, D. B. 1969. Surface electromyography frequency analysis as a diagnostic tool. J. Occup. Med. 11: 109–115.

Gottlieb, G. L., and G. C. Agarwal. 1970. Filtering of electromyographic signals. Amer. J. Phys. Med. 49: 142–146.

Kreifeldt, J. G. 1971. Signal versus noise characteristics of filtered EMG used as a control source. IEEE Trans. Biomed. Eng. 18: 16–22.

Rau, G. 1973. Der Einfluß der Elektroden- und Hautimpedanz bei Messungen mit Oberflächenelektroden (EMG). Biomed. Tech. 18: 23–27.

Rau, G., and J. Vredenbregt. 1971. Elektroden-Übergangswiderstand und Hautwiderstand für verschiedene Elektrodenmaterialien im Hinblick auf quantitative Oberflächen-Elektromyographie. Rept. no. 206, Inst.f.Perception Res., Eindhoven.

Vredenbregt, J., and G. Rau. 1973. Surface electromyography in relation to force, muscle length and endurance. In: E. Desmedt (ed.), New Developments in Electromyography and Clinical Neurophysiology. Vol. 1, pp. 607–622. S. Karger, Basel.

Author's address: Dr. Günter Rau, Forschungsinstitut für Anthropotechnik, D-5309 Meckenheim, Luftelberger Str. L1123 (West Germany)

Electromyographic analysis using metal oxide semiconductor circuitry

T. W. Roberts, M. B. Anderson, and R. F. Zernicke
University of Wisconsin, Madison

The propagated action potential is the communications code of the sensori-motor system. Recordings of action potential discharge patterns supply the researcher with clues to human and animal adaptive behavior. This is the fundamental rationale behind electromyographic analysis of movement.

Membrane depolarization currents in nerve and muscle are remarkably small, on the order of 10^{-3} amp. per cm^2 (Hodgkin and Huxley, 1952). Reliable recordings of these small currents awaited the advent of the triode electron tube (Adrian and Bronk, 1929). Even in the vacuum triode, a leakage current flows in the grid circuit, and this value often is comparable with the $10^{-6}-10^{-8}$ A sampled by the recording electrode. Furthermore, interelectrode capacitances, particularly grid to cathode, may degrade performance at higher frequencies, such as during the rising phase of a nerve action potential. To minimize these effects, degenerative voltage feedback may be applied by the cathode, reducing the charge between the two elements.

Two cathode degeneration circuits have been employed extensively in biological recording, namely, the cathode follower and the common cathode difference amplifier. The cathode follower, which has a voltage gain of less than unity (complete degeneration), has a high characteristic input impedance which may reach 10^{10} ohms in pentode electrometer configurations (Gronner, 1958). This circuit has been used principally to measure transmembrane potentials in isolated nerve-muscle preparations. The difference amplifier, on the other hand, has had broad application in recording serial activity from intact and sometimes freely moving experimental subjects. It is a push-pull circuit, a variant of the Wheatstone bridge, and has two useful advantages. First, it can drive succeeding push-pull stages of polygraphs or oscilloscopes without special coupling or intervening phase-splitting circuits. Second, and most important, is the effectiveness of the difference amplifier in rejecting extraneous electrical disturbances common to both input terminals.

Difference amplifiers using bipolar silicon transistors have been used as biological input stages. Small size, resistance to mechanical shock, and suitability to battery biasing enable solid-state devices to be carried adjacent to the recording electrodes on a moving subject (Mackay, 1968). However, base leakage currents on the order of 10^{-6} amp. in small signal transistors restrict the performance of these semiconductors in contrast to electron tubes.

More recently, however, the introduction of the metal-oxide-semiconductor (MOS) transistor has provided the physiologist with a suitable substitute for the triode while incorporating the above-mentioned advantages of conventional transistors. Gate leakage currents are as small as 10^{-14} A in some MOS devices, and, unlike bipolar transistors, they virtually are unaffected by large input signal changes. Audio spot noise figures 0.5 dB in a few types make them suitable for low-level biological recordings. The chief disadvantage of MOS transistors is their susceptibility to damage from large voltage transients applied to the control gate. In this respect, electron tubes are superior to all semiconductors. However, some MOS transistors have been provided with protective diodes which limit gate voltage swing.

MOS transistors are being used increasingly in small signal amplifiers. The input difference amplifier described here (Figure 1,a) employs RCA 40841 MOS transistors selected for matched transconductances (RCA, 1972). Gate leakage in the 40841 is on the order of 10^{-8} amp. which is satisfactory for use with surface and intramuscular electromyographic electrodes, which carry action potential currents of $10^{-6}-10^{-7}$ amp. The high forward transconductance of this MOS device permits the use of relatively small load resistors. Consequently, the amplifier is capable of driving long output cables without appreciable capacitive losses or inductive interference.

The dualgates, which are connected in order to function as a single gate, are held at ground potential by 10-megohm (or smaller) resistors and are coupled to the signal through 0.05 mfd blocking capacitors. The input coupling time constant of approximately 0.5 sec is long enough for electrocardiography. Smaller values of capacitance may be used to diminish the effects of tissue mechanical disturbances to imbedded electromyographic electrodes. Bias may be supplied by a pair of 9-volt transistor batteries (Burgess 2U6 or equivalent).

Noise referred to input over a band-pass of 10 Hz to 3 KHz is 10 μV root mean square, while common mode rejection ratio varies from 1000 to 3000:1 in the units constructed. Differential voltage gain is approximately eight.

The MOS transistors are mounted on a 1-in square printed circuit board (Figure 1, b), while the drain resistors and constant current source are located in a battery box. The two circuit boards are interconnected by a three-conductor shielded cable of desired length. For lightness and minimal cost, a suitable cable may be fabricated with a Belden 8419 phonograph tone arm cable that has been brushed with two or three coats of liquid latex of the type available at hobby shops for plaster molding. The thin latex coat remains

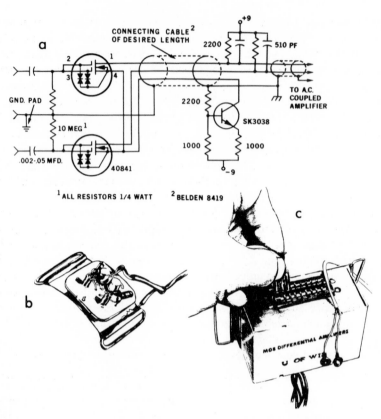

Figure 1. *a*, MOS difference amplifier circuit; *b*, circuit board mounting with ground pad; *c*, with connector tab. *GND.*, ground.

highly flexible while it insulates the braided shield and stabilizes the three conductors. Cables that are 18.3 m in length have been used with little signal decrement. The input circuit card is attached and grounded to a 1- × 1.5-in tin sheetmetal pad with a pair of 0.25-in standoff pins of no. 18 bus wire. Loops of no. 18 wire soldered to the ends of the ground pad accommodate adhesive tape for attachment to the subject. The ground pad may be bent to fit the desired contour. A pair of radio dial cord springs constitutes the input terminals when wire electrodes are employed (Basmajian, Forrest, and Shine, 1966).

Figure 2,*c* illustrates some individual action potentials recorded by intramuscular wire electrode pairs fabricated of 0.28-mm Karma alloy wire with a DC resistance of 30,000 ohms (Bigland and Lippold, 1954).

The principal merit of this circuit is that MOS transistors have an inherently high input impedance and a low output impedance suitable for driving long cables attached to subjects. At the same time, rejection of common mode interference is provided while a net voltage gain ensures that the noise contribution of the following amplifier stage will not be significant.

Several of these difference amplifiers may be assembled in a small package to facilitate recording from multiple surface electrodes on a moving subject. Figure 1,c illustrates a six-channel assembly inside a small aluminum box chassis mounted on a common ground pad. Electrode cables terminate in three-conductor printed circuit tabs which insert into printed circuit card edge connectors serving as the input receptacles. The center conductor in each tab is the cable shield and ground connection. The tab connector arrangement keeps connector bulk to a minimum and facilitates rapid change of electrode cables. The preamplifier box may be attached around the waist or thigh.

This system is designed to facilitate recording from subjects performing motor skills involving rapid movements where lead-movement artifact ordinarily might be a problem. Surface electromyography in conjunction with the system has been used to investigate the activity of selected muscles of the supporting and kicking legs in a soccer kick (Figure 2,a), and of the trunk and shoulder muscles during tennis serves (Figure 2,b). Bipolar, 8-mm silver cup

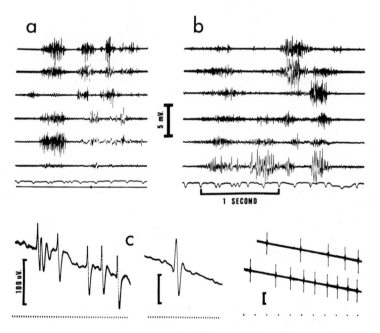

Figure 2. Electromyograms recorded with MOS amplifiers. a, Soccer place kick; b, tennis serve; c, three recordings utilizing bipolar wire electrodes: left, spontaneous fibrillation potential; center, normal muscle action potential; right, single motor unit during acceleration of firing. Time markers are 1 msec for lefthand and center records and 100 msec for righthand record.

electrodes with an average recording impedance of 5,000 ohms were used. Base line noise level remained minimal, as did base line shifts resulting from mechanical artifact.

The quick-change tabs permitted selective, simultaneous recording of any combination of muscles for subsequent analyses. Although these amplifiers are used chiefly in conjunction with an ultraviolet recording polygraph system, they may be employed effectively with ink polygraphs with an AC coupled input.

REFERENCES

Adrian, E. D., and D. W. Bronk. 1929. Discharge of impulses in motor nerve fibers. Part II. J. Physiol. 67: 119–151.

Basmajian, J. V., W. Forrest, and G. Shine. 1966. A simple connector for fine-wire electrodes. J. Appl. Physiol. 21: 1680.

Bigland, B., and O. C. J. Lippold. 1954. Motor unit activity in the voluntary contraction of human muscle. J. Physiol. 125: 322–335.

Gronner, A. D. 1958. Direct-coupled amplifiers. In: J. G. Truxal (ed.), Control Engineers Handbook. McGraw-Hill, New York.

Hodgkin, A. L., and A. F. Huxley. 1952. Currents carried by sodium and potassium ions through the membrane of the giant axon of loligo. J. Physiol. 116: 449.

Mackay, R. S. 1968. Bio-Medical Telemetry. John Wiley and Sons, New York.

RCA Solid State Databook Series. 1972. SSD-201. RCA, New York.

Author's address: *Dr. T. W. Roberts, Department of Physical Education for Women, Lathrop Hall, 1050 University Avenue, Madison, Wisconsin 53706 (USA)*

The behavior of motor units during isometric relaxation of the elbow flexors

K. L. Boon
Twente University of Technology, Enschede

Many studies in the field of biomechanics deal with the motor control of skeletal muscles during certain actions. Usually, muscle activity is studied by means of electromyographic surface electrodes. The aim of this investigation was to examine the motor control of skeletal muscle by means of needle electrodes. With these electrodes, the behavior of the different motor units can be observed.

Since the force-time curves found for voluntary isometric relaxation are very reproducible, certain averaging techniques can be used. These averaging techniques also allow the study of the cooperation of several muscles (biceps, brachialis, and brachioradialis). Besides the results of these averaging techniques, some facts concerning motor unit behavior during contraction and relaxation are described in this article.

PROCEDURES

The subject was sitting on a chair. His arm was fixed in a horizontal position, with an angle of 90° between the upper arm and forearm, which was pronated. The wrist was secured in a clamp which was mounted on two springs with strain gauges, by means of which the force exerted by the wrist could be measured.

The subject was asked to contract the elbow-flexors to a certain constant force. This force was displayed on a meter so that the subject was able to adjust the force if necessary. On command, the subject had to relax his muscles as quickly as possible. Force-time curves obtained in this manner were remarkably reproducible, and the form of the curves was nearly independent of the initial force (Boon et al., 1973). The time at which the force decreased to 50 percent of the initial force was called "half-relaxation time" (t_{hr}).

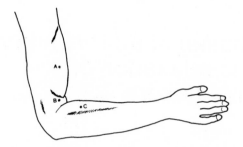

Figure 1. Places of electrode insertion in *A*. biceps, *B*. brachialis, and *C.*, brachioradialis muscles.

Bipolar needle electrodes were placed in the biceps and in the brachialis or brachioradialis muscles. The places of insertion were marked as shown in Figure 1. During each experiment, force and electrical activity were measured and recorded on magnetic tape.

To obtain average responses of the depth of electromyographic activity during isometrical relaxation, the following procedures were employed. Several recordings with the same initial force were made (see Figure 2,*A*). The standard deviations of different points on the relaxation curves are indicated in Figure 2,*A* by small bars. In this investigation, the records were compared with each other with respect to the t_{hr}. The events of motor unit activity are marked (Figure 2,*B*). Sometimes a certain discrimination level was used to separate the events from a noisy background, so that every time a spike crossed this level an event was marked. The events were counted in intervals of 12 msec (Figure 2,*C*), thus providing an impression of the pattern of activity during isometric relaxation. It also was easy to construct a distribution function of the last events (Figure 2,*D*). These events are indicated in Figure 2,*B* by longer dashes. The time of the top of this distribution function was referred to as "t_{top}." This time was the mean of the times between the last event and t_{hr}. The time ($t_{top} - t_{hr}$) was called "t_{rel}." To study the cooperation of different muscles during a relaxation, distribution functions from the three muscles were calculated. The number of experiments was designated as "n." Furthermore, an investigation was made of the behavior of different motor units. The differentiation of the motor unit type was made on the basis of the amplitude of the action potentials.

RESULTS

Results of Averaging Techniques

Figure 2 shows the results obtained from the biceps muscle. Clearly, the transition from active to passive state can be seen. The mean value of t_{rel} of

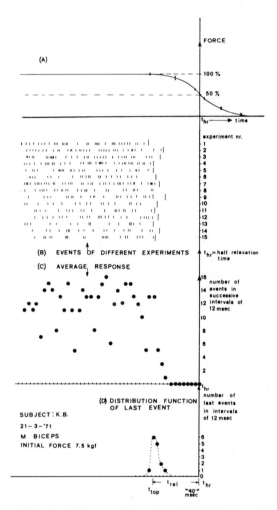

Figure 2. *(A)*, Mean isometrical relaxation curve of 15 experiments. The standard deviations of different points are indicated by *small bars*. *(B)*, events of motor unit activity of each experiment. The last event always is indicated by a *longer dash*. *(C)*, average response:number of events in successive intervals of 12 msec. *(D)*, distribution function of the last event. *nr.*, number.

the biceps muscle was 145 msec, and the standard deviation was s = 30 msec (n = 43). For the brachialis and the brachioradialis muscles, respectively, we found t_{rel} = 137 and s = 23 msec and t_{rel} = 142 and s = 23 msec (n = 22). No significant differences were found among the three distribution functions; thus, there appear to be no significant differences in the innervation process of these muscles during a relaxation.

Behavior of Different Motor Units

What happens during a gradual contraction and a gradual relaxation? During a gradual contraction, one always finds motor units with small amplitudes that become active first. Increasing contraction results in the appearance of motor units that have a greater amplitude but a lower firing frequency. During a gradual relaxation (not "as quickly as possible"), one observes the reverse, i.e., the motor units with the greatest amplitudes are the first to stop firing. The results with respect to motor unit recruitment agree with several results mentioned in the literature (Grimby and Hannerz, 1970; Henneman, Somjen, and Carpenter, 1965a, b; Person and Kudina, 1972). The fact that the observations were found to be independent of the position of the electrodes within one muscle presumably can be explained by the diffuse situation of the different muscle fibers belonging to one motor unit (Buchthal, Guld, and Rosenfalck, 1957; Edström and Kugelberg, 1968). Since the elaborate work of Henneman et al. (1965a, b), it generally has been accepted that the motor units with the small amplitudes consist of Type C muscle fiber and that motor units with the greater amplitudes consist of Type A (and B) muscle fibers.

Because motor units are homogeneous in muscle fiber type, one also can divide motor units into A-, B-, and C-type. During a sudden voluntary relaxation, one observes that Type A (and B) motor units always cease firing, whereas Type C motor units are sometimes still active during and after the relaxation period.

DISCUSSION

The results show that the activity of several motor units during a sudden voluntary isometric relaxation can be described by means of average responses and distribution functions of the last events, all with respect to t_{hr}. With this procedure, the cooperation of different muscles can be treated statistically, an important concept for the study of motor control. The author is especially interested in the voluntary isometric relaxation, inasmuch as this was found to be a good clinical diagnostic tool (Boon et al., 1973).

Only motor units of Type C sometimes remain active. It should be realized that the procedure used in the averaging techniques is at the cost of the study of the C-motor units, because the amplitudes of these units are sometimes very small. This is true especially for the distribution functions. More research must be conducted on this subject.

Because the force curves are used as a reference, it is important to know the influence of small variations of the force curves on the results. To get an impression of this influence from the records in Figure 2, some records were selected that had an identical force curve. The form of the distribution function then became more distinct.

REFERENCES

Boon, K. L., M. G. B. V. Geerdink, H. M. Greebe, and J. C. Kense. 1973. Untersuchungen des neuro-muskulären Systems des menschen durch Messungen der isometrischen Kontraktion und Relaxation. Z. EEG-EMG 4: 109.
Buchthal, F., C. Guld, and P. Rosenfalck. 1957. Multi electrode study of the territory of a motor unit. Acta Physiol. Scand. 39: 83–104.
Edström, L., and E. Kugelberg. 1968. Histochemical composition, distribution of fibers and fatigability of single motor units. J. Neurol. Neurosurg. Psychiat. 31: 424–433.
Grimby, L., and J. Hannerz. 1970. Differences in recruitment order of motor units in phasic and tonic flexion reflex in "spinal man." J. Neurol. Neurosurg. Psychiat. 33: 562–570.
Henneman, E., G. Somjen, and D. O. Carpenter. 1965a. Functional significance of cell size in spinal motoneurons. J. Neurophysiol. 28: 560–580.
Henneman, E., G. Somjen, and D. O. Carpenter. 1965b. Excitability and inhibitibility of motoneurons of different sizes. J. Neurophysiol. 28: 599–620.
Person, R. S., and L. P. Kudina. 1972. Discharge frequency and discharge pattern of human motor units during voluntary contraction of muscle. Electroencephalogr. Clin. Neurophysiol. 32: 471–483.

Author's address: *Dr. K. L. Boon, Department of Electrical Engineering, Twente Univ. of Technology, P.O. Box 217, Enschede (The Netherlands)*

Vocational electromyography: Investigation of power spectrum response to repeated test loadings during a day of work at the assembly line

R. Örtengren, H. Broman, and R. Magnusson
Chalmers University of Technology, Göteborg

G. B. J. Andersson and I. Petersen
Sahlgren Hospital, Göteborg

It has been shown that the power spectrum of the myoelectric signal responds to sustained, strong, isometric contractions with an increase of low-frequency content and a decrease of high-frequency content (Kadefors, Kaiser, and Petersén, 1968). This corresponds to a shift of the spectrum toward lower frequencies (Lindström, Magnusson, and Petersén, 1970). This shift is consistent with a decrease of the impulse conduction velocity of the muscle fibers and has been attributed to an accumulation of acid metabolites in the muscle when blood flow is restricted by the contraction (Mortimer, Magnusson, and Petersén, 1970; Lindström et al., 1970).

Jorfeldt and Wahren (1970) and Karlsson (1971) have shown that the lactate concentrations in the muscle and in the blood increase during heavy work. After the end of the work, these concentrations decline with time constants of the order of 15 min for the blood and about half that value for the muscle.

In a research program undertaken in cooperation with the Swedish car manufacturer, Volvo, in Göteborg, the described spectrum modifications have been used in the search for an objective means for the evaluation of muscular fatigue during vocational labor. Preliminary results have been reported by

This investigation was supported by grants from the Volvo AB in Göteborg and from the Swedish Board for Technical Development in Stockholm, Sweden.

Magnusson and Petersén (1971), Petersén and Magnusson (1971), and Broman et al. (1973). The aim of the present investigation has been twofold: (a) to demonstrate that it is possible to perform intricate bioelectric data acquisition from workers on the workshop floor with no or little sacrifice in signal quality in comparison with what can be obtained in the laboratory, and (b) to measure the influence of work intensity on the rate of change of the spectral levels of the myoelectric signals in certain frequency bands during a day of work by means of repeated test loadings.

METHODS

In the present investigation, signals were recorded from the left and the right biceps brachii, triceps brachii, deltoideus, and trapezius muscles of 10 experienced workers, five at each of two assembly line stations. The electrodes used were of the double-rectangle configuration type shown in Figure 1. The electrodes were fastened with rubber bands and connected to high-quality differential preamplifiers with a differential input impedance of 22 microhms and a common mode rejection of at least 90 dB. The preamplifiers were placed in a box which was strapped to the subject's chest under the clothes (see Figure 2). The signals were fed to the main amplifiers, which were placed in a separate cabinet, by a multiwire cable. The main amplifiers were connected differentially and had controls for gain and upper and lower frequency limits. The investigated frequency band was between 10Hz and 3 KHz (6-dB limits). The signals were recorded on a 14-channel FM tape recorder for subsequent analysis. An oscilloscope was used to monitor the signals continuously. Possible overloads or electrode failures were corrected as soon as noted, thus preserving signal quality.

Figure 1. Surface electrode for recording of myoelectric signals. The distance between the silver bars is 30 mm.

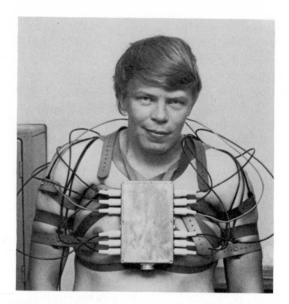

Figure 2. Location of electrodes and position of preamplifier box.

At the first station, eight different wet rubbing tasks on the car bodies were performed according to a regular scheme. The worker under study always started with the heaviest task (see Figure 3), and after 1 hr and a 12-min coffee break he proceeded with the lightest task for 1 hr. Test loads were applied before and after each task period. Then there was a 30-min break, after which the worker followed the regular scheme. The day was finished with a fifth test load series. At the second station, the work consisted of tightening nuts on brake tubings and steering stays from below the cars

Figure 3. Heavy work at the first station.

(see Figure 4). At this station, test loads were applied before and after the whole day of work and before and after each break; in all, eight test load series were performed.

During the test loading precedure, the workers were standing. The weights were held by straps, in the hands for the biceps and triceps muscles and at the wrists for the deltoideus and trapezius muscles. Both sides were tested at the same time. For the biceps and triceps muscles, the load was 50 N. The upper arms were kept vertical, with the elbow at a right angle. The hands were supinated for the biceps and pronated for the triceps muscles. The deltoideus and the trapezius muscles were tested simultaneously with a load of 30 N. The arms were kept horizontal in the frontal plane, and the hands were pronated and clenched. All loads were applied for 30 sec.

The recordings were analyzed by means of a specially designed spectrum analyzer containing four channels with octave-band filters of center frequencies 16, 63, 250, and 500 Hz and one channel for the total signal (Örtengren, 1970). The output of the analyzer was the time course of the root mean square value of each channel, all on logarithmic scales (see Figure 5). For each muscle and each filter channel, the signal level rate of change in dB per sec in relation to the total signal was calculated, except for the 63-Hz channel, which was too close to the maximum of the spectrum to respond clearly to the spectrum shifts. These values were entered into the memory of

Figure 4. Work at the second station.

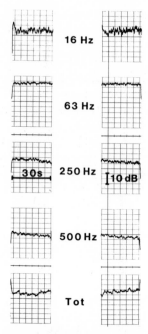

Figure 5. Example of results of test load spectrum analysis of myoelectric signal from biceps brachii muscle before and after a 12-min period of rest (Subject no. 17).

a PDP-15 computer, and statistical parameters were calculated. Single and compound tests according to the Student t test were performed at the 5-percent level of significance.

RESULTS AND DISCUSSION

The investigations which were carried out for 2 wk, were successful from the point of view of signal acquisition. There was no trouble with either reliability and signal quality or electrical safety on the workshop floor, with all of the electronic equipment. It therefore is believed that myoelectrical signals can be recorded in rather busy environments, provided that the equipment set-up and the recording procedure are planned carefully.

The recorded rates of change for signal levels are given in Tables 1 and 2. The material did not show any statistically significant differences in rates of change before and after heavy or light work, nor did the periods of rest influence the rates of change to a statistically significant degree. There was, however, a tendency toward lower rates of change after a working period and higher rates after a resting period, in comparison with the previous measurements. A possible explanation is as follows. After a period of work, an increased muscle lactate concentration shifted the power spectrum of the

Table 1. Mean rates of change of spectral levels in dB per sec in the (a) 16-Hz, (b) 250-Hz, and (c) 500-Hz bands at the first station

			Test no.		
Muscle	1	2	3	4	5
Left biceps brachii					
	0.04	0.10	0.07	0.30	0.09
(a)	(0.03)*	(0.05)	(0.04)	(0.08)	(0.06)
	−0.07	−0.13	−0.11	−0.14	−0.08
(b)	(0.04)	(0.03)	(0.02)	(0.01)	(0.02)
	−0.09	−0.04	−0.25	−0.24	−0.15
(c)	(0.03)	(0.03)	(0.03)	(0.02)	(0.03)
Right biceps brachii					
	−0.01	0.10	0.05	0.06	0.08
(a)	(0.06)	(0.07)	(0.04)	(0.06)	(0.03)
	−0.14	−0.08	−0.10	−0.12	−0.06
(b)	(0.03)	(0.06)	(0.02)	(0.01)	(0.03)
	−0.18	−0.07	−0.14	−0.18	−0.09
(c)	(0.04)	(0.09)	(0.03)	(0.01)	(0.03)
Left triceps brachii					
	0.11	0.25	0.08	0.06	0.14
(a)	(0.01)	(0.04)	(0.04)	(0.03)	(0.03)
	−0.09	−0.05	−0.08	−0.05	−0.06
(b)	(0.02)	(0.01)	(0.03)	(0.05)	(0.01)
	−0.16	−0.06	−0.09	−0.02	−0.08
(c)	(0.01)	(0.02)	(0.04)	(0.05)	(0.02)
Right triceps brachii					
	0.23	0.11	0.09	0.06	0.10
(a)	(0.04)	(0.03)	(0.01)	(0.02)	(0.01)
	−0.07	−0.02	−0.11	−0.05	−0.11
(b)	(0.01)	(0.02)	(0.02)	(0.01)	(0.01)
	−0.17	−0.11	−0.15	−0.08	−0.14
(c)	(0.02)	(0.03)	(0.02)	(0.01)	(0.01)
Left deltoideus					
	0.24	0.23	0.27	0.36	0.24
(a)	(0.03)	(0.03)	(0.03)	(0.09)	(0.01)
	−0.21	−0.24	−0.24	−0.24	−0.22
(b)	(0.02)	(0.01)	(0.03)	(0.02)	(0.02)
	−0.27	−0.34	−0.33	−0.34	−0.23
(c)	0.03	(0.03)	(0.03)	(0.02)	(0.01)
Right deltoideus					
	0.24	0.21	0.06	0.26	0.16
(a)	(0.01)	(0.03)	(0.07)	(0.02)	(0.01)
	−0.24	−0.13	−0.28	−0.29	−0.36
(b)	(0.03)	(0.11)	(0.04)	(0.01)	(0.03)
	−0.28	−0.16	−0.32	−0.33	−0.34
(c)	(0.03)	(0.13)	(0.04)	(0.02)	(0.02)
Left trapezius					
	0.23	0.20	0.15	0.15	0.73
(a)	(0.04)	(0.07)	(0.04)	(0.03)	(/)

continued.

344   Örtengren et al.

Table 1. (continued)

Muscle	Test no.				
	1	2	3	4	5
(b)	−0.17 (0.01)	−0.48 (0.15)	−0.27 (0.05)	−0.26 (0.07)	−0.05 (0.01)
(c)	−0.28 (0.02)	−0.52 (0.13)	−0.33 (0.04)	−0.23 (0.03)	−0.10 (0.02)
Right trapezius					
(a)	0.20 (/)	0.14 (0.07)	0.22 (0.03)	0.19 (0.01)	0.07 (0.02)
(b)	−0.40 (/)	−0.27 (0.05)	−0.14 (0.06)	−0.15 (0.02)	−0.03 (0.02)
(c)	−0.27 (/)	−0.21 (0.04)	−0.19 (0.05)	−0.22 (0.05)	−0.07 (0.05)

*Numbers in parentheses = S.D.

myoelectric signal toward lower frequencies. Because of this initial shift, further work has a relatively smaller effect on the spectrum. Consequently, for constant observation times of sufficient length, smaller rates of change are observed after a period of work than after a period of rest. Such findings have been reported by Örtengren (1972) for heavy, repetitive, isometrical loadings of high-duty cycle.

Statistically significant intermuscular differences were found. The test conditions for the muscles were not easily comparable, and many factors, such as load, lever arms, and agonist activity, were of importance, but the findings were in accordance with intermuscular differences in spectrum shape (Lindström, Magnusson, and Petersén, 1973).

The standard deviations of the means in the material were large, and the number of workers studied was small. It is possible that, if the material were increased, the differences would be statistically significant. A more plausible explanation of the small differences found is that this special kind of work did not affect the values of rate of change of the levels in the frequency bands investigated. This conclusion is supported by the fact that the workers were experienced and very efficient in their work, a factor which allowed them minor periods of rest after work on each car, amounting to as much as 20 percent of the 2 min devoted to every car. With these time relations, the increase of the concentration of the acid metabolites in the muscles and in the blood was small.

Table 2. Mean rates of change of spectral levels in dB per sec in the (a) 16-Hz, (b) 250-Hz, and (c) 500-Hz bands at the second station

Muscle	Test no.							
	1	2	3	4	5	6	7	8
Left biceps brachii								
(a)	0.15 (0.03)*	0.15 (0.03)	0.23 (0.02)	0.28 (0.04)	0.19 (0.02)	0.13 (0.03)	0.25 (0.01)	0.16 (0.01)
(b)	−0.13 (0.02)	−0.13 (0.01)	−0.16 (0.01)	−0.15 (0.01)	−0.16 (0.02)	−0.13 (0.01)	−0.15 (0.01)	−0.17 (0.02)
(c)	−0.25 (0.03)	−0.21 (0.02)	−0.25 (0.01)	−0.18 (0.01)	−0.26 (0.02)	−0.21 (0.03)	−0.24 (0.02)	−0.24 (0.01)
Right biceps brachii								
(a)	0.12 (0.02)	−0.01 (0.02)	0.14 (0.03)	0.17 (0.02)	0.14 (0.02)	0.16 (0.02)	0.14 (0.01)	0.15 (0.02)
(b)	−0.08 (0.01)	−0.06 (0.01)	−0.10 (0.01)	−0.11 (0.01)	−0.11 (0.01)	−0.10 (0.02)	−0.11 (0.01)	−0.13 (0.01)
(c)	−0.11 (0.01)	−0.12 (0.01)	−0.17 (0.01)	−0.14 (0.01)	−0.19 (0.01)	−0.15 (0.01)	−0.20 (0.01)	−0.18 (0.02)
Left triceps brachii								
(a)	0.15 (0.03)	0.21 (0.02)	0.12 (0.02)	0.18 (0.03)	0.23 (0.05)	0.15 (0.02)	0.13 (0.04)	0.12 (0.02)
(b)	−0.08 (0.01)	−0.05 (0.01)	−0.11 (0.01)	−0.06 (0.01)	−0.10 (0.01)	−0.09 (0.02)	−0.11 (0.02)	−0.05 (0.01)
(c)	−0.13 (0.01)	−0.17 (0.01)	−0.12 (0.02)	−0.12 (0.01)	−0.12 (0.01)	−0.13 (0.03)	−0.19 (0.02)	−0.12 (0.01)
Right triceps brachii								
(a)	0.05 (0.01)	0.16 (0.03)	0.15 (0.03)	0.16 (0.01)	0.14 (0.02)	0.13 (0.02)	0.18 (0.02)	0.12 (0.03)

continued.

Table 2. (continued)

Muscle	Test no.							
	1	2	3	4	5	6	7	8
(b)	-0.09 (0.01)	-0.10 (0.01)	-0.09 (0.02)	-0.06 (0.01)	-0.08 (0.01)	-0.05 (0.01)	-0.08 (0.01)	-0.05 (0.01)
(c)	-0.13 (0.01)	-0.15 (0.02)	-0.13 (0.01)	-0.13 (0.01)	-0.17 (0.01)	-0.13 (0.01)	-0.13 (0.01)	-0.11 (0.01)
Left deltoideus								
(a)	0.25 (0.02)	0.25 (0.02)	0.32 (0.02)	0.21 (0.01)	0.43 (0.02)	0.21 (0.01)	0.37 (0.04)	0.23 (0.01)
(b)	-0.19 (0.01)	-0.18 (0.02)	-0.25 (0.02)	-0.19 (0.01)	-0.23 (0.02)	-0.20 (0.02)	-0.25 (0.03)	-0.25 (0.01)
(c)	-0.24 (0.01)	-0.22 (0.01)	-0.30 (0.03)	-0.24 (0.02)	-0.24 (0.02)	-0.23 (0.02)	-0.28 (0.02)	-0.31 (0.02)
Right deltoideus								
(a)	0.23 (0.02)	0.18 (0.04)	0.29 (0.02)	0.19 (0.01)	0.28 (0.04)	0.30 (0.03)	0.29 (0.02)	0.29 (0.02)
(b)	-0.17 (0.01)	-0.16 (0.01)	-0.18 (0.01)	-0.22 (0.02)	-0.13 (0.01)	-0.20 (0.01)	-0.18 (0.01)	-0.15 (0.02)
(c)	-0.18 (0.01)	-0.19 (0.01)	-0.24 (0.02)	-0.26 (0.02)	-0.17 (0.01)	-0.23 (0.01)	-0.23 (0.01)	-0.17 (0.01)

	1	2	3	4	5	6	7	8
Left trapezius								
(a)	0.14 (0.04)	0.17 (0.01)	0.31 (0.05)	0.24 (0.02)	0.23 (0.03)	0.17 (0.02)	0.15 (0.01)	0.17 (0.02)
(b)	−0.11 (0.01)	−0.09 (0.01)	−0.22 (0.03)	−0.10 (0.01)	−0.10 (0.01)	−0.11 (0.01)	−0.16 (0.02)	−0.09 (0.02)
(c)	−0.17 (0.02)	−0.22 (0.03)	−0.32 (0.05)	−0.22 (0.02)	−0.19 (0.03)	−0.17 (0.04)	−0.25 (0.03)	−0.14 (0.04)
Right trapezius								
(a)	0.22 (0.02)	0.15 (0.03)	0.14 (0.03)	0.18 (0.04)	0.15 (0.03)	0.11 (0.03)	0.18 (0.03)	0.15 (0.04)
(b)	−0.07 (0.01)	−0.07 (0.02)	−0.13 (0.01)	−0.09 (0.01)	−0.08 (0.01)	−0.13 (0.01)	−0.15 (0.01)	−0.13 (0.01)
(c)	−0.17 (0.02)	−0.14 (0.01)	−0.23 (0.02)	−0.16 (0.02)	−0.16 (0.02)	−0.21 (0.01)	−0.18 (0.02)	−0.15 (0.02)

*Numbers in parentheses = S.D.

348   Örtengren et al.

REFERENCES

Broman, H., R. Magnusson, I. Petersén, and R. Örtengren. 1973. Vocational electromyography: Methodology of muscle fatigue studies. *In:* J. E. Desmedt (ed.), New Developments in EMG and Clinical Neurophysiology. Vol. 1, pp. 656–664. S. Karger, Basel.

Jorfeldt, L., and J. Wahren. 1970. Human forearm muscle metabolism during exercise. V. Quantitative aspects of glucose uptake and lactate production during prolonged exercise. Scand. J. Clin. Lab. Invest. 26: 73–81.

Kadefors, R., E. Kaiser, and I. Petersén. 1968. Dynamic spectrum analysis of myo-potentials with special reference to muscle fatigue. Electromyography 8: 39–74.

Karlsson, J. 1971. Muscle ATP, CP, and lactate in submaximal and maximal exercise. *In:* B. Pernow and B. Saltin (eds.), Muscle Metabolism during Exercise, pp. 383–393. Plenum Press, New York.

Lindström, L., R. Magnusson, and I. Petersén. 1970. Muscular fatigue and action potential conduction velocity changes studied with frequency analysis of EMG signals. Electromyography 10: 341–356.

Lindström, L., R. Magnusson, and I. Petersén. 1973. Muscle load influence on myoelectric signal characteristics. Scand. J. Rehabil. Med., In press.

Magnusson, R., and I. Petersén. 1971. Vocational Electromyography. Proceedings of the Second Nordic Meeting on Medical and Biological Engineering, Oslo, Norway.

Mortimer, J. T., R. Magnusson, and I. Petersén. 1970. Conduction velocity in ischemic muscle. Effect on EMG frequency spectrum, Amer. J. Physiol. 219: 1324–1329.

Petersén, I., and R. Magnusson. 1971. Vocational Electromyography. Methodology of Muscle Fatigue Studies. Digest of the Ninth International Conference on Medical and Biological Engineering, p. 163. Melbourne, Australia.

Örtengren, R. 1970. A Measurement System with Simultaneous Read-Out for Spectrum Analysis of Myoelectric Signals. Proceedings of the First Nordic Meeting on Medical and Biological Engineering, Otaniemi, Finland, pp. 139–141.

Örtengren, R. 1972. Spectrum Analysis of Myoelectric Signals from the Brachial Biceps Muscle during Isometric Dynamic Loading. Digest of the Third International Conference on Medical Physics, Including Medical Engineering, Göteborg, Sweden.

Author's address: *Mr. R. Örtengren, Department of Applied Electronics, Chalmers University of Technology, S-402 20 Goteborg 5 (Sweden)*

An electromyographic study of competetive road cycling conditions simulated on a treadmill

M. Desiprés

University of the Orange Free State, Bloemfontein

Houtz and Fischer (1959) analyzed muscle action and joint movement during exercise on a stationary bicycle. Hamley and Thomas (1967) experimented with skilled and unskilled cyclist-athletes to determine the effect of postural factors on the physiological efficiency on a bicycle ergometer. In both studies, stationary cycle ergometers were used. The purpose of this project was to study the effect of saddle height and load on the pattern of surface muscle interaction and accompanying joint ranges under competitive road riding conditions, which were simulated on a treadmill.

METHODS

The subjects of this investigation were three experienced male cyclists, who, although still junior competitors, had represented their province regularly at the national level. Anthropometric measurements were made to determine body mass, standing pubic symphysis height, and limb segment lengths.

Each series of experiments consisted of the subject riding the bicycle with the seat at 95 percent of the pubic symphysis height, i.e., measured from saddle surface to pedal axle surface in a straight line along saddle pillar and crank, and also with the seat elevated to 105 percent of the pubic symphysis height.

The bicycle used by each subject throughout the experiment had a gear ratio of 48:21. Each subject cycled with the seat at both heights at 0, 2, and 4° incline for 30 sec, with 30 sec of rest between each ride. The pedaling cadence, which was set at 90 cycles per min (± 27 kph), was controlled with a metronome. The experiments were repeated twice on different days. On the first day, each subject started at the low saddle height, and on the second day each started at the higher saddle height. The saddle height was readjusted following the first ride each day. During the exercise, the heart rate was monitored continuously and recorded at the end of each ride. This heart rate was used later to predict the work load as determined on a Monark cycle ergometer. This predicted work load was found to be in the vicinity of 1,200–1,500 kg per m per m. During a 5-min rest period after the first ride each day, the recovery pulse was taken and recorded.

Cinematographic recordings were made with a Bolex H16, 16-mm camera set at 24 frames per sec, which was modified so that a microswitch on a sprocket in the camera produced a pulse. This pulse was recorded on the polygram so as to be able to synchronize foot positions with the electromyograms and electrogoniograms. Eastman Plus X negative film (7231 PXN 449) was used to produce positive prints and thus eliminated projection analysis. Electrogoniometer arms were fixed in-line on foot, lower leg, and thigh segments, with potentiometers on the knee and ankle joint centers.

Muscle action potentials were picked up by surface electrodes and, after appropriate amplification, recorded on two four-channel Elema Schönander Mingograph jet ink recorders. Two channels were used for cinematic frame pulses, two for electrogoniograms, and four for electromyograms. Prior to each experiment, the electromyograph was calibrated with the same setting of amplification for all recordings and for all channels (10 mm = 1,000 μV, 700 Hz, 0.006 s). The placement of surface electrodes was standardized in advance according to the recommendations of Battey and Joseph (1966). The 12 superficial muscle groups sampled were: rectus femoris, sartorius, gluteus maximus, tensor fasciae latae, hamstrings, i.e., biceps femoris and semitendinosus, quadriceps femoris, i.e., vastus medialis and vastus lateralis, tibialis anterior, soleus, pectoralis major, rectus abdominis, erector spinae, and flexor digitorum superficialis.

The surface electrodes were Siemens 7-mm diameter silver chloride cupped discs with adhesive rings. These were placed 2–2.5 cm apart equidistant from the motor point on a line parallel with the muscle fibers. The exceptions were the vasti, hamstrings, and soleus. Skin abrasion and use of Aquasonic transmission gel ensured that the source impedance of electrodes ranged from 3,000 to 15,000 ohms which, according to O'Connell and Gardner (1963), is sufficient for reliable recordings. Electromyograms were recorded (paper speed, 50 cm per sec) during the last 5 sec of the exercise at each incline and for the two alternate saddle heights. Muscle action potential patterns for one pedal cycle were quantified by measuring the amplitude with a Wildt measuring magnifier in conjunction with a Siemens Cardiometer and by summing the microvolt output and frequency which was multiplied by the duration of activity to give units in mV-sec. An analysis of variance was computed to determine whether the influences of saddle height and treadmill incline on electrical activity in muscles were statistically significant.

RESULTS

The relationship between the ranges of movement in the ankle and knee joints varied with the changing position of the extremity. Saddle height significantly influenced plantar flexion in the ankle but did not significantly affect the extension of the knee joint. Range of movement in the ankle at low saddle height was $37°$ ($66-103°$), as opposed to a range of motion of $53°$ ($73-126°$) at the higher saddle height. The range of movement in knee was from 64 to $137°$ or $73°$ at the higher saddle height. It was found that the pattern of muscle excitation of the three subjects during cycling under the different conditions was consistent with respect to the leg muscles.

In accordance with the findings of Houtz and Fischer (1959), it was found that electrical activity was minimal and variable in the rectus abdominis and erector spinae muscles. This also was found to be true for the pectoralis major and flexor digitorum superficialis muscles. For this reason, the patterns of these four muscles were not considered in the overall analysis.

With respect to the leg muscles, the following observations and conclusions were made. On the average, increased saddle height apparently led to earlier and longer duration of muscle activity throughout, but the magnitudes of muscle action potentials were not significantly greater when the subject pedaled at 105 percent than that at 95 percent of public symphysis height. The activities of individual leg muscles were as follows. The rectus femoris, the basic function of which is to flex the thigh in reference to the hip and to extend the lower leg in reference to the knee joint, was found to be more active at the lower saddle height. The sartorius, which is basically a flexor of both hip and knee joints as well as a lateral rotator, was active at a very early stage during the 105-percent saddle height. The muscle activity also was quantitatively larger at this saddle height. The gluteus maximus, an extensor of the hip, was not significantly more active than the rectus femoris and the sartorius. The tensor fasciae latae, which assists in flexing the thigh in reference to the hip joint, was of major importance during extension of the knee because it tensed the facial sheath of the thigh muscles through the iliotibial band, i.e., it is possible that this muscle performed a synergistic function. The hamstrings apparently played a major role in extending the hip during cycling under these conditions, more so than in flexing the knee. Saddle height also influenced electrical muscle activity in this muscle group more than in the other leg muscles. The quadriceps femoris, an extensor of the knee was also very active, as expected, but not as active as the hamstrings. Its action seemed to be more that of a prime mover to extend the knee.

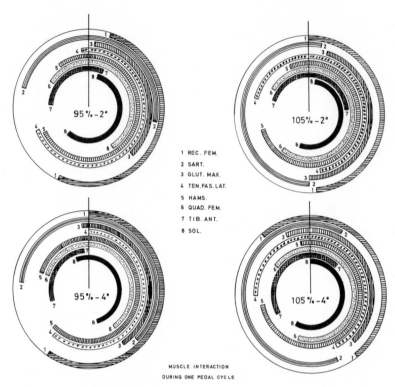

Figure 1. Coordinated muscle activity for one pedal cycle, i.e., from top dead center clockwise to top dead center. *REC. FEM.*, rectus femoris; *SART.*, sartorius; *GLUT. MAX.*, gluteus maximus; *TEN. FAS. LAT.*, tensor fasciae latae; *HAMS.*, hamstrings; *QUAD. FEM.*, quadriceps femoris; *TIB. ANT.*, tibialis anterior; *SOL.*, soleus.

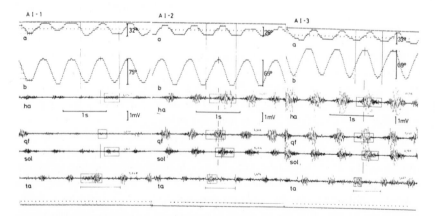

Figure 2. Electrogoniograms and electromyograms at 95 percent of pubic symphysis height. *a*, ankle goniogram; *b*, knee goniogram; *ha*, hamstrings; *qf*, quadriceps femoris; *sol*, soleus; *ta*, tibialis anterior; *1mV*, 1 cm, 700 Hz, 0.006 tc; *1 s*, 50 mm.

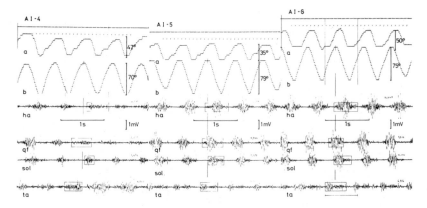

Figure 3. Electrogoniograms and electromyograms at 105 percent of pubic symphysis height.

Lower saddle height resulted in greater activity in this muscle group. The tibialis anterior acts as an inverter as well as dorsiflexor of the foot. It was interesting to note considerably more activity in this muscle at the zero incline. This could have been the result of a "breaking effect," utilized in order to maintain equilibrium of the machine on the belt. The activity of the soleus, which is basically a plantar flexor, increased with increased incline, but its activity was not significantly greater at lower saddle height. Rank order of muscle activity in units of electrical action potentials liberated during cycling under these conditions were: hamstrings, quadriceps femoris, soleus, tibialis anterior, and tensor fasciae latae.

CONCLUSIONS

During all rides, the coordinated muscle interaction was orderly and of cyclic nature, and, the pattern of muscle activity was highly reproducible in each of the three cyclists. It was found that the higher height of saddle increased the range of movement in plantar flexion of foot significantly but did not influence knee extension by a significant amount. Furthermore, varying the height of the saddle did not influence muscle activity significantly. An increase in treadmill incline did intensify the electrical activity in the muscles of the leg. The hamstrings and quadriceps femoris were influenced to the greatest degree. The tibialis anterior was the only muscle which was electrically more active at zero incline. Finally, it was found that the flexor digitorum superficialis, pectoralis major, rectus abdominis, and erector spinae muscles apparently fire randomly during movements which help to maintain equilibrium. Their electrical activity patterns were not consistent.

Table 1. Action potentials μV-sec during one pedal cycle

Height (%)	Incline (degrees)	Muscles											
		rf*	sar	gl	tfl	ha	qf	ta	sol	pec	ra	es	fds
		1	2	3	4	5	6	7	8	9	10	11	12
95	zero	589	643	208	1,924	3,937	1,264	6,357	1,706	540	702	9	1,131
95	2 (3.5%)	1,244	602	1,224	2,817	13,949	4,978	1,639	4,109	217	346	82	59
95	4 (7 %)	3,567	2,696	2,345	4,153	16,986	8,935	2,034	5,498	646	420	264	903
105	zero	576	852	455	2,185	7,825	1,459	4,686	1,274	1,094	2,037	10	3,498
105	2 (3.5%)	901	1,483	1,245	3,162	14,440	5,917	2,310	3,939	1,711	1,109	554	1,132
105	4 (7 %)	2,547	3,052	2,670	4,790	15,771	7,007	3,575	5,853	842	570	508	1,945

*Abbreviations: rf, rectus femoris; sar, sartorius; gl, gluteus maximus; tfl, tensor fasciae latae; ha, hamstrings; qf, quadriceps femoris; ta, tibialis anterior; sol, soleus; pec, pectoralis major; ra, rectus abdominis; es, erector spinae; fds, flexor digitorum superficialis.

REFERENCES

Battey, C. K., and J. Joseph. 1966. An investigation by telemetering of the activity of some muscles in walking. Med. Biol. Eng. 4: 125–135.

Hamley, E. J. and V. Thomas. 1967. Physiological and postural factors in calibration of the bicycle ergometer. J. Physiol. 191: 55–57.

Houtz, S. J., and F. J. Fisher. 1959. An analysis of muscle action and joint excursion during exercise on a stationary bicycle. J. Bone Joint Surg. 41-A: 123–131.

O'Connell, A. L., and E. B. Gardner. 1963. The use of electromyography in kinesiological research. Res. Quart. 166–184.

Author's address: *Dr. Marius Desiprés, Department of Physical Education, University of the Orange Free State, P.O. Box 339, Bloemfontein (Republic of South Africa)*

Muscle contraction mechanisms: An EMG study of the staircase phenomenon

K. Hainaut

Université Libre de Bruxelles, Bruxelles

In man, staircase potentiation of the isometric twitch can be elicited by delivering supramaximal electrical pulses to the ulnar nerve at the wrist and by recording the isometric myogram of the abductor pollicis muscle (Desmedt, 1958). Repetitive stimulation of a rested muscle at low frequencies (1–3 per sec) shows first a slightly progressive decrease of the force in the first 10–20 twitches (negative staircase) and thereafter a steady increase (positive staircase). When the series of stimuli is interrupted, the force returns to control value within a few minutes. This type of potentiation is unique because the time course of the potentiated twitch is accelerated (Desmedt and Hainaut, 1967). In previous studies, we reported the effect of temperature on this phenomenon in man (Hainaut, 1968) and discussed the modification of the relationship between force and speed in human muscle (Hainaut, 1971). We came to the conclusion that, contrary to the current view that the active state would be maximal in the single twitch, the staircase potentiation depends on an intensification of the electromechanical activation of the myofilaments early in the contraction cycle (Desmedt and Hainaut, 1968).

One possible mechanism for this effect could be a lowering of the muscle fiber mechanical threshold (Hodgkin and Horowicz, 1960), whereby the muscle action potential would trigger an earlier and larger intracellular release of Ca^{++} from the sarcoplasmic reticulum that would increase the number of cross-bridges formed between actin and myosin and thus the tension produced at any given time in the twitch cycle. The shortening of relaxation, observed in the potentiated twitch, could be related to an acceleration of the subsequent process of uptake of myoplasmic Ca^{++} by the sarcoplasmic reticulum sink (Desmedt and Hainaut, 1968). Indeed, it has been shown that the active state can be graded in intensity by the concentration of myoplasmic

Ca^{++} and that, on the other hand, Ca^{++}-uptake kinetics affect the twitch time course (Ebashi and Endo, 1968). The purpose of the present study was to test the mechanical threshold modification induced by repeated activity.

METHODS

Single muscle fibers were dissected from the semitendinosus muscle of *Rana temporaria* immediately after decapitation of the frog and kept in Ringer's solution at 20° C. The isolated muscle fiber was stretched to 125 percent of its slack length and the diameter was estimated by the mean of 10 measurements along the fiber (90–175 μ). The fiber then was tested by electrical stimulation and transferred to the experimental chamber. The latter was made of Perspex (Hodgkin and Horowicz, 1959), and solutions of different composition can be introduced by a multiple tap with four inlets. The tendon at one end of the fiber was hooked to a metal support, and the other end was attached to an RCA 5734 mechanical transducer. Action potentials can be recorded with external electrodes (Lüttgau, 1965). The mechanical and electrical activities of the fiber, elicited by electrical stimulation, were studied, as well as the tension produced when different external potassium concentrations were applied briefly. The isometric myogram and its first derivative were recorded simultaneously on a slow-speed polygraph and on a cathode-ray oscilloscope.

Under these conditions, a series of experiments were performed in order to test the modification of the mechanical threshold induced by the staircase potentiation. The test used was the brief application of a Ringer's solution with increased potassium concentration, so that a known contracture plateau, obtained before the staircase series, was compared with the plateau obtained a few seconds after the series for the same potassium concentration. We studied the following parameters: (a) influence of the number of stimuli, (b) the stimulation frequency, (c) the interval between the staircase series and the time of test application, and (d) the effect of temperature on the observed modification of the mechanical threshold.

RESULTS

The results showed clearly that, during the staircase phenomenon, the electro-mechanical threshold was not lowered, but increased. Indeed, the contracture induced by the testing solution containing an increased potassium concentration showed a reduced amplitude 4 sec after the staircase series (Figure 1). The recovery to control value occurred progressively over the next 15 min. These results were consistent and they indicated that the working hypothesis

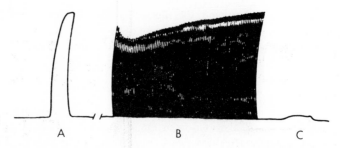

A   B   C

Figure 1. Illustration of mechanical threshold increase as a result of the repeated activation of an isolated muscle fiber from *Rana temporaria* (Hodgkin and Horowicz, 1959, 1960). *A*, submaximal potassium contracture at rest; *B,* after recovery from A, the fiber was stimulated repetitively and showed successively an initial decrease of force (negative staircase) followed by a steady increase (positive staircase); *C,* 4 sec after the staircase series, a testing solution containing the same potassium concentration as in A induced only a very limited contraction plateau.

must be modified. We think that the staircase potentiation probably involves intracellular modifications in the electromechanical activation process instead of muscle fiber membrane modifications.

REFERENCES

Desmedt, J. E. 1958. Méthodes d'étude de la fonction neuromusculaire chez l'homme. Myogramme isométrique, électromyogramme d'excitation et topographie de l'innervation terminale. Acta Neurol. Belg. 58: 997–1017.

Desmedt, J. E., and K. Hainaut. 1967. Modifications des propriétés contractiles du muscle strié au cours de la stimulation électrique répétée de son nerf moteur chez l'homme normal. C. R. Acad. Sci. 264: 363–366.

Desmedt, J. E., and K. Hainaut. 1968. Kinetics of myofilament activation in potentiated contraction: Staircase phenomenon in human skeletal muscle. Nature 217: 529–532.

Ebashi, S., and M. Endo. 1968. Calcium ion and muscle contraction. Progr. Biophys. Mol. Biol. 18: 123–183.

Hainaut, K. 1968. Effect of temperature on muscle potentiation in man. *In:* J. Wartenweiler, E. Jokl, and M. Hebbelinck (eds.), Biomechanics, pp. 255–257. S. Karger, Basel, and University Park Press, Baltimore.

Hainaut, K. 1971. Force and speed in human muscle. 1971. *In:* J. Vredenbregt and J. Wartenweiler (eds.), Biomechanics II, pp. 168–169. S. Karger, Basel, and University Park Press, Baltimore.

Hodgkin, A. L., and P. Horowicz. 1959. The influence of potassium and cloride ions on the membrane potential of single muscle fibers. J. Physiol. (Lond.) 148: 127–160.

Hodgkin, A. L., and P. Horowicz. 1960. Potassium contractures in single muscle fibers. J. Physiol. (Lond.) 153: 386–403.

Lüttgau, H. C. 1965. The effect of metabolic inhibitors on the fatigue of the action potential in single muscle fibres. J. Physiol. (Lond.) 178: 45–67.

Author's address: *Dr. Karl Hainaut, Brain Research Unit, Université Libre de Bruxelles, Faculte de Medccine, 115 Boulevard de Waterloo, 1000 Brussels (Belgum)*

Orthopedics and rehabilitation

Axes of joint rotation of the lumbar vertebrae during abdominal strengthening exercises

B. F. LeVeau

University of North Carolina, Chapel Hill

Movements between individual lumbar vertebrae occur because of compression and tension on the intervertebral discs which lie between adjacent vertebral bodies. Schmorl and Junghanns (1971) call these individual areas "motor segments," that is, the motor spaces arranged between the bony vertebrae. The motor segments that maintain their own posture through tension from the nucleus pulposus make up a single joint. The nucleus pulposus is the articular cavity; the anulus fibrosus makes up the articular ligaments; and the cartilaginous plates on the joint surface of the vertebral bodies comprise the articular cartilage.

Allbrook (1957) stated that selection of a satisfactory fixed point from which the angle of movement can be assessed is essential in the investigation of the vertebral movement. However, Charnley (1951) considered a "bending axis" for the spine as a whole more important than locating the axis between each pair of vertebral bodies. He contended, along with Kester (1969), that as the spine moves from flexion to extension, the intervertebral space enlarges anteriorly and diminishes posteriorly. The opposite motion was said to occur in the movement of extension to flexion. Charnley located the bending axis on a line where the least motion appeared to occur. This axis was in the posterior quarter of each intervertebral disc. From his measurements, he placed the axis at about 5 mm anterior to the front of the spinal theca.

Elward (1939) and Armstrong (1965) disagreed with the idea that a single center of motion of the lumbar spine as a whole existed. They believed that a center at each intervertebral level with a limited range of motion was present. However, Elward and Armstrong, as well as others (Clayson et al., 1962; Steindler, 1955) agreed that the axis between adjacent vertebrae at each level

The data for this study were collected while the author was a graduate student at the Biomechanics Laboratory at The Pennsylvania State University.

was located within the intervertebral disc, with the posterior space closing and the anterior space opening from flexion to extension and with the anterior space narrowing and the posterior space widening from extension to flexion.

Contrary to this, Dittmar (1930) found that movements were not around a single axis, but around several variously situated transitory axes. He did not find elevation of the intervertebral discs anteriorly to correspond to their decline posteriorly, and vice-versa. Wiles (1955) also suggested that the intervertebral axis is not fixed.

METHODS

In this investigation, the movements of the lumbar vertebrae were studied during the execution of four abdominal strengthening exercises performed by 18 volunteer subjects. Lateral view roentgenographs were filmed during the six test positions: (a) hook-lying, (b) long-lying, (c) double leg raising, (d) hook-lying sit-up, (e) long-lying sit-up, and (f) V-sit.

The axis of rotation for the separate vertebral levels was determined by observing the changes in the anterior and posterior distances between the bodies of adjacent vertebrae. If one distance increased and the other decreased, the axis of motion between the adjacent vertebrae would be located within the interposed disc. Depending on the ratio of change, the axis was located posteriorly, anteriorly, or near the center of the disc. If both distances increased or decreased together, the axis of motion between the adjacent vertebrae would be located outside the disc, either posteriorly or anteriorly. Table 1 provides the anterior and posterior distance differences for selected position changes.

RESULTS

The measured changes in anterior and posterior disc spaces revealed that at all five lumbar levels the adjacent vertebrae were flexed when the position changed from the rest position to either the hook-lying sit-up, the long-lying sit-up, or the V-sit. The movement from the long-lying rest position to the double leg raising position showed flexion at the lumbosacral joint, but extension at the other four levels.

Analysis of the values in Table 1 provides the approximate joint center locations for the individual motor segments during the changes in position.

DISCUSSION

The results of the present study showed agreement with those of Dittmar (1930) and Wiles (1955). Each lumbar intervertebral level did not have the

Table 1. Change in the anterior and posterior disc spaces (in cm) at the five lumbar levels for selected changes in the test positions

Position changes	Lumbar levels $L_5 S_1$		$L_4 L_5$		$L_3 L_4$		$L_2 L_3$		$L_1 L_2$	
	Anterior	Posterior	Anterior	Posterior	Anterior	Posterior	Anterior	Posterior	Anterior	Posterior
Hook-lying to long-lying	0.72	−0.51	0.56	−0.63	0.08	−0.38	0.40	−0.22	0.44	0.56
Hook-lying to hook-lying sit-up	−4.38*	−0.29	−4.33*	1.23	−2.83*	1.19	−2.37*	1.23*	−2.19*	1.57*
Hook-lying to V-sit	−3.55*	0.26	−3.64*	0.92	−1.89*	−0.09	−2.09*	0.53	−1.58*	1.56*
Long-lying to long-lying sit-up	−5.27*	0.95	−3.31*	1.44*	−1.48*	0.62	−2.03*	0.38	−2.25*	0.54
Long-lying to double leg raising	−2.43*	0.29	0.22	−0.62	0.80	−0.63	0.82*	−0.59	1.13*	−0.79

*Measures significantly different at the 0.05 level.

same relative location for the axis of motion between adjacent vertebrae, as indicated by Charnley (1951). Change from one position to another indicated a transitory axis that not only moved within the disc, but at times was located externally to the disc either anteriorly or posteriorly. In some instances the posterior and anterior spaces both increased or decreased together, although not the same amount. For example, from the hook-lying position to the hook-lying sit-up position at the level of the fifth lumbar vertebra and the sacrum, both anterior and posterior spaces narrowed, with the anterior narrowing to a greater extent than the posterior. Thus, the axis of rotation between the two vertebrae was located posterior to the inter-vertebral disc. Cinefluoroscopic studies of the vertebral motion should reveal the change of axis of movement from one position to another. The results of this study did indicate that the axis of motion is not stationary as the lumbar vertebrae move in relationship to one another. They agree with Armstrong (1965) that flexion and extension for these test positions takes place by serial movements between individual vertebrae rather than by simultaneous move-ment among all vertebrae concerned.

REFERENCES

Allbrook, D. 1957. Movements of the lumbar spinal column. J. Bone Joint Surg. 39B: 339–345.
Armstrong, J. R. 1965. Lumbar Disc Lesions. 3rd Ed. Williams & Wilkins, Baltimore.
Charnley, J. 1951. Orthopedic signs in diagnosis of disc protrusion. Lancet 266: 186–192.
Clayson, S. J., I. M. Newman, D. F. Debevec, R. W. Anger, H. V. Skowlund, and F. J. Kotte. 1962. Evaluation of mobility of hip and lumbar vertebrae of young women. Arch. Phys. Med. Rehabil. 43: 1–8.
Dittmar, O. 1930. Die Saggittal-und Lateral flexorische Bewegung der Menschlichen Lendenwirbelsaule im Rontgenbild. Z. Anat. Gesch. 92: 644–667.
Elward, J. E. 1939. Motion in the vertebral column. Amer. J. Roentgenol. 42: 91–99.
Kester, N. 1969. Evaluation and medical management of low back pain. Med. Clin. North. Amer. 53: 525–540.
Schmorl, G., and H. Junghanns. 1971. The Human Spine in Health and Disease, translated by E. F. Besemann. Grune & Stratton, New York.
Steindler, A. 1955. Kinesiology of the Human Body. Charles C Thomas, Springfield, Ill.
Wiles, P. 1955. Movements of the lumbar vertebrae during flexion and extension. Proc. Roy. Soc. Med. 28: 647–651.

Author's Address: *Dr. Barney F. LeVeau, Division of Physical Therapy, The University of North Carolina, Chapel Hill, North Carolina 27514 (USA)*

Biomechanics in lumbar lordosis

J. Polster, U. Spieker, H. R. Hoefert and J. U. Krenz
Westfälischen Wilhelms University, Munster

For the purpose of analysis, lumbar lordosis, being a product of functional adaptation to dynamic stress, always should be compared with its opposite, i.e., the flattened lumbar lordosis which from the functional point of view must be regarded as the response to static stress. However, lumbar lordosis can fulfill its specific role only when all of the elements responsible for spinal movement are cooperating properly. If one element, for example, a vertebral body, joint, or muscle, is pathologically affected, the results are functional variations of an increased or flattened lumbar lordosis that are subjected to the laws of biomechanics, as the following three clinical examples demonstrate.

EFFECTS OF DISEASE

The first example illustrates the reaction of the lumbar spine when it is affected by a basic inflammatory disease, as shown in the two silhouettes (Figures 1 and 2) of patients suffering from ankylosing spondylitis. From evaluation of the lateral radiographs of 274 patients with ankylosing spondylitis, it became apparent that we can differentiate between two types of manifestation: (a) the flat, so-called "ironing-board spine" with practically no lumbar lordosis, and (b) the spine with a very pronounced thoracic kyphosis. In the latter condition, the patient's head usually is stretched forward and sometimes held sideways, and a lumbar lordosis is present at least in parts. Both types also are related to the clinical progression of the disease. The flat, ironing-board spine shows ascending ankylosis starting at the iliosacral joints, whereas in the other type of spine, the ankylosis descends in the craniocaudal direction. The flat shape of the lumbar spine with caudal beginning of ankylosis is the result of the tendency to immobilize the lumbosacral joint. Furthermore, a flat spine with a tilted pelvis needs less muscle power to support an upright posture because the line of gravity passes very close to the

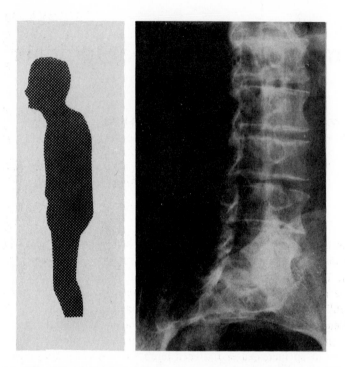

Figure 1. Silhouette of a patient suffering from ankylosis spondylitis and radiograph of his lumbar spine with reduced lordosis and caudal beginning of ankylosis.

spinal axis and practically parallel to a flat one, as has been shown by the investigations of many authors. Groh, Thös, and Baumann (1967), among others, mentioned that there is a definite dependence of the muscle power required for body support on the shape of the spine. Flattening of the lumbar spine in connection with a tilted pelvis thus facilitates the immobility of the iliosacral joint, and the need for supporting muscle power is reduced to a minimum. On the other hand, the missing curvature of the spine means decreased or even eliminated spring effect of the spine and, consequently, little shock absorbtion, so that vertical impacts are felt more intensely by patients suffering from ankylosis spondylitis. We often hear these patients complain about the pain from such impacts.

EFFECTS OF HIP ARTHRODESES

The problem of a pathologically increased lumbar lordosis is explained by the second example. Following surgery for hip arthrodesis, we find increased lumbar lordosis, depending on the hip flexion angle of the fixed leg. To analyze the compensation mechanism of increased lumbar lordosis, we exam-

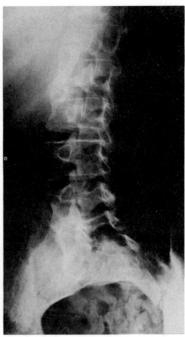

Figure 2. Silhouette of a patient suffering from ankylosis spondylitis with cranial beginning of ankylosis. The radiograph shows that lumbar lordosis and motility are preserved.

ined 12 patients with hip arthrodesis, utilizing still photography, motion pictures, and radiographs. The positions of the spine, pelvis, and the fixed leg were studied during walking movements and in the sitting position. Comparative data were obtained from four healthy subjects.

Figure 3 shows a schematic drawing of the straightening up movement of the lumbar spine. From left to right, the hip flexion angle referred to as "β" is increasing from a neutral position to $40°$. For determination of the straightening up movement, the angle between the superior and inferior vertebral surface and the horizontal axis was measured. The angle α corresponds to the neutral position between the adjacent segments as measured in the upright position of a healthy control person. The angle $\Delta\alpha$ indicates the additional movement of the vertebrae required for erect position of a patient with the respective hip flexion angle. $\Delta\alpha^1$ is the range of motion that still would be possible if the total physiological range of motion as reported by Bakke (1931) is accepted as a basic norm. The angles indicated in Figure 3 refer only to a lumbar spine with a specific hip flexion angle of $40°$. The results show that the straightening up movement of a lumbar spine with a hip flexion angle of up to $15°$ lies within the normal range of motion of L5/S1 as established

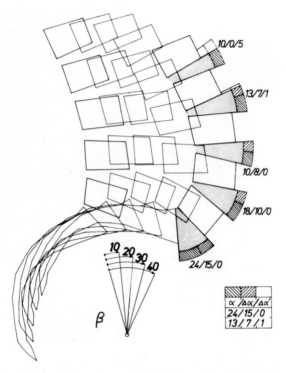

Figure 3. *Left,* the neutral position of a lumbar spine. *Following from left to right* are four lumbar spines with increasing lordosis which correspond to an increasing hip flexion angle from 10 to 40°. See text for detailed explanation.

by Bakke (1931). Even a hip flexion angle of up to 30° can be compensated for in the lumbar spine by additional movement in the higher segments. We have found that the spine motion for extension starts in the lowest segment, L5/S1, and, if necessary, spreads to L4/5 and L3/4. Up to a hip flexion angle of 15°, adjustment is reached completely in L5/S1. For larger angles, movements of the higher segments in ascending sequence are required. An extreme angle of 40° no longer can be compensated for in the lumbar spine because it exceeds the total range of motion of the lumbar segments.

If, contrary to the physiological process of motion, we distribute the required range of motion equally over all the segments in order to achieve an upright position for the various hip flexion angles, the result is rather different (Figure 4). In the lowest segment, the maximal range of motion is not utilized, whereas, in the higher segments, the total physiological range is reached and even exceeded. The important difference between the upright position, which is achieved mainly by an adjusting movement in Segment L5/S1, and the theoretical one, in which all the segments participate equally, is in the shifting of the center of gravity. If the upright position is achieved by

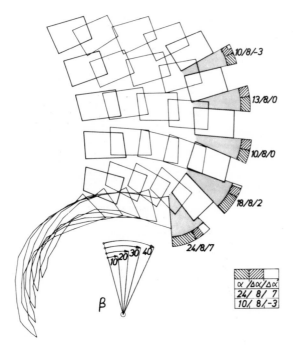

Figure 4. The arrangement of the diagram is identical with that in Figure 3. The necessary straightening up movement has been distributed equally over all of the segments, according to theoretical calculations. See text for detailed explanation.

a large movement in L5/S1, erect body posture is reached at an earlier stage; however, if it is distributed over all of the lumbar segments in equal parts, the whole lumbar spine—and with it, the center of gravity—is shifted ventrally for at least one-half the width of the vertebral body, as shown in Figure 5. Thus, the concentration of the straightening up movement in the lowest segment has economical advantages, because only a minimum of supporting muscle force is needed for the upright posture, but at the same time it leads to an increased shearing stress in the pars interarticularis of L5/S1 that probably is incompatible with the physiological material.

EFFECT ON NERVE INFLAMMATION

Also subject to the laws of biomechanics, although its pathological feature is developed on different grounds, is the so-called "reflexly induced relief lordosis." This relief lordosis is found in an inflammation process with adhesion of the dura or the nerve roots in the spinal canal. It may be noticed also in a tumor starting from the myelin, specifically of the dura, when the inflammation is located in the region of the lumbar spine. The pronounced

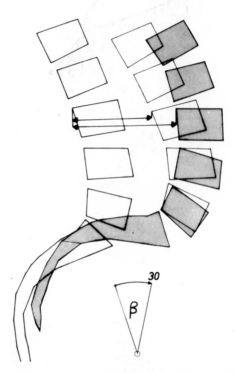

Figure 5. *Left,* the neutral position of a lumbar spine; *right,* the light-shaded spine shows the physiological extension with maximal movement in L5/S1, representing the values measured in a patient with a hip flexion angle of 30°. The superimposed *dark-shaded spine* indicates the theoretical form when the total straightening up movement is shared by all of the segments. The difference between the *arrows* indicates the shorter distance of ventral shifting of the third lumbar vertebral body when maximal straightening up movement of the spine takes place in Segment L5/S1.

lumbar lordosis is a relief posture when the gliding capability of the dural sac of the efferent nerve roots in the region of the spinal canal and the interverte- bral foramina are reduced or even eliminated. An increased lumbar lordosis shortens the gliding path considerably from maximum to 25 percent of maximum, according to our measurements (Figure 6). The clinical manifesta- tion is seen in the patient with pronounced lumbar lordosis, retroversion of the thorax, and tilting of the pelvis which leads to the typical "pushed gait." This posture is illustrated in Figure 7.

These examples indicate that under normal conditions, i.e., when the interacting elements such as vertebral joints, musculature, nerves, etc. are intact, the shape and extent of the lumbar lordosis are determined by the various forces acting on the body. However, when the elements are affected by pathological changes, this principle is inoperable, and the lumbar lordosis then is determined by the character and grade of the pathological process.

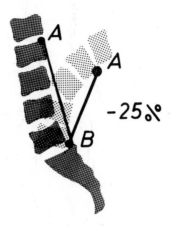

Figure 6. The difference between the two lines *A* and *B* indicates the possible shortening of the gliding path in the spinal canal attributable to maximal lordosis.

Figure 7. Typical "pushed gait" with pronounced lumbar lordosis in a patient with meningioma and adhesion of the dura.

REFERENCES

Bakke, S. N. 1931. Röntgenologische Beobachtungen über die, Bewegungen der Wirbelsäule. Kungl. Boktyckerirt. P. A. Norstedt & Söner, Stockholm.

Groh, H., F. R. Thös, and W. Baumann. 1967. Die statische Belastung der Wirbelsäule durch die Sagittalkrümmungen. Internatl. Z. Angew. Physiol. 24: 129–149.

Author's address: *Dr. J. Polster, Orthopadische Klinik, Westfälischen Wilhelms-Universitat, 44 Munster (West Germany)*

Transmission of stresses in the vertebral column

R. Hernandez-Gomez
Madrid, Spain

It is the opinion of this author that the normal vertebral column acts as an empty cylinder that is formed by three equivalent points of support. It is known from Mutel (1922) and Gallois and Japiet (1925), as cited by Perez-Casas (1965), that the special triangular configurations of individual vertebrae architecturally transmit the forces of stress on the vertebral column. Attempts have been made to study different theoretical and practical aspects of the human spine from this point of view, i.e., by analyzing the spine as a continuous empty cylinder formed at each vertebral level by three main contact points which are connected by the fasciculus obliquus, the U-fascicles, and the intertransverse fascicles, respectively.

THEORETICAL ASPECTS

Distribution of Stress

With the viscera positioned in front of the spine and attached to it, one can speculate about the importance of vertebral bodies. De Seze's scheme showed the intervertebral posterior joints as simple units for motion, able to move independently of one another. Nevertheless, this is possible only in the thoracic region and not in the lumbar area. Therefore, it can be deduced that weight is supported more directly by each vertebra as the angle between the vertebrae and a vertical line is decreased. It is possible to express the idea that the amount of stress is inversely proportional to the sine of the angle. With an angle of $0°$, most of the stress is transmitted through the vertebral bodies, but this is possible only at the initial vertebrae of the three spinal regions and then only in the resting positions (Figure 1).

At the lumbosacral level, Strasser demonstrated (Steindler, 1955) that an angle of $45°$ produced a distribution of stress which is equally divided

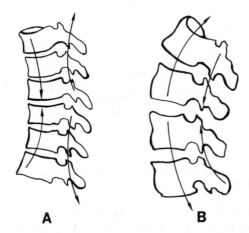

A   **B**

Figure 1. In kyphotic situations (*A*), maximal support is provided by the vertebral body. In lordotic situations (*B*), on the other hand, maximal support is given by the intervertebral posterior joints. In A, mobility occurs in the posterior area of the vertebrae; in B, it occurs in the anterior portion.

between the anterior and posterior parts of the lumbosacral junction. This indicates that the intervertebral posterior joints not only are points of motion, but that, in many cases, they are the main supporting points, especially when support or stressful forces are being transmitted. This same problem can be approached from a different point of view.

Vertebral Beams

One may assume that each vertebra acts mechanically as a beam. If it is further assumed that if there are three transmission points for every vertebral beam, then that beam can be separated into three component beams, as shown in Figure 2. *A* and *B* are the intervertebral posterior joints and the *C* nucleus pulposa A-B, A-C, and B-C are then the component beams. At the cervical and lumbar levels, this is the arrangement, but at the dorsal level, the ribs extend the vertebral beams in lateral and backward directions. The A-C beam then is transformed into the A'-C' beam, with four supporting points (continuous beam). The A-B beam becomes the B-D and the B-C beam becomes the B-E (Figure 2). The result of this alteration is a change in the mechanical situation. First of all, the A'-C' beam is very long; it has four points of support, but the external ones (A' and C') are quite moveable because they are fixed only by the muscles and the long elastic ribs. This beam needs a medial point of support, which is the nucleus pulposus (B) situated very close to it. Mechanically, A'-C' provides twice the amount of support.

Second, the auxiliary beams B-D and B-E have only three support points and no real support at D and E. These beams therefore are transformed into

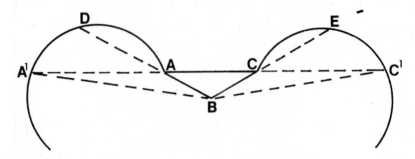

Figure 2. Scheme of the vertebrae and the ribs as a system of beams (see text for detailed explanation).

cantilever beams. There is a good solution to the architectural problem of these cantilever beams, namely, to interlock the proximal points of support by fastening them to a wall or other structure (Morley, 1921). This interlocking of the nucleus pulposus occurs in the vertebral column at the dorsal level by pressing the adjacent vertebral bodies against each other. In order to do this, a kyphotic curve must be formed at the dorsal level.

On the other hand, because the cervical and lumbar regions move freely, they do not need to interlock the proximal or distal support points. They are in a lordotic position, which increases the requirements of the intervertebral posterior joints for support and, finally, they have greater mobility, rolling over the nucleus pulposus. This kind of mobility in the anterior part and support in the posterior area means a great deal in relation to the biomechanical and pathological behavior of the human spine. Animals have a different mechanical structure. Their spines act as a single shaft and therefore have no need for support provided by large vertebral bodies. Therefore, they have a different kind of articulation at the posterior level. Moreover, they have no constant curves, and the muscles are relaxed because they do not need to be under constant tension. A similar relaxation occurs in human beings when they are lying down or swimming.

PRACTICAL ASPECTS

Equilibrium of the Spine

The need to maintain a vertical position creates many problems relative to the human spine. Increased curves, rib deformities, and spinal rotations all must be looked on as normal compensatory reactions. Studying the spine with a radioscopic closed-circuit television, one can find a number of abnormalities. We usually try to correct these with supplementary risers under one of the feet, trying them in each shoe and using different sizes. Many times, a simple riser is best for obtaining maximal progress in the correction of deformities. It is essential, nevertheless, to disregard the lengths of the extremities. The major problem is in the vertebral column. Sometimes we even put the riser in the shoe of the longer leg.

The riser must bring the pathological spine into equilibrium, thus producing the correct forces. It is necessary to make the riser uniform, with the same thickness in the heel and the ball of the foot, so that it produces an equivalent effect. A riser underneath only the heel increases the lordotic component at lumbar level as well as the kyphotic one at the dorsal level. It is also well known that certain forms of risers will increase scoliotic curves. One can demonstrate these changes in the spine very easily with radiographic, teleradiologic, and radiocinematographic aids. Filming is especially important in the evaluation of postural changes in different positions, particularly in the upright position (Hernandez-Gomez, 1971, 1973).

Correction of Deformities

It is the opinion of this author that the best method for correcting a deformity of the spine in the developing adolescent is usually the elongation method. The use of risers, physical therapy, and elongation plaster is the best method before the spine is developed completely (at about 19–21 years of age). With the use of plaster casts, we also attempt to correct the thoracic deformities, creating a series of forces in attempts to obtain the same result in different ways. To correct deformities of ribs performing contrary to their requirement to act as extended beams, small pillows of felt are employed.

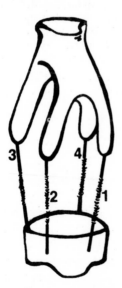

Figure 3. Scheme of an original plaster corset. Between the two main pieces, the upper cervical one, and the lower pelvic one, are four adjustable screws. These can be elongated and, at the same time, they can produce an important effect of derotation by adjusting the screws according to the established rotation of the vertebral column.

These are similar to the derotating kind of plaster casts. We have introduced a new model of elongation plaster corset, schematically shown in Figure 3. This plaster corset has upper and lower portions, similar to those in the Donaldson model, which are separated by four traction screws (Hernandez-Gomez, 1973). These screws are used to elongate, of course, but the main idea is to create the desirable component of derotation by putting each screw in an optimal position to accomplish this purpose. Generally, traction is applied with the aid of a radiologic television to obtain maximal accuracy of rotation either counter-clockwise or clockwise, according to the degree of rotation in the vertebral column. Numbers are utilized as shown in Figure 3, but revisions are made every week. This corset is used for 2 months on each patient. In our experiences with only four cases up to the present time, very good results have been achieved and patients have shown a surprisingly good tolerance for the corset. Felt for one-rib deformities and, of course, risers also are used. This seems to be a very good nonoperative approach at the present time.

REFERENCES

Hernandez-Gomez, R. 1971. El Raquis, Unidad Funcional. Monografias Medi-
cas LIADE, Madrid.
Hernandez-Gomez, R. 1973. Biomecánica y patomecánica del raquis. Rev.
Iberoam. Rehabil. Med. 9: 2.
Morley, A. 1921. Resistencia de Materiales. Editorial Labor S.A., Barcelona.
Perez-Casas, A. 1965. Anatomia Funcional del Aparato Locomotor y de la
Inervación Periférica. Editorial Bailly-Bailliere, Madrid.
Steindler, A. 1955. Kinesiology of the Human Body. Charles C Thomas,
Springfield, Ill.

Author's address: *Dr. Ricardo Hernandez-Gomez, General Peron 13, Madrid 20 (Spain)*

A dynamic concept for the diagnosis of idiopathic scoliosis

J. K. Ober
Medical Academy, Poznan

Idiopathic scoliosis is manifested clinically as a disturbance in the static characteristics of the spinal column. The diagnosis of scoliosis also involves the examination of its static aspects. However, these static methods do not permit detection of scoliosis before readiologic and clinical symptoms appear, that is, in the prescoliotic phase.

New diagnostic techniques are possible, assuming that the primary factor in the development of idiopathic scoliosis is dynamic in nature and can be identified by investigating the dynamic behavior of the neuromuscular system which stabilizes the spinal column. Consequently, methods of examination should contain dynamic as well as static components.

How does dynamic dysfunction of the spine stabilizing system result in the static appearance of scoliosis?

The reflex contractions of the spinal column muscles compensate for the bending of the spinal column. This is a characteristic behavior of the spine stabilizing system during human movement. As a result, even small compensatory differences between the right and left sides cause permanent, asymmetric, dynamic overloadings of the soft tissues (Figure 1). These overloadings are accumulated and after some time result in the partial loss of elasticity in the soft tissues on one side of the spine. This loss results in modification of the previously neutral position of the intervertebral linkage, and scoliosis appears.

The curvature of the spinal column increases the asymmetry of the load and further alters the function of the spine stabilizing system. This may account for the rapid progression of the idiopathic scoliosis.

The natural regeneration of the elastic properties of the soft tissues of the spine may accelerate or moderate the effects of the primary dynamic dysfunction, depending on the effectiveness of the metabolic processes and the age of the patient.

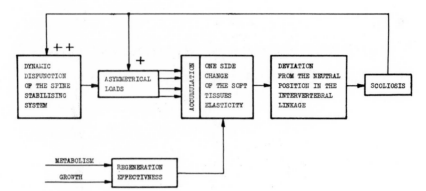

Figure 1. Hypothetical model of the development of the idiopathic scoliosis.

This hypothetical model for the development of the idiopathic scoliosis indicates that a dynamic approach to the diagnosis of idiopathic scoliosis is needed.

METHOD

To obtain information about the dynamic behavior of the spine stabilizing system, an impulse response method was used. The stabilizing system is activated by a short-duration bending of the spinal column in the frontal plane, a movement which causes a reflex contraction of the spine muscles. The muscular activity received from both sides of the spine on the Th_9 and

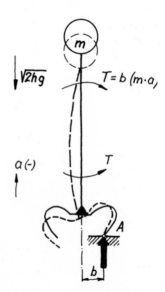

Figure 2. Diagram of bending forces applied to the spinal column.

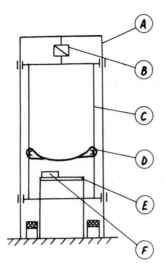

Figure 3. Scheme of devices generating an impulse of torque.

L_1 level is recorded electromyographically. The inertial forces of the mass of the trunk, appearing at the moment of the sudden stopping of the patient's free fall, was used to generate the impulse of bending torque. Inertial force tilts the pelvis, which is supported on one side at the point of ischial tuberosity, and thereby bends the spinal column (Figure 2).

To assure relaxation of the back muscles before excitation of the spine system, it was decided that the patient should assume a sitting position during examination. A special device was built to meet these conditions (Figure 3).

The diagram in Figure 4 presents the complete experimental setup used to record the exact time of excitation as well as the reaction of the stabilizing

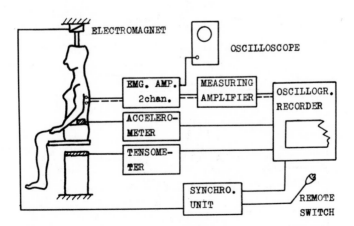

Figure 4. Diagram of the experimental setup for dynamical examination of the spine.

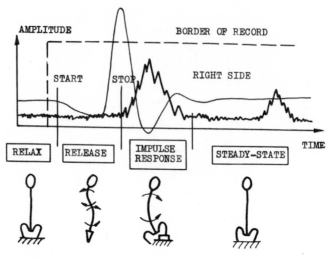

Figure 5. Simplified diagram of a recording /upper half/.

system. The examination process starts when the electromagnet is deactiv-
ated, thus releasing the free-fall mechanism and simultaneously triggering the
recording device.

RESULTS

A typical record is presented in Figure 5. The smooth, thin line represents the
vertical component of the acceleration of the patient's body. The heavy,

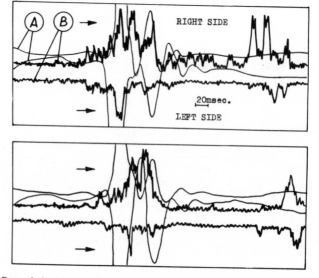

Figure 6. Record showing acceleration and bioelectrical activity during dynamical test.

zig-zag line is the record of the rectified and integrated bioelectrical activity of the spine muscles. Figure 6 shows an example of two original records, with Trace A indicating acceleration and Trace B the bioelectrical activity. The asymmetry of reaction between the right and left sides of the spine should be noted. This dynamic examination, if applied to large groups of children, may be very important in the establishment of new criteria for definition of the prescoliotic state.

Author's address: *Dr. Jan K. Ober, Institute for Orthopedics and Rehabilitation, Medical Academy, Poznan-61-545, ul. Dzierzynskiego 135/147 (Poland)*

Evaluation of treatment success for spastic paralyzed children by means of quantitative electromyography

M. Rotzinger and H. Stoboy

Institut für Leistungsmedizin, Berlin

The evaluation of neuromuscular performance in children with cerebral palsy often is performed completely subjectively by investigators (Freeland and Vorweg, 1973; Roth, 1973; Steen, 1969). Sometimes the effect of rehabilitation is estimated by comparing the productivity of normal patients with that of patients with cerebral palsy (c.p.) or other disabling diseases (McLeod, 1969). Spira (1972) measured the improvement during a sports program by having subjects climb and descend five stairs. He stated that there was an improvement in the speed of the various performances. Levine, Jossmann, and DeAngelis (1972) and Bergamini et al. (1972) tried to determine the effect of muscle relaxants (Lioresal) by means of integrated muscle tone and amplitudes of the H-reflex.

In our investigations, the measurement of H-reflex was excluded because of the individual discrepancies in thresholds. Therefore, we preferred to measure the integrated electrical activity (E.A.) of subjects in both supine and standing positions.

Independently of our investigation, and without communication with us during the test period, another team of researchers (Tepper, Soeffky, and Rauterberg, 1973) judged the effect of rehabilitation by means of subjective ratings and muscular tests.

METHODS

Investigations were carried out using two groups: (a) 70 healthy children ranging in age from 6 to 8 yrs. and (b) five children (6–8 yrs. old) with the diplegic type of cerebral palsy. An achillotenotomy was performed on one subject before the investigation. Before and/or during this period of investigation, neuromuscular exercises like those of Bobath were carried out. In addition, a trampoline exercise program was developed (Soeffky, 1971) which included only simple jumps in an upright position.

The E.A. was obtained by use of surface electrodes on the tibialis anterior and gastrocnemius muscles of both sides. The original and the average electromyogram were recorded simultaneously. The integrated electromyogram could be read on a digital scale.

The measurements were obtained with the subjects in both relaxed supine and standing positions. The noise level of amplifiers was subtracted from the E.A. The E.A. also was registered during dorsiflexion and plantar flexion of the feet.

Both groups of children were investigated before training while the c.p. children were also tested four times during a period of 1 yr, with an interval of approximately 3 months between each evaluation. Because of the high mental drive of c.p. patients (Hubermann, 1973; Struppler, 1972) with frequent spastic bursts, all subjects were investigated in a nearly noise-sheltered and half-darkened room with an indifferent environmental temperature.

Differences between groups were evaluated statistically by means of t tests.

RESULTS

In a first approach, the healthy children were classified into five age groups. Because of statistically nonsignificant differences, all were considered as one group.

Muscle tone of healthy subjects in a supine position ranged between 64 and 82 μV_{sec}. No statistically significant differences could be detected between the particular muscle groups. Therefore, the E.A. and the standard deviation of all investigated muscles were used as reference values.

In comparison with normal subjects, a significantly increased muscle tone was detected in the left tibialis anterior muscle of c.p. children in the first and second investigations ($p < 0.01$). In the third investigation, the tone of all four muscle groups increased significantly ($p < 0.01$). In the fourth investigation, no significant difference could be detected. The fifth investigation showed again a highly increased tone of the left gastrocnemius and tibialis anterior muscles ($p < 0.05$) (see Figure 1).

The value of muscle tone in c.p. patients was compared among the particular investigations. Only between the third and fourth investigations could a significant decrease of the muscle tone be detected in both gastrocnemius muscles ($p < 0.05$).

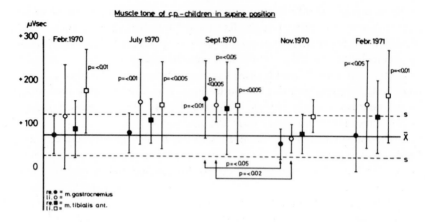

Figure 1. Muscle tone of c.p. children compared with that of normal subjects (*ant.,* anterior).

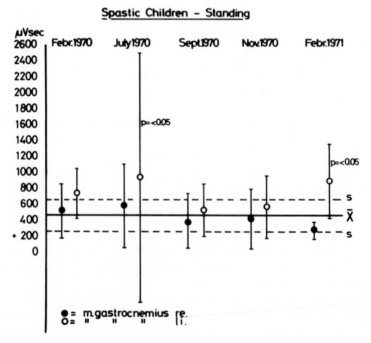

Figure 2. E.A. (gastrocnemius muscle) of c.p. children compared with that of normal subjects in standing position.

In an upright position, the E.A. of healthy subjects was significantly higher in the gastrocnemius than in the tibialis anterior muscle ($p < 0.005$). For this reason, the E.A. of the gastrocnemius and tibialis anterior muscles of c.p. patients and normal subjects were compared separately. No significant difference could be found between the right gastrocnemius muscles of c.p. patients and those of normal subjects. In the second investigation, the E.A. of the left gastrocnemius muscle was markedly higher in c.p. patients than in normal subjects ($p < 0.05$). In the fourth and fifth investigations, this difference disappeared, but in the fifth investigation, the E.A. of c.p. patients rose again ($p < 0.05$) (Figure 2).

In the right and left tibialis anterior muscles a highly significant difference with $p < 0.001$ was detected in all five investigations with subjects in the standing position (Figure 3).

Spastic Children – Standing

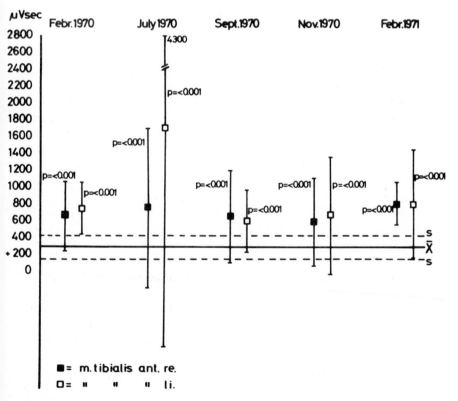

Figure 3. E.A. (tibialis anterior muscle) of c.p. children compared with that of normal subjects in standing position.

DISCUSSION

Tepper et al. (1973) judged the effect of rehabilitation by subjective impressions and by some muscular tests. At the end of trampoline exercises, more skilled jumps were observed, i.e., spread-eagle jumps, and an improvement of approximately 25 percent was observed in skipping and tapping. Also, the gait patterns seemed to be more regular. In the number of steps, walking of a certain distance, and step length, no improvements were detected. Also, in aimed ball rolling, no improvement was found in hitting of the target. On the average, the duration of one-leg standing increased by 78 percent. From these findings, it was concluded that there was an improvement in motor performance.

The measurement of muscle tone in c.p. children is very complicated. Low noises, sudden movements of the investigator, etc., trigger bursts of E.A. In spite of attempts to avoid all of these factors they may be partially responsible for the completely irregular pattern. Hubermann (1973) stated that spasticity may increase enormously before sport competitions, so that performance may become very low.

Levine et al. (1972) tried to quantify the effect of a muscle relaxant (Lioresal) by measurement of muscle tone. A difference with $p < 0.05$ was found between the values before treatment and the sum of all values during therapy. The graphs clearly showed a large fluctuation of muscle tone, so that in some cases at different times it was higher or nearly as high as before treatment. If we had stopped our measurements after the fourth investigation, in which all values for muscle tone were in the normal range, we would have been in accordance with clinical judgment. But the values of the fifth investigation showed again a marked increase (left gastrocnemius and left tibialis anterior muscles), so that no correspondence between clinical judgment and muscle tone could be found. As suspected, the E.A. of gastrocnemius muscles in normal subjects was significantly higher than in the tibialis anterior muscles. During the rehabilitation period, the E.A. of left gastrocnemius muscle showed an inconsistent pattern. In both tibialis anterior muscles, the E.A. was markedly higher in comparison with normal range in all five investigations. The reason for these findings is probably that c.p. patients with talipes equinus have to bend their knees in order to touch the ground with the entire sole of the foot. As a result of this maneuver, the gastrocnemius muscles are relaxed. In this position, the center of gravity is projected behind the ankle joint so that the tibial anterior muscles have to counteract the effect of gravity.

In only three clinical tests out of six could an average improvement be detected, and the duration of one-leg standing and the frequency of skipping were enhanced noticeably in only 50 percent of the subjects. The children showed a marked increase in frequency of tapping with the upper extremity, which was less spastic than the lower extremities. In the electromyographic investigations of only one c.p. patient was the muscle tone in a supine position consistently diminished, and in only one other patient did the coordination pattern of dorsiflexion and plantar flexion of the foot approach that of normal standards.

Thus, the subjectively rated evaluations of gait and trampoline skills as well as motor tests led to positive judgment of the effect of rehabilitation exercises, which could not be substantiated by our electromyographic investigations.

REFERENCES

Bergamini, L., F. Birgnolio, R. Fariello, L. Fra, G. Quatrocoilo, R. Mutani, and A. Riccio. 1972. Elektromyographische (H-Reflex) und kinetometrische Studien über die antispastische Wirkung. *In:* CIBA 34'647-Ba. Aspekte der Muskelspastik. Verlag Hans Huber, Bern-Stuttgart-Wien.

Freeland, S. P., and F. M. T. Vorweg. 1973. Cerebral palsied adolescents in South Africa. Paper presented at the International Symposium on the Disabled Adolescent, Tel Aviv, Israel, June 24–28.

Hubermann, G. 1973. Sport activities for cerebral palsied adolescents. Paper presented at the International Symposium on the Disabled Adolescent, Tel Aviv, Israel, June 24–28.

Levine, I. M., P. B. Jossmann, and V. DeAngelis. 1972. Quantitative Wirkungsbeurteilung. *In:* CIBA 34'647-Ba. Aspekte der Muskelspastik. Verlag Hans Huber, Bern-Stuttgart-Wien.

McLeod, N. 1969. Vocational programming for the cerebral palsied. *In:* Proceedings of the XIth World Congress of Rehabilitation. International Community, Responsibility for Rehabilitation, The National Rehabilitation Board, Dublin.

Roth, V. G. 1973. The adolescent with meningomyelocele. Paper presented at the International Symposium on the Disabled Adolescent, Tel Aviv, Israel, June 24–28.

Soeffky, E. 1971. Möglichkeiten der Beeinflussung des spastischen Bewegungssyndroms durch Leibesübungen. Diplomarbeit, Institut für Leibeserziehung, Freie Universität Berlin.

Spira, R. 1972. Influence of sport activities in rehabilitation of paralytic subjects. *In:* Rehabilitation Research in Israel. GOMEH Scientific Publications, Bezalel Tcherikover, Publishers Ltd., Tel Aviv.

Steem, M. 1969. Unusual treatment procedures in cerebral palsy—vibration therapy. *In:* Proceedings of the XIth World Congress of Rehabilitation. International Community, Responsibility for Rehabilitation, The National Rehabilitation Board, Dublin.

Struppler, A. 1972. Zur Physiologie und Pathophysiologie des Skelettmuskeltonus. *In:* Aspekte der Muskelspastik. Verlag Hans Huber, Bern-Stuttgart-Wien.

Tepper G., E. Soeffky, and E. Rauterberg. 1973. Trampolinspringen mit Kindern, die an einer spastischen Hemiplegie leiden. Z. Heilpadagog., In press.

Author's address: *Dr. M. Rotzinger, Freien Universität Berlin, Clayallee 229/233, Berlin (West Germany)*

Electromyographic and phonomyographic patterns in muscle atrophy in man

M. Marchetti, A. Salleo, F. Figura, and V. Del Gaudio
University of Rome, Rome

Since the report by Buchthal and Clemmesen (1941) on the electromyographical (EMG) changes caused by disuse atrophy, the only article published on this subject has been that by Wolf, Magora, and Gonen (1972). The most important feature described in this article was a reduction of frequency and amplitude of EMG waves during maximal voluntary effort. Furthermore, the interference pattern is not as markedly modified as in other diseases affecting the neuromuscular apparatus. Therefore, EMG generally is not employed in orthopedics and rehabilitation practice in cases of disuse atrophy. In past years, we have adopted the Fourier frequency analysis to obtain a better characterization of the EMG patterns during maximal effort in athletes and have gained some useful information on the effects of training (Cerquiglini et al., 1973). Thus, it seemed appropriate to apply the same technique to the study of changes in EMG produced by disuse atrophy. Moreover, in our previous work, phonomyography (PMG) was utilized with EMG. This method reveals by means of suitable microphone the microvibration produced by muscles during contractions. These vibrations can be considered as a mechanical counterpart to the bioelectrical signal (Gordon and Holbourn, 1948); and, in our experience, PMG reveals mechanical activity in muscle just as well as EMG reveals the excitatory process.

METHODS

This investigation was carried out in 14 orthopedic patients previously submitted to immobilization of one or more joints for treatment of bone fracture. Neither the muscles explored nor their motoneurons were affected by anatomical damage. Soon after the plaster restraint was removed, the hypotrophic muscle, as well as the corresponding muscle of the contralateral limb, were explored during maximal isometric voluntary contractions. The surface EMG signal was amplified by means of an electromyograph (DISA Model 14A30) and conveyed to an electronic frequency analyzer (Kay Sona-Graph Model 7029-A). In two cases, needle electromyography also was used.

The PMG was picked up simultaneously through a piezoelectric microphone (Hewlett-Packard Model 21050A, Hewlett-Packard Co., Palo Alto, Calif.) fixed on the skin between the two EMG surface electrodes. The sound signal was amplified (Hewlett-Packard Heart Sound Model 350-1700A) and sent to the frequency analyzer, as was the EMG signal. The analyses were performed in the frequency band of 5 to 500 Hz. In the display obtained from the analyzer, the frequency of the harmonics was plotted on the ordinate against real time on the abscissa. The intensity of the harmonics was expressed by the degree of blackening of the paper. The profile of the frequency spectrum was obtained by means of a photoelectric device reading the spectrum display along the ordinates. The mean profile of each contraction was obtained by superimposing five readings carried out at prefixed time intervals (40 msec).

In four cases, the exploration was repeated during the course of the muscle rehabilitation. In two cases, EMG and PMG signals were explored during electrical stimulation of the motor nerve (single pulses).

RESULTS

Immediately after the plaster restraint was removed, the comparison between the EMG and PMG patterns of the hypotrophic muscle and those of the corresponding normal muscle for all subjects showed that the former produced a much sparser interference pattern, as well as waves of lesser amplitude (Figure 1). It must be emphasized that such changes hardly can be evaluated in quantitative terms without the aid of frequency analysis.

In all subjects explored during the rehabilitation period, changes in both EMG and PMG patterns were observed. The nature of these changes was a progressively larger number of waves with greater amplitudes that finally become similar to those exhibited by the control muscles. Such results have been confirmed by needle EMG. The EMG and PMG changes described were followed immediately by an improvement of muscle strength.

In five subjects, we found an unusual aspect of the EMG and PMG, namely, the appearance of an easily recognizable rhythmic pattern of the waves. Observations in one of these subjects (Figure 2) are described. Even during the first exploration, a phonomyographical pattern characterized by rhythmic waves with repetition frequency of approximately 13 cycles per sec was observed. The frequency analysis of the PMG revealed the presence of a fundamental harmonic, with the same frequency as the waves observed in the

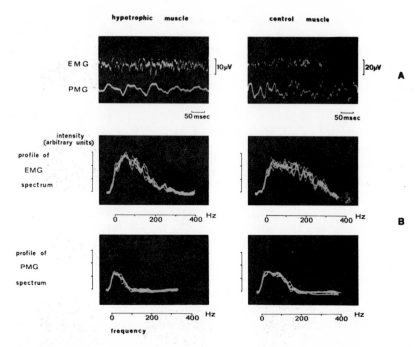

Figure 1. *A*, EMG and PMG recordings from hypotrophic and control muscle during voluntary maximal isometrical contraction. *B*, profiles of frequency spectra of the tracings shown in A.

tracing. However, the inspection of EMG obtained during the same session did not reveal a regular repetitive frequency. Nevertheless it was possible in the EMG frequency spectrum to recognize a fundamental harmonic similar to that present in the PMG. In the following session, the phonomyographical tracing showed an accentuated rhythm, whereas the EMG pattern appeared extremely modified. In fact this pattern consisted of large, rhythmic waves with the same repetition frequency as in the PMG. A sudden improvement of muscle strength occurred with the appearance of EMG rhythmic pattern.

Finally, the muscle potential obtained in response to electrical stimulation of the motor nerve in hypotrophic muscles was quite different from that in the control muscle; the former showed a lesser amplitude and a longer duration (Figure 3). Such differences progressively disappeared during rehabilitation.

DISCUSSION AND CONCLUSION

Our results show that when a muscle is hypotrophic as a result of immobilization, the EMG and PMG waves are composed of harmonics of lower frequency than those of homologous unaffected muscle. These results are in

VASTUS LATERALIS

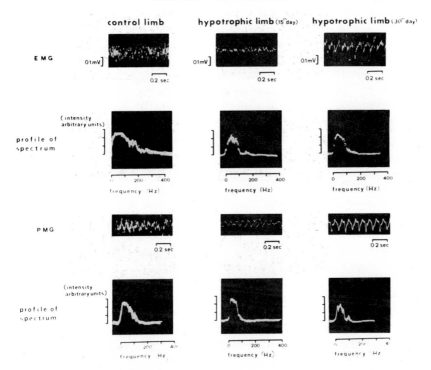

Figure 2. Confront between normal and hypotrophic muscle (vastus lateralis at different times during rehabilitation period. *EMG*, surface electrodes).

agreement with those obtained in our previous research. In fact, it was observed then that training caused changes in the processes of muscle excitation and activation that appeared in the EMG and PMG tracings as an increase in the frequency of the harmonic components (Cerquiglini et al.. 1973; Marchetti et al., 1972a, b). It is not surprising that a condition that is diametrically opposed to that of training—in this case, a prolonged motor inactivity—produces changes that also are diametrically opposed to those produced by training.

The suggestion is made that the EMG and PMG patterns observed by us in disuse atrophy could be the result of some deterioration of the ability of the motor centers to recruit motor units and/or to fire them at higher frequencies. An alternate explanation is that the decreases in harmonic components may depend on some change in conduction of muscle fibers. Indeed, the results obtained through motor nerve stimulation suggest that immobilization is accompanied by an impairment of the peripheral components of the neuromuscular system. Nevertheless, the observed differences between hypotrophic and control muscles in both electrical and mechanical activities are greater in voluntary contraction than in artificial stimulation (Figure 3). Therefore, the hypothesis that the motor centers are involved seems more tenable. Consequently, the relearning of normal activity by motoneurons seems of primary importance in the process of motor rehabilitation. In our opinion, the central and peripheral structures of the neuromuscular apparatus are so intimately interdependent in regard to trophism and function that separate damage to one of them hardly is conceivable.

Synchronous activity of motoneurons has been observed widely under conditions with one aspect in common; that is, the maximal muscle strength output. This is the case also in muscle fatigue, in which there is the reduction in the number of recruitable motor units following an impairment of the common final pathway that has been induced by pathological or other means. Moreover, it has been established by others (Stepanov, 1959) and ourselves (Cerquiglini et al., 1973; Marchetti et al., 1972a) that very fit athletes are able to obtain this kind of motor unit synchronization when performing maximal effort.

The present evidence clearly shows that the same effect also occurs in disuse atrophy during maximal contraction. According to a hypothesis suggested by Lippold (1971) in regard to muscle fatigue, this synchronization can be considered as a compensatory mechanism produced by the central nervous system to overcome the relative weakness of the effector with respect to the task to be accomplished. Such a hypothesis is supported by the sudden enhancement of muscle strength that, in our experiments, coincided with the appearance of synchronization.

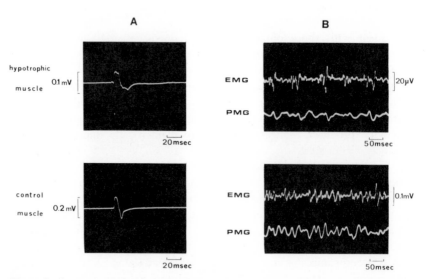

Figure 3. A, muscle potentials induced by electrical stimulation of the motor nerve. B, EMG and PMG recordings from the same muscles during voluntary maximal isometrical contraction.

Finally, it can be inferred from our results that muscle sound can supply a good deal of information about muscle capabilities. As a matter of fact, the degree of its expressiveness is comparable in every respect to that of the EMG. Furthermore, muscle sounds can be appreciated qualitatively by a practiced ear and pose less problems from the point of view of electronic assembly and economy than do EMG techniques. The utilization of this simpler method in clinical practice therefore is recommended.

REFERENCES

Buchthal, F., and S. Clemmesen. 1941. On differentiation of muscle atrophy by electromyography. Acta Psychiat. Neurol. 16: 143.

Cerquiglini, S., F. Figura, M. Marchetti and A. Salleo. 1973. Evaluation of athletic fitness in weight-lifters through biomechanical, bioelectrical and bioacoustical data. In: S. Cerquiglini, A. Venerando, and J. Wartenweiler (eds.), Biomechanics III, pp. 189–195. S. Karger, Basel, and University Park Press, Baltimore.

Gordon, G., and H. S. Holbourn. 1948. Sound from single motor units in a contracting muscle. J. Physiol. 107: 456.

Lippold, O. 1971. Physiological tremor. Sci. Amer. 224: 63.

Marchetti, M., A. Salleo, F. Figura, and E. Calistroni. 1972a. Aspetti fonomiografici ed elettromiografici della massimizzazione della forza muscolare in atleti sollevatori di peso. Arch. Fisiol. In press.

Marchetti, M., A. Salleo, F. Figura, and M. Mondino. 1972b. Analisi elettromiografica e fonomiografica in cultori di atletica leggera. Arch. Fisiol., In press.

Stepanov, A. S. 1959. Electromyogram changes produced by training in weight lifting. Physiol. J. USSR 45: 115.

Wolf, E., A. Magora, and B. Gonen. 1972. Disuse atrophy of the quadriceps muscle. Electromyography 11: 479.

Author's address: Dr. Marco Marchetti, Instituto di Fisiologia Umana, Università di Roma, 00100, Roma (Italy)

Force-time characteristics of reflexive muscle contractions

G. L. Soderberg
University of Iowa, Iowa City

C. A. Morehouse
The Pennsylvania State University, University Park

Studies of the force-time properties of the myotatic reflex have been conducted in a medical, physiological, and biophysiological context. There have been few, however, dealing with the variability of these reflex characteristics. The purpose of this investigation was to analyze these components of the response of the human quadriceps muscle when phasic stretch was applied to the patellar tendon of the right leg.

METHODS

A specially constructed pendular hammer elicited the reflex according to a previously described method (Soderberg and Morehouse, 1973). Quadriceps electrical activity (electromyogram) and force at the ankle were recorded (Figure 1). Fifteen male subjects, ranging in age from 18 to 29 yrs., with a mean age of 23.5 yrs., were each tested for 3 consecutive days. Test trials were completed at a defined and predetermined threshold level and also at a suprathreshold level. Five suprathreshold level trials provided the data for this study. High-speed converters and a computer provided an on-line data collection and reduction system.

FORCE–TIME CURVE ANALYSIS

Of the total 204 force data points recorded at 2-msec intervals, only those on the tension development portion of the curve were considered. Slope, point

This study was supported in part by Research Trainee Grant No. 0EG–0–70–3966 from the United States Office of Education.

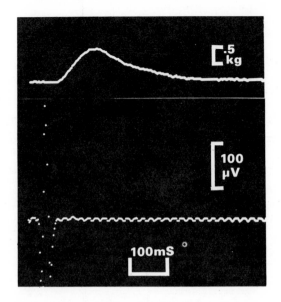

Figure 1. Computer oscilloscope display of typical force and electromyographical response.

of inflection, and tangent angles (at point of inflection) were calculated by means of a special computer program.

RESULTS

Mean tangent angles at the point of inflection were calculated for the group for each day. These values were: Day 1, 34.2°; Day 2, 31.1°; and Day 3, 32.6°. This showed that there were no systematic changes in the group of 15 subjects during the 3 days of testing. Taking into account both consistency and the fact that each of these values included five trials for each subject, it appeared that tangent angles of the force-time curve of 30–35° could be considered representative of this level of applied force.

Means and standard deviations of the tangent angles for the five suprathreshold trials on each day for each subject also were calculated and are presented in Table 1. Standard deviations showed marked inter- and intrasubject differences. There appeared to be no significant trends to the changes in the tangent angles during the 3 test days.

Group means for time to point of inflection for each of the 3 days were 37.9, 38.3, and 39.1 msec, respectively. Although these figures showed an increase over the 3 days, this range of times to point of inflection could be considered representative of the responses of this level of applied force. Means and standard deviations of the time to inflection points for the five suprathreshold trials of each day for each subject are shown in Table 2. As with the tangent angles, there appeared to be no significant trends in the times to point of inflection over the 3 test days.

DISCUSSION

Analysis of the characteristics of the rate of tension development showed that the tangent angles at the point of inflection were a more variable measure than the time to point of inflection. Data indicated that the standard deviations for the within-subject time to point of inflection comparisons tended to be less variable, and generally lower, than those reported for the tangent angle data (Subjects 1–4, 6–8, 10, and 13). In addition, standard

Table 1. Means and standard deviations for tangent angles* (degrees)

Subject	Day	$\bar{X}$	S	Subject	Day	$\bar{X}$	S
1	1	54.0	1.5	9	1	65.0	3.2
	2	44.4	6.3		2	63.1	4.9
	3	43.1	11.4		3	67.9	5.5
2	1	17.5	7.5	10	1	21.2	11.2
	2	23.7	12.8		2	11.3	8.0
	3	41.0	9.3		3	6.8	5.1
3	1	32.2	6.4	11	1	9.2	4.5
	2	21.9	11.1		2	10.7	9.7
	3	47.5	17.9		3	7.4	6.8
4	1	43.6	17.0	12	1	32.8	22.5
	2	26.0	3.9		2	44.9	18.2
	3	36.3	10.7		3	28.1	13.6
5	1	17.4	16.4	13	1	53.0	9.6
	2	17.3	8.8		2	45.0	11.4
	3	15.8	6.2		3	38.4	7.9
6	1	27.7	6.3	14	1	15.7	14.8
	2	24.3	5.0		2	10.7	5.9
	3	19.9	5.4		3	7.5	5.9
7	1	42.4	10.5	15	1	51.8	13.4
	2	39.2	5.4		2	54.8	21.0
	3	29.8	5.9		3	62.2	3.2
8	1	28.9	8.3				
	2	30.8	3.2				
	3	38.0	5.0				

*Five suprathreshold trials administered on each day.

Table 2. Means and standard deviations for time to point of inflection* (msec)

Subject	Day	$\bar{X}$	S	Subject	Day	$\bar{X}$	S
1	1	36.8	4.3	9	1	43.8	9.0
	2	30.8	3.7		2	47.6	3.9
	3	26.3	10.9		3	47.9	2.9
2	1	23.5	6.0	10	1	16.6	1.6
	2	28.5	6.4		2	15.2	5.5
	3	28.3	5.8		3	16.7	6.1
3	1	19.2	5.1	11	1	39.4	12.3
	2	17.5	5.2		2	44.3	5.5
	3	22.1	6.4		3	51.6	30.3
4	1	29.2	11.7	12	1	42.9	3.9
	2	25.6	8.8		2	89.8	29.8
	3	28.0	6.4		3	68.3	27.3
5	1	53.9	14.4	13	1	41.9	8.8
	2	48.7	11.1		2	48.4	9.7
	3	33.5	11.9		3	50.0	7.2
6	1	34.9	3.5	14	1	35.8	4.6
	2	37.6	3.2		2	27.1	11.4
	3	30.3	4.4		3	54.8	40.8
7	1	30.4	1.6	15	1	94.6	31.7
	2	31.8	2.5		2	76.4	17.7
	3	29.9	2.5		3	64.2	6.9
8	1	25.8	5.5				
	2	27.1	4.1				
	3	35.1	4.4				

*Five suprathreshold trials administered on each day.

deviations computed from other data suggested that the times to peak force were less deviant than the force measures. However, there is no further statistical evidence in this study that supports the low variability and the subsequent possible importance of this time factor. The maximal force and rate of tension developed by muscle undergoing a myotatic reflexive contraction theoretically should approach constant values. However, because the state of the muscle spindle is altered by the previous strike and influenced by other neurophysiological events, all conditions present for two consecutive force measurements would not be identical. This study presents evidence to support the idea that time factors are less variable than force measures. With adequately controlled testing conditions, however, variability of the maximal force and rate of tension development should be minimal.

An additional factor that should be considered is the low level of tension elicited by the method of testing used in this study. It is feasible that the rate of tension and maximal force developed would be less variable at higher levels of applied force. It was noted that Subject 9, who had the greatest mean values for the tangent angles (65.0, 63.1, and 67.9°, respectively) in the present study, also had standard deviations that were all under 5.5°.

The means of the tangent angles at the point of inflection cannot be compared with those from previous research because the angle is dependent on the scale of units used. In addition, no studies were found in the literature that evaluated force-time curves of reflex muscle contractions. Sukop, however, reported values for the slope of maximal voluntary tension development of 143–439 kg per sec (J. Sukop, personal communication, 1970). The range of values obtained in the present study of involuntary contraction was from 2.5 to 32.0 kg per sec for the suprathreshold level of testing. The times to point of inflection were also much lower for the present study (37–39 msec) in comparison with those reported by Sukop (50–120 msec).

It may be inferred from these data that, as a muscle contraction progresses from an involuntary to a maximal contraction at the greatest possible speed, there would be a large increase in rate of tension development. Although there also would be an increase in the time required to reach the point of inflection the increase would be within the range of 26–67 percent, in comparison with a 93–98 percent increase for rate of tension development. Again, it would be interesting to use higher levels of force to elicit the reflex in order to follow the changes in the rate of tension development and the time to point of inflection.

SUMMARY

The analyses of the force-time curves must be regarded as preliminary information. No research was located which reported force-time data on muscle contracting as the result of a reflex mechanism. Testing needs to be completed with a much larger number of trials and subjects in order to clarify the results reported. This study has provided pertinent information, but further study of the characteristics of this reflex is both indicated and necessary if the reflex mechanism is to be understood completely.

REFERENCES

Soderberg, G. L., and C. A. Morehouse. 1973. Selected electrical, force and time characteristics of the human myotatic reflex. Phys. Ther. 53: 961–969.

Author's address: *Dr. Gary L. Soderberg, Department of Physical Therapy, University Hospitals, University of Iowa, Iowa City, Iowa 52242 (USA)*

Application of the classical contact theory to deformation of the skeletal joints

H. K. Huang and R. S. Ledley
Georgetown University Medical Center, Washington, D.C.

It is well known that when two bodies come into contact under an applied load, the body surfaces will deform at the point of contact and an area of contact will appear. The size and the shape of this contact area depend on the nature of the applied load, the original shape, and the material properties of the contact bodies. From the size and the shape of this contact area, the stress distributions of the bodies caused by the applied load also can be determined.

In the case of the skeletal structure, the most important contact areas are the knee joint, where the contact bodies are the femur and the tibia, and the hip joint, where the femur and the acetabulum make contact. In a normal standing position, these joints support 86 and 65 percent of the body weight, respectively, which are enormous loads in comparison with the other skeletal joints. It is therefore important to study the contact area of these two joints to determine the stress distribution under such loads.

Walker and Hajek (1972) tried to measure the load-bearing (contact) area in the knee joint. Fresh intact knee joints were used as material, and the contact areas at different flexion angles under a constant load were measured by the casting method. Walker and Hajek formulated a simple geometric theory to estimate the contact area on the basis of the measurement of the total deflection of the articular cartilage. Although the comparison was reasonable, the theory did not take into consideration the nature of the load. Greenwald and O'Connor (1971) studied the transmission of the load through the human hip joint. They measured contact areas corresponding to different

This work was supported by Grant RR-05681 from the National Institutes of Health to the National Biomedical Research Foundation.

points on a walking cycle by first assembling the joint components on a loading frame, applying a desired load, and injecting a dye solution. The area of contact then was identified as the nonstained area on the joint components. A simple mechanical model of the hip joint also was derived on the basis of spherical symmetry.

In the present study, the classical theory of two bodies in contact is introduced, and the results obtained by applying this theory to the knee and hip joints are compared with the work of other investigators.

THE CLASSICAL THEORY OF TWO BODIES IN CONTACT

If two elastic bodies of arbitrary shape come into contact, a small surface of contact is formed as a result of local deformation. Certain parameters describing this surface can be obtained by applying the classical theory of two bodies in contact, sometimes called Hertz's theory (Love, 1944). These parameters include the major and minor diameters of the contact surface, the total deflection of the bodies, the area of the contact surface, the maximal contact pressure, and the stress distribution within the contact surface.

This theory requires a geometric description of the contacting bodies at the point of contact in the form of the radius of curvature of each body and the angle between the planes normal to each radius of curvature. Then, knowing Young's modulus and the Poisson ratio for the contact bodies, one can predict these parameters on the basis of the following simple assumptions: (a) the surface of contact is an ellipse; (b), the principal radii of curvature at the point of contact are much larger than the major and minor axes of the ellipse; and (c) the total compressive load is distributed over the entire surface of contact, with the maximal pressure acting at the point of contact.

The semiaxes, a and b, of the elliptical contact surface are computed as:

$$a = m \sqrt[3]{\frac{3\pi}{4} \cdot \frac{P(k_1 + k_2)}{(C + D)}} \tag{1}$$

$$b = n \sqrt[3]{\frac{3\pi}{4} \cdot \frac{P(k_1 + k_2)}{(C + D)}}$$

where

P = the total compressive force; $k_i = \frac{1 - \nu_i^2}{\pi E_i}$, $i = 1, 2$; ν_i = the Poisson ratio of the ith body; and E_i = Young's modulus for the ith body.

$(C + D)$ accounts for the geometrical configuration of the contacting bodies and is determined from the equation:

$$C + D = \frac{1}{2} \left[\frac{1}{R_1} + \frac{1}{R_1'} + \frac{1}{R_2} + \frac{1}{R_2'} \right]$$

where R_i, R_i' are the principal radii of curvature[1] of the ith body. The coefficients m and n are transcendental functions of the auxiliary angle $\cos^{-1} \left[\frac{D - C}{D + C} \right]$, expressed in terms of elliptical integrals. $(D - C)$ is defined as:

$$(D - C) = \frac{1}{2} \left\{ \left[\frac{1}{R_1} - \frac{1}{R_1'} \right]^2 + \left[\frac{1}{R_2} - \frac{1}{R_2'} \right]^2 + 2 \left[\frac{1}{R_1} - \frac{1}{R_1'} \right] \cdot \left[\frac{1}{R_2} - \frac{1}{R_2'} \right] \cos 2\alpha \right\}^{1/2}$$

where α is the angle between the normal planes containing curvatures $\frac{1}{R_1}$ and $\frac{1}{R_2}$. The total deflection (d) at the point of contact is determined as:

$$d = \frac{t}{2} \sqrt[3]{\left[\frac{3\pi}{4} \right]^2 p^2 (k_1 + k_2)^2 (C + D)} \tag{2}$$

where t, like m and n, is a transcendental function of $\cos^{-1} \left[\frac{D - C}{D + C} \right]$. The numerical values of m, n, and t have been computed (Whittemore and Petrenko, 1921) and are shown in Table 1.

The contact area (A) is found according to the equation:
$$A = \pi ab \tag{3}$$

where a and b are the semiaxes of the elliptical surface of contact.

The maximal pressure at the center of the contact surface (q_o) is determined as:

$$q_o = \frac{3}{2} \cdot \frac{P}{a} \tag{4}$$

where P = the total compressive force and A = the contact surface area.

Finally, aligning the x and y axes with the elliptical semiaxes a and b and the z axis with the direction of the compressive force, then the principal stresses at the center of the contact surface are:

$$\sigma_x = -2\nu q_o - (1 - 2\nu) q_o \left[\frac{b}{a + b} \right]$$
$$\sigma_y = -2\nu q_o - (1 - 2\nu) q_o \left[\frac{a}{a + b} \right] \tag{5}$$
$$\sigma_z = -q_o$$

Furthermore, it has been shown by application of these equations (Timoshenko and Goodier, 1951) that the maximal shearing stress (τ_{max}) is along

[1]The principal curvatures are the maximal and the minimal curvatures which are in the planes at right angles.

Table 1. Coefficients m, n, and t in terms of the auxiliary angle

$\cos^{-1} \dfrac{[D-C]}{[D+C]}$	30°	35°	40°	45°	50°	55°	60°	65°	70°	75°	80°	85°	90°
m	2.73	2.40	2.14	1.93	1.75	1.61	1.49	1.38	1.28	1.20	1.13	1.06	1.00
n	0.49	0.53	0.57	0.60	0.64	0.68	0.72	0.76	0.80	0.85	0.89	0.94	1.00
t	1.45	1.55	1.63	1.71	1.77	1.83	1.88	1.91	1.94	1.97	1.99	2.00	2.00

Table 2. Location of the maximal shearing stress

b/a	z_1	τ_{max} $(\nu = 0.3)$
1	0.47a	0.31 q_0
0.34	0.24a	0.32 q_0

the z axis at a small distance, z_0, from the surface of contact. Table 2 shows values of τ_{max} and z_1 for several ratios of the semiaxes a and b.

Hirsch (1944) was the first to use the classical contact theory to predict joint deformation. He applied this theory to the patella of the knee, where he employed spheres to simulate the joint. This allowed a slight simplification of the more generalized theory.

APPLICATION OF THE CLASSICAL CONTACT THEORY TO THE KNEE JOINT

The condyles of the femur and of the tibia join at the knee. Each condyle is covered with a layer of articular cartilage; it is assumed that the cartilage has the same shape as the condyle. Also, because the elasticity modulus of bone is 100–1,000 times greater than that of cartilage (Greenwald and O'Connor, 1971), the deformation of the knee joint is assumed to correspond to the deformation of the articular cartilage between the condyles. These assumptions ignore the medial and lateral menisci between the condyles for the time being to allow comparison with the results of Walker and Hajek (1972). It also should be noted here that the articular cartilage is not an ideal elastic material but rather viscoelastic. Therefore, the predictions of the classical contact theory will not be valid over the continuous time deformation of the knee joint. However, the deformation at discrete points in time still can be predicted accurately.

The results of applying the classical contact theory to predict the area of contact at the knee joint and semiaxes of the elliptical contact surface at various angles of flexion are shown in Table 3. The radii of curvature of the femoral and tibial condyles used in these calculations were taken from the measurements of Walker and Hajek (1972). A compressive load of 150 kgf, which was assumed to be divided equally between the lateral and medial contact surfaces, was used. Finally, the results were computed for several values of Young's modulus and Poisson's ratio within the range of values for articular cartilage (Hayes et al., 1972). The comparison of these results with the casting method of Walker and Hajek appears very reasonable, as shown in Table 3 and Figure 1.

Table 3. Comparison between contact areas at the knee joint obtained by the classical theory and by other methods (total compressive load = 150 kgf)

Degree of flexion	Measured from cast by Walker and Hajek (1972)	Computed by Walker and Hajek (1972)	Contact area medial condyles (cm²)								
			E = 1100 N/cm²			Classical theory ($\nu = 0.42$) E = 890 N/cm²			E = 680 N/cm²		
			Area	Semiaxes (cm) a	b	Area	Semiaxes (cm) a	b	Area	Semiaxes (cm) a	b
0	2.45	1.6	1.14	(0.65,	0.56)	1.31	(0.70,	0.60)	1.57	(0.77,	0.65)
15	1.95	1.3	0.97	(0.58,	0.53)	1.12	(0.62,	0.56)	1.34	(0.68,	0.62)
30	1.9	1.4	1.22	(0.64,	0.61)	1.41	(0.68,	0.65)	1.68	(0.75,	0.72)
60	1.35	1.5	1.09	(0.67,	0.51)	1.25	(0.72,	0.55)	1.50	(0.79,	0.60)
90	2.0	1.8	1.28	(0.65,	0.62)	1.47	(0.70,	0.67)	1.76	(0.77,	0.73)
120	1.4	1.4	1.19	(0.62,	0.62)	1.37	(0.66,	0.66)	1.64	(0.72,	0.72)

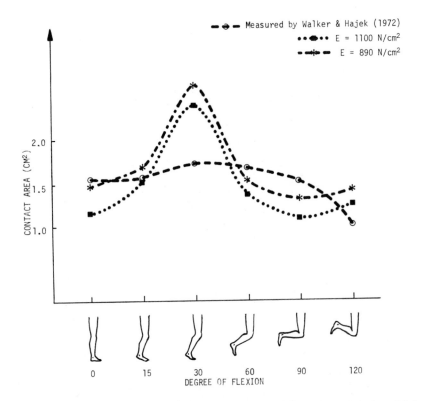

Figure 1. Comparison of the values of contact area in lateral condyles of the knee joint obtained by direct measurement and computed by classical theory.

The other parameters predicted by the classical contact theory are shown in Table 4. These include the principal stresses (σ_x, σ_y, σ_z), the maximal pressure at the point of contact, the maximal shearing stress (τ_{max}), and the point of maximal shearing stress. In addition, two sets of values for Young's modulus and the Poisson ratio were used in the computation, and the results compare rather well with those estimated by Hayes and Mockros (1971).

APPLICATION OF THE CLASSICAL CONTACT THEORY TO THE HIP JOINT

Unlike the knee joint, the hip joint does not have distinct contact points at the medial and lateral condyles. However, the application of the single contact theory to this joint may have some value in predicting the general characteristic of the joint during the walking cycle, as defined by Paul (1966, 1967).

Table 4. Principal stresses and maximal shearing stress at the point of contact in the knee joint (total compressive load = 150 kgf; unit for stresses N/cm²)

Contact Area (cm²)	Medial (0° flexion)		Lateral (0° flexion)	
	$\nu = 0.42$ E = 1,100 N/cm²	$\nu = 0.3$ E = 680 N/cm²	$\nu = 0.42$ E = 1,100 N/cm²	$\nu = 0.3$ E = 680 N/cm²
q_0	1.14	1.68	1.29	1.90
σ_z	98.6	66.9	86.9	59.0
σ_x	-98.6	-66.9	-86.9	-59.0
σ_x	-90.1	-52.5	-76.4	-41.1
σ_y	-91.3	-54.6	-83.5	-53.3
τ_m	30.6	20.7	26.9	18.3
z_1 (cm)	0.27	0.33	0.26	0.32

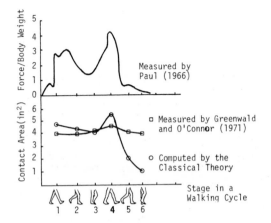

Figure 2. Magnitude of the resultant force transmitted through the hip and the corresponding contact area during a walking cycle.

The results of the classical contact theory and its comparison with the results of Greenwald and O'Connor (1971) are shown graphically in Figure 2. It can be seen that the comparison breaks down for the last two points of the walking cycle. This breakdown may occur because the conditions of the experiment may be in error. However, it is also possible that the initial assumption of a single contact point may not be true in the case of light loads.

DISCUSSION

The classical contact theory predicts the contact surface in the knee joint quite well. For the hip joint, the application does not seem to apply under light loads, and it is quite possible that this may be due to the characteristics of the joint, such as multiple contact areas and stiffness of the contact surface. It is hoped that through a noninvasive measurement technique, e.g., ultrasonic waves (Kendall and Tabor, 1971), these contact areas can be determined in vivo. This would allow in turn the prediction of the mechanical properties of the articular cartilage by reapplying the classical contact theory.

Although the classical contact theory is applicable only to compressive loading, many revised contact theories which deal with small tangential forces and a small torsion couple have been developed (Mindlin, 1949). These would aid in the analysis of the contact areas and the stress distributions of the knee joint under different loading conditions. In addition, work has been done on the problem of more than two bodies in contact (Tu and Gazis, 1964) that will encourage the inclusion of the menisci in future studies of the computation of the contact surfaces at the knee joint.

REFERENCES

Greenwald, A. S., and J. J. O'Connor. 1971. The transmission of load through the human hip joint. J. Biomech. 4: 507–528.

Hayes, W. C., L. M. Keer, G. Herrmann, and L. F. Mockros. 1972. A mathematical analysis for indentation tests of articular cartilage. J. Biomech. 5: 541–551.

Hayes, W. C., and L. F. Mockros. 1971. Viscoelastic constitutive relations for human articular cartilage. J. Appl. Physiol. 31: 562–568.

Hirsch, C. 1944. A contribution to the pathogenesis of chondromalacia of the patella. Acta Chir. Scand. Suppl. 83.

Kendall, K., and D. Tabor. 1971. An ultrasonic study of the area of contact between stationary and sliding surfaces. Proc. Roy. Soc. Lond. A 323: 321–340.

Love, A. E. H. 1944. A Treatise on The Mathematical Theory of Elasticity. 4th Ed. Dover Publications, New York.

Mindlin, R. D. 1949. Compliance of elastic bodies in contact. Trans. ASME 71: 259–268.

Paul, J. P. 1966. The biomechanics of the hip-joint and its clinical relevance. Proc. Roy. Soc. Med. 59: 943–948.

Paul, J. P. 1967. Forces transmitted by joints in the human body. Proc. Inst. Med. Engng. 181: 8–15.

Timoshenko, S., and J. H. Goodier. 1951. Theory of Elasticity. McGraw-Hill, New York.

Tu, Y.-O., and D. C. Gazis. 1964. The contact problem of a plate pressed between two spheres. J. Appl. Mech. 31: 659–666.

Walker, P. S., and J. V. Hajek. 1972. The load-bearing area in the knee joint. J. Biomech. 5: 581–589.

Whittemore, H. L., and S. N. Petrenko. 1921. Friction and carrying capacity of ball and roller bearings. Tech. Paper of NBS No. 201.

Author's address: *Dr. H. K. Huang, National Biomedical Research Foundation, Georgetown University Medical Center, 3900 Reservoir Rd., N.W., Washington, D.C. 20007 (USA)*

Measurement of moments in the knee-ankle orthosis of ambulating paraplegics

C. G. Warren, J. F. Lehmann, and G. S. Kirkpatrick
University of Washington, Seattle

Paraplegics customarily are fitted with bilateral orthoses that stabilize their ankles and knees during ambulation (Edburg, 1967; Heiser, 1967).

There are many factors that influence the success of a paraplegic's ambulation; several of these can be evaluated biomechanically (Lehmann and Warren, 1969). The objective of the program of which this research is a part is to increase the utility of lower extremity orthoses to the paraplegic patient.

Measurements of moments in the uprights of the orthosis were taken to determine data on dynamic forces that could be used in the design of a phasic knee lock mechanism and in the evaluation of new materials for orthoses and their joints. The phasic knee lock would allow bending of the knee during swing phase and would provide stability during the stance phase, thus contributing to a more efficient gait. Many new applications of materials that need to be evaluated for maximal strength and durability are being incorporated into new designs of orthoses. Therefore, the objectives of this study were to determine the magnitudes and phasic relationship of moments carried by the brace uprights, specifically at the knee and ankle joints, and also to determine the effect of design and adjustment changes on these moments.

METHOD

Five paraplegic patients with lesion levels ranging from T-9 through L-1 were selected. They ranged in age from 17 to 46 yr and in weight from 45 to 99 kg. Each patient had been fitted clinically with a pair of double upright orthoses with fixed knee joints and double stopped ankle joints with sole plates extending to the metatarsal heads. The orthoses were adjusted to give the patient maximal standing balance. The orthosis for each patient was instrumented with a strain gauge transducer that was clamped across each knee joint and that measured moments about the joint axis. The ankle moment was obtained from moment and shear forces measured with transducers installed on the uprights just above the ankle joint. The output of the transducers was input to a digital data acquisition system that was on-line to a Honeywell DDP 516 computer. Wire mesh was placed over the heel and sole of the shoe and indicated the phases of stance when in contact with a grounded floor surface. This information also was input into the data acquisition system.

The subjects ambulated in the parallel bars using a swing-to and swing-through gait. Data were recorded on the second of three consecutive steps, and six steps for each configuration of the orthosis were recorded. Each channel of data was processed with its calibration, and the force data were printed out for verification of instrumentation. The data subsequently were stored on a magnetic disc for further processing at the completion of the experiment. Each of the six steps varied slightly in duration. They therefore were normalized by interpolation to a common scale of percentage of stance phase, and the curves then were averaged.

The evaluation of maximal moments occurring in the uprights of the orthosis included three conditions in which the orthoses would be used. First, they were evaluated under normal clinical fitting with anterior stops; second, in the worst case condition of fitting, with 8° of anatomical knee flexion in the orthosis; and third, without the anterior ankle stops. Evaluation of the data showed that moments were supported symmetrically between the left and right extremities. Therefore, illustrations are shown only for the right side. It also was noted that the moments were supported symmetrically between the medial and lateral uprights. Therefore, these moments were summed to give total moment on the orthosis.

The knee moment data in Figure 1 are a composite of six steps for each of five subjects. The maximal moment occurred shortly after heel strike and shortly before midstance. As the sole plate of the brace is loaded, the reactive force from the floor moves anterior to the axis of the ankle. The magnitude of the moments were in proportion to the weight of the patient. Figure 2 shows the knee moment data from the evaluation in which the patient had an increased anatomical knee flexion of 8° within the orthosis. This did not cause an increase in the maximal moment carried by the brace; however, the duration of the flexion moment was increased markedly. The areas under the positive segments of the curves are statistically different. Figure 3 shows the knee moment data from the evaluation with a free ankle dorsiflexion and the knee held in maximal extension. The maximal moments immediately after heel strike were similar to those with the fixed ankle. However, there was no marked drop-off in the knee moment after midstance. This moment was

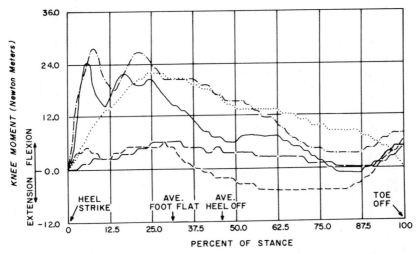

Figure 1. Knee moment in the right orthosis with leg maximally extended, and swing-through gait. *A VE.*, average.

maintained until the extremities were being unweighted in preparation for the next swing-through.

Figures 4 and 5 present the ankle moment data for evaluations with and without an anterior stop at the ankle. Again the magnitudes of the moments were in proportion to the body weights of the patients. From heel strike to foot flat, the moment from the heel lever arm caused the moment tending to

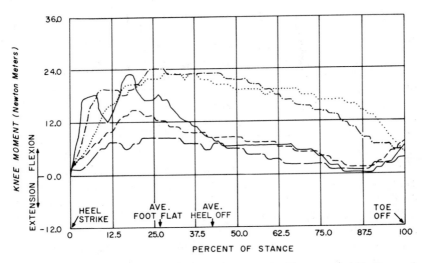

Figure 2. Knee moment in right orthosis with leg in 8° anatomical flexion, and swing-through gait.

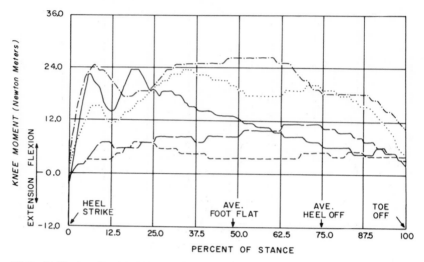

Figure 3. Knee moment in right orthosis with free ankle dorsiflexion, and swing-through gait.

plantarflex. After foot flat, the toe lever is loaded. Without anterior stops, the toe lever is ineffective.

DISCUSSION

Increasing the anatomical knee flexion in the orthoses increases the applied moment of rotation about the anatomical knee axis during stance phase, as

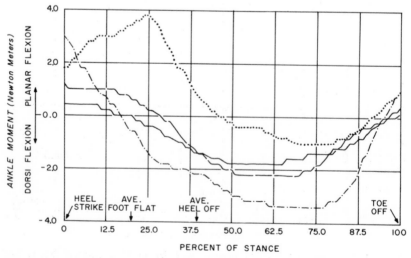

Figure 4. Ankle moment in right orthosis with fixed ankle, and swing-through gait.

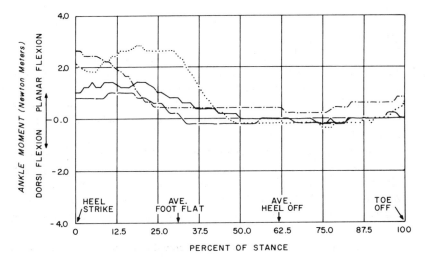

Figure 5. Ankle moment in right orthosis with free dorsiflexion, with swing-through gait.

load is borne through the skeletal system. Therefore, more reactive moment is required from the orthosis to maintain the stability of the knee joint.

When an anterior stop is used at the ankle, the sole plate in the shoe of the orthosis is loaded shortly after midstance. This moves the reactive force from the floor forward on the toe lever, resulting in a moment's being generated about the ankle that is opposite to that at the knee. Thus, the moment at the ankle contributes to the reactive forces required to maintain knee stability and effectively reduces the moment at the knee joint. This effect is demonstrated in the comparison of the knee moment measured for fixed and free ankle dorsiflexion. During the latter part of the stance phase with free dorsiflexion at the ankle, there is no reduction of the knee flexion moment because the sole plate cannot be loaded to provide a counter-moment in the upright. This is further demonstrated by comparing the ankle moment for the free and fixed ankles. During the late stance, no dorsiflexion moment is generated with the free ankle.

The maximal moments during the stance phase were compared with the moments required to cause yielding in the uprights and in the knee joint assembly. The upright material was 2024-T4 aluminum, 5/8 x 1/4 in. These yield points were measured on a Tinius-Olson tension test apparatus. The maximal moments measured during ambulation ranged between 11 and 15 percent of the moment measured at the yield point of the upright and knee joint assemblies. The yield occurred in the upright material, but the rivets and knee joint assembly remained intact. On the basis of the minimal guaranteed specifications of the manufacturer, the moments conceivably could be as high as 30–35 percent of a moment, thus causing failure in the upright.

SUMMARY AND CONCLUSIONS

In the evaluation of knee moments, the range of maximal values has been defined for a representative sample of paraplegics and related to the ultimate strength of the upright material. The effect of malalignment on the knee moment has been demonstrated, as was the effect of utilizing a fixed ankle and sole plate.

The phasic relationship of the knee moment varies considerably between the patients; this would indicate that an automatic knee lock mechanism would have to be tailored to each application or at least be programmable. Only in limited instances was the brace knee moment present in extension during stance phase, indicating that stabilization was required throughout the stance phase.

A comparison between the maximal moments measured in the orthoses and those causing the structures to yield indicates considerable overdesign. In redesigning these orthoses, one should consider using structures of lesser yield strength, providing that durability can be maintained. The application of new materials and designs can contribute significantly to the utility of lower extremity orthoses to the paraplegic patient.

REFERENCES

Edburg, E. 1967. Bracing for patients with traumatic paraplegia. J. Amer. Phys. Ther. Assoc. 47: 818—823.

Heiser, D. 1967. Long leg brace designs for traumatic paraplegia. J. Amer. Phys. Ther. Assoc. 47: 824—826.

Lehmann, and C. G. Warren. 1969. Biomechanical evaluation of braces for paraplegics. Arch. Phys. Med. 50: 179—188.

Author's address: *C. Gerald Warren, Department of Rehabilitative Medicine, University of Washington School of Medicine, Seattle, Washington 98195 (USA)*

Research
on
muscle
contractions

A model of skeletal muscle suitable for optimal motion problems

H. Hatze

University of Stellenbosch, Stellenbosch

The question of what meaning to give the term "optimization" in connection with human motions has been discussed by Hatze (1973a). A certain set of ordinary second-order differential equations has been shown to provide an adequate mathematical representation of the dynamic behavior of the model of the human body. It has been pointed out that the muscle torques ($F_{q_r}^M$) present the key problem for the optimization. Recent attempts (Chow and Jacobson, 1971; Townsend and Seireg, 1972) to optimize bipedal locomotion via optimal programming have had some limited success. However, a proper model of muscle, suitable for treatment by methods of optimal control is still missing. The present study attempts to fill this gap partially.

THE RELATIONSHIP BETWEEN THE MUSCLE TORQUES ($F_{q_r}^M$) AND THE FORCES PRODUCED BY THE MUSCLES

The theory is presented for one-joint muscles but can be extended easily to two-joint muscles and muscle groups.

The righthand side of the equations of motion of the dynamical body model reads: $F_{q_r}^L + F_{q_r}^E + F_{q_r}^M$ (Hatze, 1973a). These are the applied generalized forces corresponding to the generalized coordinate, q_r. Only the muscle torques ($F_{q_r}^M$) are subject to nervous control and hence are of interest for the optimization procedure.

Figure 1 shows a joint with three degrees of freedom (described by the Euler angles Ψ, θ, and ϕ, as defined by Wells, 1967) and a fixed instantaneous center of rotation (an idealization, which can be realized for only a few joints).

Now, the muscle force ($\underline{F}^M$) produces a torque ($\underline{N}^M$) in the system xyz, given by the vector cross-product $\underline{N}^M = \underline{s} \cdot \underline{F}^M$. This can be written as:

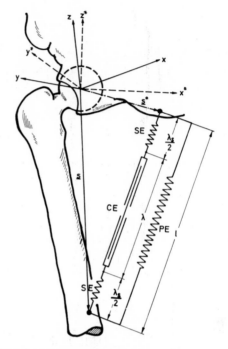

Figure 1. Three-degree-of-freedom joint spanned by a muscle of Length 1. The position vector s*, measured relative to System x*y*z* (fixed to upper limb), defines the origin of the muscle, while s, measured relative to the moving system xyz (fixed to lower limb), defines its insertion. The muscle model comprises the series elastic element SE (*Length* λ_S), the parallel elastic element PE (*Length l*), and the contractile element CE (*Length* λ). Slight dampings of SE and PE are neglected.

$$\underline{N}^M = |\underline{F}^M| \cdot (\underline{s} \cdot X^{-1}\underline{s}^*)/[(\underline{s} - X^{-1}\underline{s}^*) \cdot (\underline{s} - X^{-1}\underline{s}^*)]^{1\!/\!2} \qquad (1)$$

where X^{-1} is the inverse of the transformation matrix (in terms of Euler angles) from System *xyz* into System *x*y*z** (Wells, 1967). The components N_x^M, N_y^M, and N_z^M of the torque vector $\underline{N}^M$ still have to be transformed so that they correspond to the Euler angles $\overline{\Psi}$, Θ, and Φ. This can be accomplished by the transformation:

$$\begin{bmatrix} F_\Psi^M \\ F_\Theta^M \\ F_\Phi^M \end{bmatrix} = \begin{bmatrix} \sin\Theta\sin\Phi & \sin\Theta\cos\Phi & \cos\Theta \\ \cos\Phi & -\sin\Phi & 0 \\ 0 & 0 & 1 \end{bmatrix} \bullet \begin{bmatrix} N_x^M \\ N_y^M \\ N_z^M \end{bmatrix}, \qquad (2)$$

thus establishing the required relationships among the muscle torques $F_{q_r}^M$ (for $q_r = \Psi$, Θ, Φ) appearing in the differential equations and the actual contractive force $\underline{F}^M$ produced by the muscle.

THE CONTROL MODEL OF THE MUSCLE

According to previous works (Bahler, 1968; Jewell and Wilkie, 1958; Hatze, 1973b), a skeletal muscle can be modeled as indicated in Figure 1. The following relation follows directly from Figure 1:

$$1 - \hat{\imath} = [(\underline{s} - X^{-1}\underline{s}^*) \cdot (\underline{s} - X^{-1}\underline{s}^*)]^{1/2} - \hat{\imath} = (\lambda - \bar{\lambda}) + (\lambda_s - \bar{\lambda}_s) \qquad (3)$$

where $\hat{\imath}$, $\bar{\lambda}$, and $\bar{\lambda}_s$ are the resting lenghts of the muscle, the contractile element (CE), and the series elastic element (SE), respectively.

It is seen easily that an additional degree of freedom is introduced for each muscle in the system, and the corresponding coordinate for the jth muscle will be defined by $x_{2n+j} = \lambda_j - \bar{\lambda}_j$. Here, n is the number of positional generalized coordinates of the dynamic model of the body.

It has been shown (Jewell and Wilkie, 1958) that the load-extension relations for the parallel elastic element (PE) and SE are approximately exponential in the physiological range of movements, which may be expressed by:

$$|\underline{F}^{PE}| = 0.01155 \, \bar{F} \, [\exp(k_{PE}(1-\hat{\imath})) - 1] \, , \text{and} \qquad (4)$$

$$|\underline{F}^{SE}| = -0.2985 \, \bar{F} \cdot \ln(1 - (\lambda_s - \bar{\lambda}_s)/k_{SE}l_o) \qquad (5)$$

for the jth muscle. The constants k_{PE}, k_{SE}, $\bar{F}$, $\hat{\imath}$, l_o, and $\bar{\lambda}_s$ must be determined experimentally on the living muscle in situ.

From Figure 1, it follows that

$$|\underline{F}^M| = |\underline{F}^{PE}| + |\underline{F}^{SE}|, \text{and} \qquad (6)$$

$$|\underline{F}^{SE}| = |\underline{F}^{CE}| \, . \qquad (7)$$

Now, the contractile element has some unique features. Its force output $|\underline{F}^{CE}|$ depends nonlinearly on the muscle temperature, the time t, the velocity of shortening $\dot{x}_{2n+j}$, the position x_{2n+j}, and on the degree of stimulation (to be defined later). The time dependence is twofold. First, the active state (and hence the contraction force) of the CE does not reach its full intensity immediately on stimulation but rises as a function of several variables (Close, 1972). Second, if $\dot{x}_{2n+j}$ is positive, i.e., if the active muscle is stretched, the contractive force is a complicated function of t, x_{2n+j}, $\dot{x}_{2n+j}$, and the degree of stimulation (Sugi, 1972).

However, it is not desirable to include the time explicitly in the modeling equations because this complicates the optimization procedure (Boltyanskii,

1971). So far, the author has succeeded only in eliminating this complication for the time function of the active state but not for the very complex behavior of the CE in stretching. For this reason, the present model is suitable only for concentric contractions. It is not difficult to show that the delayed rise of the active state can be accounted for by the introduction of an additional degree of freedom for each muscle. Labeling the corresponding coordinate x_{2n+m+j} and its derivative $x_{2n+m+j+1}$ (for the jth muscle), one obtains the following differential system:

$$\dot{x}_{2n+m+j} = x_{2n+m+j+1}$$

$$\dot{x}_{2n+m+j+1} = a_{2j} [sech(k_{vj} [v_j^{-1} -1]) - x_{2n+m+j}] - a_{1j} x_{2n+m+j+1} , \qquad (8)$$

for $j = 1, \ldots, m$. The relative frequency of stimulation v_j (defined by: stimulation frequency/optimal stimulation frequency) is the *first control parameter*, while k_{vj} is a constant and a_{1j} and a_{2j} are both certain functions of x_{2n+j} .

Now, a graded response to stimulation in muscle is achieved mainly by varying the number of active motor units and, to a lesser extent, by altering the stimulation frequency. Defining the relative number of active motor units in the jth muscle to be the *second control parameter* u_j, the degree of stimulation becomes a function of u_j and v_j $(0 < (u_j,v_j) \leqslant 1)$.

It can be shown that an appropriate function describing the contractile force of CE is given by:

$$|F_j^{CE}| = u_j \cdot x_{2n+m+j} \cdot k_j \cdot \bar{F}_j [(exp(A_j \cdot \dot{x}_{2n+j}) + (exp(A_j \cdot \dot{x}_{2n+j}) - 1)$$

$$/(exp[-D_j \cdot k_{1j} \cdot x_{2n+m+j}^{0.75}] - 1)] , \qquad (9)$$

including the length-tension relation $k_j = 1-4.29(x_{2n+j}/\bar{\lambda}_j)^2$, for $0.76 \leqslant \lambda_j/\bar{\lambda}_j \leqslant 1.24$, the velocity-length dependence $k_{1j} = 1-3.5(x_{2n+j}/\bar{\lambda}_j)^2$, three characteristic constants of the jth muscle $(\bar{F}_j, A_j, \text{ and } D_j)$, and the influence of the active state function x_{2n+m+j} on the contractive force.

Substituting Equation 5 into Equation 7 and then Equation 7 into Equation 9, one can solve for $\dot{x}_{2n+j}$ to obtain a second differential system, namely:

$$x_{2n+j} = (1/A_j) \cdot ln(-0.2985 .ln [(1 - [l_j-\bar{l}_j-x_{2n+j}]$$

$$/k_{SE} \cdot l_o)(1 - exp[D_j \cdot k_{1j} \cdot x_{2n+m+j}^{0.75}])$$

$$/k_j \cdot \bar{F}_j \cdot u_j \cdot x_{2n+m+j} + exp(D_j \cdot k_{1j} \cdot x_{2n+m+j}^{0.75})], j=1, \ldots ,m , \qquad (10)$$

where $l_j - \bar{l}_j$ is given by Equation 3. By putting $q_1 = x_1, \dot{q}_1 = x_2, q_2 = x_3, \dot{q}_2$

= $x_4, \ldots$, the original set of second-order differential equations for the dynamic model (Hatze, 1973a) can be written as the first-order system

$$\dot{x}_i = f_i(x_1, \ldots, x_{2n}, x_{2n+1}, \ldots, x_{2n+m}), i=1, \ldots, 2n. \quad (11)$$

This system then, taken together with the systems of Equations 10 and 8, constitutes the required set of $(2n+3m)$ first-order differential equations for optimal programming.

AN EXAMPLE: THE ACTIVE STATE FUNCTION FOR HUMAN MUSCLE

System 8 can be solved for the jth muscle for fixed values of x_{2n+j} and v_j. Its solution is given by:

$$x_{2n+m+j} = U_v - [([U_v - x_{2n+m+j}(0)] / [p_{1j} - p_{2j}])(p_{1j} \cdot \exp[p_{2j} \cdot t]$$
$$- p_{2j} \cdot \exp[p_{1j} \cdot t]) + (x_{2n+m+j+1}[0] /$$
$$(p_{1j} - p_{2j})(\exp[p_{1j} \cdot t] - \exp(p_{2j} \cdot t)] \quad (12)$$

where U_v = sech $[k_{vj}(v_j^{-1} - 1)]$; $p_{1,2j} = -a_{1j}/2 \pm (\frac{1}{4}a_{1j}^2 - a_{2j})^{\frac{1}{2}}$, $p_{1j} = \bar{p}_{1j}[1 - 0.65.\text{sech}(1.62 - 3.24.x_{2n+j}/\bar{\lambda}_j)]$; and $\bar{p}_{1j}, p_{2j}, k_{vj}$ as well as the optimal frequency f_{opt} (the stimulation frequency at which the rate of tetanic tension rise is maximal) are constants, depending on the type of muscle under consideration (fast or slow). Their values can be estimated from known properties of human muscle at 37°C (e.g., Eberstein and Goodgold, 1968) and are for fast and slow muscles, respectively: $\bar{p}_{1j} = (-170, -81), p_{2j} = (-519, -790)$, $k_{vj} = (0.46, 0.21)$, and $f_{opt} = $ (approximately 100.60).

It may be seen that the delayed decay of the active state in the relaxation phase of a tetanus is not accounted for. This feature of the muscle, however, is thought to be a result of a stretching effect (SE stretches, thus relaxing CE) and hence is not included in the present model (concentric contractions only). On the other hand, the dependence of the decay of the active state on the muscle length and the increase in the maximal rate of rise of the active state, with the stimulating frequency, have been included.

It should be noted that the active state, as defined here, is an overall property of the total muscle in situ and thus will never show twitch-like fluctuations, not even at low stimulation frequencies. This is because the individual motor units usually are fired asynchronously, thus producing a smoothing effect.

The present control model should be regarded as a theoretical basis for experiments on the control behavior of human muscle. It most probably will have to be modified when more experimental evidence accumulates on the properties of human muscle in situ.

REFERENCES

Bahler, A. S. 1968. Modeling of mammalian skeletal muscle. IEEE-BME 15: 249.

Boltyanskii, V. G. 1971. Mathematical Methods of Optimal Control. Holt, Rinehart and Winston, New York.

Chow, C. K., and D. H. Jacobson. 1971. Studies in human locomotion via optimal programming. Math. Biosci. 10: 239–306.

Close, R. I. 1972. Dynamic properties of mammalian skeletal muscle. Physiol. Rev. 52: 129–197.

Eberstein, A., and J. Goodgold. 1968. Slow and fast twitch fibers in human skeletal muscle. Amer. J. Physiol. 215: 535–541.

Hatze, H. 1973a. Optimization of human motions. *In:* S. Cerquiglini, A. Venerando, and J. Wartenweiler (eds.), Biomechanics III, pp. 138–142. University Park Press, Baltimore, and S. Karger, Basel.

Hatze, H. 1973b. A theory of contraction and a mathematical model of striated muscle. J. Theoret. Biol. 40.

Jewell, B. R., and D. R. Wilkie. 1958. An analysis of the mechanical components in frog's striated muscle. J. Physiol. 143: 515.

Sugi, H. 1972. Tension changes during and after stretch in frog muscle fibres. J. Physiol. 225: 237–253.

Townsend, M., and A. Seireg. 1972. The synthesis of locomotion. J. Biomech. 5: 71.

Wells, A. D. 1967. Lagrangian Dynamics. Schaum, New York.

Author's address: *Dr. Herbert Hatze, Department of Physical Education, University of Stellenbosch, Stellenbosch (South Africa)*

Simulated muscle contraction utilizing a spring model

V. Karas and S. Otáhal
Charles University, Prague

Fung (1970), having applied Hill's three-element model of a muscle contraction (Hill, 1938; cited by Fung) and having discussed the paper by Parmley and Sonnenblick (1967), presented the equation expressing the relation of the stiffness of a series elastic element to its tensile stress S:

$$\partial S/\partial \eta = \alpha\,(S + \beta).$$

An integration of that equation yields

$$S = (S^+ + \beta)\,e^{\alpha(\eta - \eta^+)} - \beta \tag{1}$$

where η is the extension of the series elastic element, S^+, η^+ is the integration constant, and α, β are the constants.

The above equations have been considered in relation to the heart or papillaris muscles. The authors questioned whether Fung's equation might apply also to the description of the activity of a group of agonistic muscles controlling motion of an extremity about a joint of the human body. Once that question was solved, the next step was to determine the integration constants η^+ and S^+ for such a state of the muscle when

$$\tau^{(S)} = \tau^+ = O = S(\eta^+, \Delta)$$

where $\tau^{(S)}$ is the stress in the series elastic element and Δ is the insertion. Overlap of actin-myosin fibers occurs when a muscle takes Length L^+ and the extension of the series elastic element $\eta^+ = O$. Under these conditions, the tensile stress of the series elastic element $S^+ = O$ and Equation 1 becomes

$$S = \beta(e^{\alpha\eta} - 1). \tag{2}$$

Furthermore, Equation 2 included for the entire increase of the tensile stress a constant value of the actin-myosin filaments insertion (Δ = constant). In other words, the stimulation of the muscle was supposed to be constant for each S. Under these conditions it is possible from the geometrical changes of the sliding-filament model (Fung, 1970) to determine that the muscle, developing the output force S, takes instantaneous positional length $L = \eta + L^+$. Therefore, the elastic extension of the series element is $\eta = L - L^+$. Equation 2 then will become

$$S = \beta(e^{\alpha(L-L^+)} - 1). \tag{3}$$

It is possible to say that the relation of the tensile stress S to the positional muscle length L (when L^+ = constant and L in Equation 3 is varying) represents the invariant characteristics of the behavior of an activated muscle, because it has been obtained during the constant stimulation corresponding with the value Δ of the insertion (respective to L^+). For different levels of stimulation, Z, for various values of Δ (or L^+), one gets diverse and contrasting invariant characteristics $S(L)$ $\Delta = L^+$, Z = constant (Figure 1, a).

Therefore, it is possible to interpret Equation 3 as the behavior of a muscle during constant stimulation as well as the behavior of a spring with a

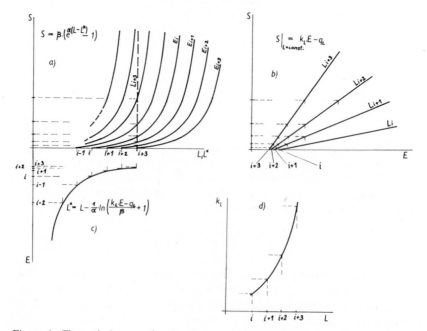

Figure 1. Theoretical curves for description of the muscle-spring model (*const.*, constant).

certain initial "zero-length" L^+ (controlled spring). The tensile stress S then will be exponentially increasing with the increase of the positional length L. When the muscle already has taken a positional length L = constant, the tensile stress S will be increasing exponentially with the decrease of the forced "zero-length" L^+.

The controlled spring described by Equation 3 has a linearly varying stiffness with the developed tension S. An equation analogous to 3 but for the group of elbow-flexors was reported by Asatrjan and Feldman (1965) and by Feldman (1966). This equation was:

$$F = a(e^{\alpha(x-\eta)} -1).$$ (4)

Comparing the notation used by Fung (1970) and by Feldman (1966) in Equation 3 and 4, it is evident that $\beta \sim a$, $\alpha \sim \alpha$, $L \sim x$, and $L^+ \sim \lambda$. On the basis of these it followed that Fung's equation (3) was applicable also to the description of the behavior of a whole group of agonistic muscles. Equations 3 and 4 allow simulation of the usually unsolvable situation in mechanical analysis of a group of agonistic muscles by means of a simple idea of one spring with controlled "zero-length" $\lambda \sim L^+$ and instantaneous positional length $x \sim L$.

The muscle force gradation is done by quantum summation of the contractions of muscle fibers. The muscle output tension S (without extra- and intraarticular resistance tension), which is developed by a muscle during an isometrical contraction, is for each positional muscle length L given by:

$$\partial$$
$$S_L \ (L = \text{constant}) = n_L \cdot Z - P_L \cdot$$ (5)

where Z is stimulation proportional to the number of muscle fibers contractions and to the value of tension S, and n_L and p_L are the slope and the shift of the linear relation $S(Z)$.

The level of the stimulation Z, activating a human muscle group during natural motion, is impossible to determine. Therefore, for practical use, it is better to substitute for Z in Equation 5 the integrated electrical muscle activity (E), which can be determined electromyographically:

$$S|L = \text{constant} = k_L \cdot E - q_L.$$ (6)

The relations $S(E)|L$ = constant are linear, as was found experimentally by several authors. One of the straight lines that is valid for the length L_{i+3} is shown in Figure 1, b.

According to Equations 3 and 6, $L^+(E)$ is given by (see Figure 1, c):

$$L^+ = L - \frac{1}{\alpha}\ln(\frac{k_L \cdot E - q_L}{\beta} + 1).$$ (7)

The relation S(E) in Equation 6 for other positional lengths (L_{i+2}, L_{i+1}..) could be obtained by the diagram in Figure 1, a, and by converse procedure from the diagram in Figure 1, c.

In the diagram (Figure 1, b) for S(E)|L = constant, it is apparent that the muscle has a different initial E depending on the value L even though its output tension S = O. With respect to Hill's three-element concept, this state is expressed by Equation 3 $S(\eta, \Delta)$ = O, when η = O, $S(\eta, \Delta)$ > O, when η > O. Then each invariant characteristic i in Figure 1, a, has its own constant activity, E_i. From the diagram in Figure 1, b, d it is further evident that the slopes of the relations S(E)|L = constant are decreasing exponentially with the decreasing of the muscle length L_i. It results in the fact that a muscle that is in a smaller positional length must be stimulated more intensively to produce a specific tension S than the same one in a greater positional length. The shift of the invariant characteristics in Figure 1, a, along the abscissa L is from the left limited by minimal value of the "zero-length" L^+ and from the right by maximal excursion, of the muscle attachment in a joint system. The amplitude of the tension S is for each L limited by isometrical maximum. The output isometrical tension of an activated muscle (group of agonistic muscles) is the function of three variables:

1. The level of of exciting processes stimulating a muscle. In theory, these are in "controlled spring-muscle model" adequate to the "zero-length" of the spring L^+ (respective to muscle electrical activity E).

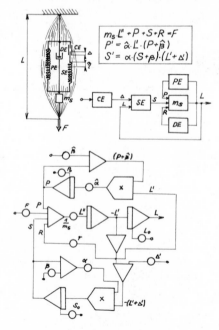

Figure 2. Analog computer model of muscle contraction.

2. The instantaneous positional length L of the stimulated muscle. (It is in the model analogous to the extension of the spring with "zero-length" L⁺).

3. The magnitude of the relevant constants.

The problem also was modeled by analog computer (Figure 2). Exploiting the analog model, the reactions of a muscle evoked by changes of an external load (during fixed levels of stimulation), which were stimulated by the size of filament insertion (Δ), as well as the reactions on changes of stimulation during loaded with constant external load, the influence of the constants α, β on the shape of the invariants characteristics (Figure 1, a), etc., can be observed.

The analog model also permitted study of the activity of the parallel element (PE), including its hysteresis, series element (SE), contractile element (CE), and also the viscose-elastic character of a muscle group of agonists.

REFERENCES

Asatrjan, A. G., and A. G. Feldman. 1965. O funkcionalnoj nastrojke nervnoj sistemy pri upravlenii dviženiem ili sochranenii stacionarnoj pozy. Biofizika (Moscow) 10: 837–846.

Feldman, A. G. 1966. O funkcionalnoj nastrojke nervnoj sistemy pri upravlenii dviženiem ili sochranenii stacionarnoj pozy. II. Reguliruemye parametry myšc. Biofizika (Moscow) 11: 498–508.

Fung, Y.-C. 1970. Mathematical representation of the mechanical properties of the heart muscle. J. Biomech. 3: 381–404.

Paramley, W. W., and E. H. Sonnenblick. 1967. Series elasticity in heart muscle: Its relation to contractile element velocity and proposed muscle models. Circ. Res. 20: 112–123.

Author's address: *Dr. V. Karas, Faculty of Physical Education and Sport, Charles University, Ujezd 450, Prague 1 (Czechoslovakia)*

Mechanical and electrical behavior of human muscle during maximal concentric and eccentric contractions

P. V. Komi

University of Jyväskylä, Jyväskylä

The force-velocity relationships of human muscle (e.g., Asmussen, Hansen, and Lammert, 1965; Komi, 1973a) characterize the main mechanical differences in concentric and eccentric work. Some information also has been reported on the interrelationship between neural input and mechanical output during sub-maximal concentric and eccentric contractions. The slope of the regression line representing the relationship between IEMG and muscle tension is greater when muscle shortens at a constant velocity than when it lengthens at the same velocity (Bigland and Lippold, 1954). When recordings were made with a greater number of velocities, then a family of curves was obtained, a result which emphasizes the importance of the contraction velocity in determining the pattern of IEMG-muscle tension (kg) relationship (Komi, 1973b).

In maximal contractions, the IEMG activity seems to be the same in both concentric and eccentric work, and its value may be independent of the contraction velocity (Komi, 1973b). In this study, the IEMG analysis was restricted to the midportion of the movement. Because it is expected that not only the contraction velocity, but also the length of the muscle, influences the IEMG-tension relationships, this study was designed to investigate further the mechanical and electrical behavior of human muscle during maximal contractions of concentric and eccentric work.

METHODS

Electromyogram (EMG) Measurements

The biceps brachii, brachioradialis, and triceps brachii muscles were selected for the EMG analysis. EMG activity was monitored using the wire electrode technique with a wire 50 μm in diameter. We used our earlier method (Komi and Buskirk, 1970) to prepare the electrode. Both Karma Alloy (Driver-Harris Co., Harrison, N.J.) and Evanohm wires (Wilbur & Driver Co., Newark, N. J.) were used. The recording technique for the brachioradialis muscle was the same as that used by Basmajian and Stecko (1962); however, for the biceps and triceps brachii muscles, the two uninsulated ends (2 mm) were separated over the muscle belly by 4 cm. All the electrode pairs had a common ground electrode placed medial to the midpoint between the two biceps electrodes.

EMG activity was amplified with Tektronix 122 amplifiers (Tektronix, Inc., Beaverton, Ore.) using the voltage gain of 1,000 and lower and upper cut-off frequencies of 8,000 and 10,000, respectively. The amplified signals first were full-wave rectified, then electronically integrated, and finally displayed on an oscillograph.

Dynamometer and Testing

An electromechanical dynamometer (Komi, 1973a) was used to load the forearm flexors of eight male students with the following constant velocities: -5.3, -1.9, -0.7, $+0.7$, $+1.9$, and $+5.3$ cm per sec. The minus and plus signs denote eccentric contraction (negative work) and concentric contraction (positive work), respectively. The velocities were calculated for the biceps brachii muscle, which, on the average, changed its length by 7 cm during the movement range of $120°$ ($50-170°$). The order of the contraction velocities was selected randomly. In all movements, the forearm was kept supine by a special wrist cuff on which the strain gauges were installed.

RESULTS AND DISCUSSION

The results of the elbow flexors are compiled in Figure 1, which shows the obtained force-length and IEMG-length relationships. The form of the force-length curves is in agreement with those reported previously, thus confirming that maximal eccentric tension is greater than maximal concentric tension in all muscle lengths (e.g., Doss and Karpovich, 1965; Jörgensen and Bankov, 1971; Komi and Buskirk, 1972; Singh and Karpovich, 1966) and that the magnitude of this difference is determined by the force-velocity characteristics of the muscle in concentric and eccentric work (e.g., Asmussen et al., 1965; Komi, 1973a; Levin and Wyman, 1927). It must be emphasized that Figure 1,A does not show the force-length relationship for a specific flexor muscle but for all flexors involved in the movement. It is not easy to evaluate the influence of each flexor muscle on the total force-length curve, although calculations have been made to estimate that the contributions of the biceps brachii and brachioradialis muscles to elbow flexion. These contributions were 34 and 19 percent, respectively, to the torque output of the flexors (Jörgensen and Bankov, 1971).

430   Komi

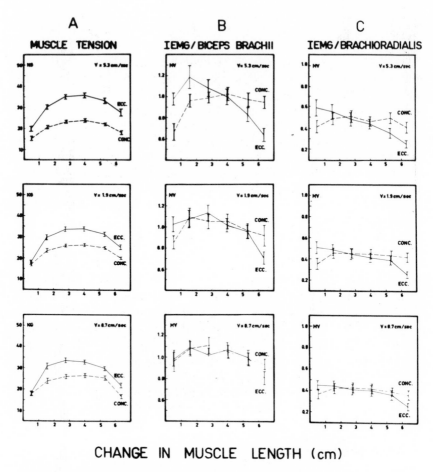

Figure 1. Comparison of the force-length relationship for elbow flexors (*A*), and the IEMG-length relationship for biceps brachii (*B*) and brachioradialis (*C*) muscles in different constant velocities of concentric (*conc.*) and eccentic (*ecc.*) work.

The IEMG activity of the brachioradialis was approximately 50 percent of that recorded from the biceps brachii. Although it is difficult to interpret this difference exactly, it is in concordance with the function of this muscle in elbow flexion (Basmajian and Latif, 1957) and with other reported values (Komi, Aunola, and Eloranta, 1972).

In both muscles, the IEMG values during the midportion of the movement were, on the average, the same for both contraction types and for all velocities used (Figure 1,*B* and *C*). This seems to be in agreement with other reports on the same muscles (Jörgensen and Bankov, 1971; Komi, 1973b). In slow contraction velocity (0.7 cm per sec), IEMG-length curves were rather flat, and a slight, although questionable, parallelism with the IEMG and force curves was observed for the biceps brachii muscle. The positions of the average eccentric and concentric IEMG curves were changed slightly in

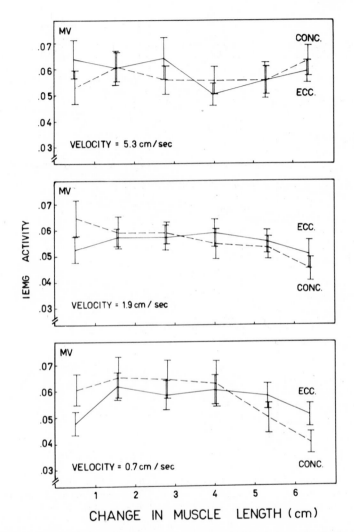

Figure 2. IEMG-length relationship for the anatagonist (triceps brachii) muscle in concentric (*conc.*) and eccentric (*ecc.*) work of elbow flexors. The velocity conditions are the same as those in Figure 1.

moderate velocity (1.9 cm per sec) and clearly at the highest velocity (5.3 cm per sec) of contraction. In eccentric work, the curve rotated clockwise so that less IEMG-activity was recorded in longer muscle lengths and more in shorter muscle lengths. In concentric work, the velocity did not have an appreciable effect on the IEMG-length relationship, except perhaps for the biceps brachii muscle, from which a slightly reduced IEMG-activity was recorded during the fast contraction (Figure 1,*B*).

Vredenbregt and Koster (1966) have shown in isometric elbow flexion that the maximal IEMG-activity of the biceps brachii is independent of muscle length. Komi and Buskirk (1972) obtained similar results for slow-velocity concentric and eccentric work. The slight tendency for parallelism with the IEMG (biceps brachii muscle) and force curves in the present study is not as clear as that reported by Liberson, Dondey, and Asa (1962). The differences in the recording techniques and especially in the volumes of muscle covered by the EMG electrodes make it difficult to compare the present results with other reports.

The observed effect of the contraction velocity on the IEMG-length curve can be difficult to explain. It is likely that the elastic elements in the muscle are at least partly responsible for the differences in mechanical tension between eccentric and concentric work. It cannot be deduced from the present results how the possible changes in the mechanical behavior of muscle at different working lengths affect the IEMG-length curve. The decrease of IEMG activity during the longer muscle lengths of the fast eccentric contraction may be explained by the increased inhibitory effect from the Golgi tendon organs; however, because the patterns of the force curves were similar for all conditions studied, some other mechanisms also must be involved. It seems logical to suggest that the observed changes in the IEMG-length relationships are caused by differences in the overall behavior of the proprioceptive control mechanisms. The similarity in the observed activation of the antagonist muscle in all measured conditions, however, challenges this hypothesis (Figure 2).

REFERENCES

Asmussen, E., O. Hansen, and O. Lammert. 1965. The relation between isometric and dynamic muscle strength in man. Communication No. 20 from the Testing and Observation Institute of the Danish National Association for Infantile Paralysis.

Basmajian, J. V., and A. Latif. 1957. Integrated actions and functions of the chief flexors of the elbow: A detailed electromyographic analysis. J. Bone Joint Surg. 39-A: 1106–1118.

Basmajian, J. V., and G. Stecko. 1962. A new bipolar electrode for electromyography. J. Appl. Physiol. 17(5): 849.

Bigland, B., and O. C. J. Lippold. 1954. The relation between force, velocity and integrated electrical activity in human muscles. J. Physiol. 123: 214–224.

Doss, W. S., and P. V. Karpovich. 1965. A comparison of concentric, eccentric, and isometric strength of elbow flexors. J. Appl Physiol. 20(2): 351–353.

Jörgensen, K., and S. Bankov. 1971. Maximum strength of elbow flexors with pronated and supinated forearm. In: J. Vredenbregt and J. Wartenweiler (eds.), Biomechanics II. pp. 174–180. S. Karger, Basel, and University Park Press, Baltimore.

Komi, P. V. 1973a. Measurement of the force-velocity relationship in human muscle under concentric and eccentric contractions. In: S. Cerquiglini, A.

Venerando, and J. Wartenweiler (eds.), Biomechanics III, pp. 224–229. S. Karger, Basel, and University Park Press, Baltimore.

Komi, P. V. 1973b. Relationship between muscle tension, EMG, and velocity of contraction under concentric and eccentric work. *In:* J. E. Desmedt (ed.), New Developments in Electromyography and Clinical Neurophysiology. Vol. 1, pp. 596–606. S. Karger, Basel.

Komi P. V., S. Aunola, and V. Eloranta. 1972. Recruitment of the elbow flexor muscles in different positions of the forearm. Presented at the 6th International Congress of Physical Medicine, Barcelona.

Komi, P. V., and E. R. Buskirk. 1970. Reproducibility of electromyographic measurements with inserted wire electrodes and surface electrodes. Electromyography 4: 357–367.

Komi, P. V., and E. R. Buskirk. 1972. Effect of eccentric and concentric muscle conditioning on tension and electrical activity of human muscle. Ergonomics 15(4): 417–434.

Levin, A., and J. Wyman. 1927. The viscous elastic properties of muscle. Proc. Roy. Soc. 101B: 218–243.

Liberson, W. T., M. Dondey, and M. M. Asa. 1962. Brief repeated isometric maximal exercises. Amer. J. Phys. Med. 41(1): 3–14.

Singh, M., and P. V. Karpovich. 1966. Isotonic and isometric forces of forearm flexors and extensors. J. Appl. Physiol. 21(4): 1435–1437.

Vredenbregt, J., and W. G. Koster. 1966. Some aspects of muscle mechanics in vivo. *In:* Proceedings of the International Symposium on Attention and Performance, pp. 94–100. Soesterberg/Driebergen, Holland.

Author's address: *Dr. Paavo V. Komi, Kinesiology Laboratory, University of Jyväskylä, 40100 Jyvaskyla, 10 (Finland)*

The relation between length and the force-velocity curve of a single equivalent linear muscle during flexion of the elbow

A. E. Chapman

Simon Fraser University, Burnaby

Numerous authors have verified the following equation describing the relationship between the force and velocity of shortening (F–V curve) of the contractile component (CC) in a variety of isolated muscles

$$(F + a) (V + b) = (F_{max} + a)b. \qquad (1)$$

A series elastic component (SEC) also has been identified and its compliance studied (Jewell and Wilkie, 1958). The relationship between the force developed isometrically and the length of the CC (F_{max} –length curve) has been investigated in whole muscle and single fibers (Gordon, Huxley, and Julian, 1966). More recently, approximately parallel F-V curves have been obtained, each curve being appropriate to a given length of the CC (Bahler, Fales, and Zierler, 1968).

Wilkie (1950) made the first significant attempt to apply this information to human muscles and showed the applicability of a two-component model (CC plus SEC) to human muscular contraction. He developed the concept of a single equivalent muscle at the hand to represent the combined action of the muscles producing flexion of the elbow. From measurements made at one joint-angle during dynamic and static contractions, he demonstrated a single characteristic F-V curve for the CC and deduced the compliance of the SEC. Cavanagh and Grieve (1970) used Wilkie's concept of a fixed F-V relationship to calculate the energy stored in the SEC during a dynamic pull, although

This work was completed with the financial support of President's Research Grants 714-038 and 714-040, Simon Fraser University, and with the technical aid of Mr. D. Sanderson.

no account was taken of the effect of changes in muscle length on the F-V curve, as has been suggested by the work of Bahler et al. (1968) and Wilkie (1950).

There is controversy concerning the total change in length of any one muscle over the full range of joint-movement. However, Wilkie (1950) stated that a change of 13 percent of maximal isometric force (F_{max}) of the equivalent muscle accompanied a joint-excursion of $60°$, although his data also showed that, over a $90°$ excursion, a change in force of 20 percent was obtained.

The F-V curve of the CC and the stress/strain characteristics of the SEC are nonlinear. However, Houk (1963) linearized both of these relationships and deduced a transfer function for isometric contraction that gave an exponential output, the input being maximal instantaneous activation of the CC. If the first few milliseconds of the contraction were ignored, he obtained close agreement between the observed and the calculated rise of force during supination of the forearm. The initial slow rise of force was rejected as being due to the compliance of soft tissue at the hand that was not part of the model. Assuming that the properties of the SEC are independent of muscle length, any change in the time constant (T) of the exponential rise of force against time is related directly to the slope (B) of the linearized F-V curve.

This analysis, and the foregoing assumptions, were used to investigate the possibility of obtaining relative changes in the F-V curve of the CC of a single equivalent muscle at the hand during elbow flexion.

METHODS

Twenty-one young adult male subjects performed maximal isometric contractions with the humerus horizontal, which was aligned in the frontal plane and supported from below at a point proximal to the elbow joint. A rigid board placed against the lateral aspect of the thorax prevented shoulder adduction, and the hand was fixed in a fiber-glass cast in a supine position with the wrist extended and fingers slightly flexed. In all cases, torque produced about the elbow joint (T_{max}) was measured and subsequently converted into horizontal force (F_{max}) at the hand by use of simple trigonometric relationships.

Group 1, comprised of 10 subjects, exerted two contractions at each of the five joint angles shown in Figure 1,A. The cast was fixed to a lever incorporating a torque transducer and an angular potentiometer. The output from each transducer was recorded separately on an FM tape recorder, and both parameters were digitized through an A-D converter (torque at 200 per sec; angle at 100 per sec) interfaced with a PDP-8e (12 K) minicomputer on which subsequent analysis was performed. The data were submitted to an exponential curve-fitting routine after correction for the compliance of the system and removal of the initial slow rise of force (Houk, 1963). Data

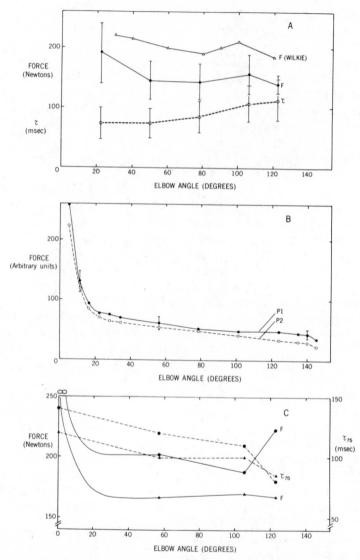

Figure 1. *A*, mean forces (*solid line*) and mean time-constants (Υ, *broken line*) in relation to elbow angle ($0°$ = full extension) with vertical bars representing ± 1 standard deviation ($N = 8$); the results obtained by Wilkie (1950) are shown for comparison. *B*, mean force ($N = 9$) against angle for P1 (*solid line*) and P2 (*broken line*) with 3 vertical bars showing ± 1 standard deviation for P1 at 3 elbow-angles. *C*, forces (*solid lines*) and Υ_{75} (*broken lines*) against elbow-angle for Subject 1 (▲) and Subject 2 (●).

demonstrating a poor fit were rejected (resulting in the rejection of data from two subjects). F_{max}, T_{max}, and the time-constant Υ (time to 63 percent of F_{max}) of the most rapid rise of force for eight subjects were determined in this way.

Group 2, comprised of nine subjects, performed two contractions at the 13 joint-angles shown in Figure 1,B. The apparatus and posture (P1) were identical to those used for Group 1, but the second contraction of the two was performed with the humerus aligned in the sagittal plane (P2). From this group, only F_{max} and T_{max} were obtained in arbitrary units.

Group 3, comprised of two subjects, contracted isometrically against a linear voltage differential transformer, the sensitive axis of which was placed to register tangential force. All other procedures were identical to those used for Group 1. As the position of the transformer was maintained constant for each subject, the tangential force was proportional to torque. Each subject exerted 10 contractions at joint angles shown in Figure 1,C. F_{max} and the time to reach 75 percent of F_{max} (T_{75}) were obtained from those contractions which showed a smooth rise of force against time. Any curves deviating from this criterion were judged to be due to postural changes during the contraction and were rejected from the analysis.

RESULTS AND DISCUSSION

The F-V curve (Equation 1) can vary as a result of variations in F_{max}, V_{max} (when F = 0), and a/F_{max}. Changes in F_{max} and V_{max} result in a change in the value of B in the linearized F-V curve of Houk (1963). In this study, it was not possible to obtain values of a/F_{max} and V_{max}, but T_{75} and T, the latter of which is related to B, were obtained in addition to F_{max}. Figure 2 shows how changes in F_{max} and V_{max} affect the value of T_{75} for any given value of a/F_{max}.

The results from the three groups of subjects are shown in Figure 1. For Group 1, the large standard deviations of F_{max} at any joint angle are a consequence of both the large variation in F_{max} among subjects and the different-shaped curves relating F_{max} to joint-angle within each subject (Figure 1,A). The data from Group 2, P1 and P2 (Figure 1,B), and from Group 3 (Figure 1,C) show similar results. For Group 2 (Figure 1,B), it is probable that each part of the biceps brachii muscle is set at a different length in Postures P1 and P2. The consistent effect of this postural change plus the large individual variation in the shape of the T_{max}–joint-angle curves suggests that each subject is unique in terms of where, in the ranges of joint-angles, his isometric F_{max}–length curve lies.

In all groups, the values of F_{max} obtained within the range of 0–25° are higher than could be accounted for by the characteristic F_{max}–length curve (Gordon et al., 1966). Thus, it is concluded that a simple geometric correction is inapplicable over the full range of joint movement if it is based on the premise that all the flexor muscles are parallel to the bones (Wilkie, 1950). The probable reason for the high values of F_{max} is that the tendons of both biceps brachii and brachialis muscles pass over the capitulum and

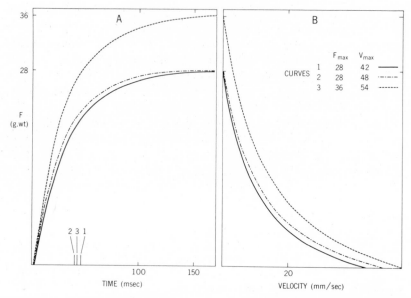

Figure 2. *A*, rise of force against time computed using the technique and some data of Jewell and Wilkie (1958) concerning muscle from frog. A constant load-extension curve (SEC) was maintained, and the rise of force against time was computed from the formula:

$$t = \int_{F=0}^{F=F_t} \frac{1}{V_{(CC)} \frac{dF}{dx}} \cdot dF$$

using each corresponding F-V curve shown in *B*. The values of Υ_{75} for each curve are shown on the abscissa. *B*, F-V curves computed from the formula $[(F + a) (V + b) = (F_{max} + a)b]$ for different values of F_{max}, $a/F_{max} = b/V_{max} = 0.25$ being maintained constant for all three curves.

trochlea of the humerus, respectively, at angles between 0 and 25°. Any or all of the above factors which lead to gross changes in F_{max} make the use of a single equivalent muscle at the hand inapplicable over approximately the first 25° of flexion from full extension.

No significant differences were found among mean values of Υ at different joint angles for Group 1 (Figure 1,*A*), the mean values being comparable with those reported by Houk (1963) for supinators of the forearm. Because the value of Υ depends on both maximal and immediate muscular activation, it was probable that two contractions were insufficient to ensure these conditions, particularly in the light of the results from Group 3 (Figure 1,*C*). For both subjects in Group 3, the value of Υ_{75}, which decreased with practice at any joint angle, showed a consistent decrease with increasing joint flexion. For Subject 1, F_{max} was the same at 57 and 123°, so that the relative F-V curves would approximate Curves 1 and 2, respectively, in Figure

2,*B*. For Subject 2, whose F_{max} at $123°$ is above that at $57°$, the two relative F-V curves would approximate Curves 3 and 1, respectively, in Figure 2,*B*. As Bahler et al. (1968) demonstrated, taking into account the parallel nature of curves with similar values of a/Po at different muscle lengths, F_{max} changing proportionately, it is unlikely that the single equivalent muscle at the hand can be considered as acting like a real muscle over a range of muscle lengths and joint angles.

In analyses involving measurement of the shape of an isometric force-time curve, it appears necessary to build into the model some measure of the rate of onset of activation in order to minimize the number of contractions required. The quantified electromyogram may prove useful in this respect.

REFERENCES

Bahler, A. S., J. T. Fales, and K. L. Zierler. 1968. The dynamic properties of mammalian skeletal muscle. J. Gen. Physiol. 51: 369–384.

Cavanagh, P. R., and D. W. Grieve. 1970. The release and partitioning of mechanical energy during a maximal effort of elbow flexion. J. Physiol. (Lond.) 210: 44–45.

Gordon, A. M., A. F. Huxley, and F. J. Julian. 1966. The variation in isometric tension with sarcomere length in vertebrate muscle fibres. J. Physiol. (Lond.) 184: 170–192.

Houk, J. C. 1963. A mathematical model of the stretch reflex in human Muscle systems. M.S. thesis, MIT, Cambridge, Mass.

Jewell, B. R., and D. R. Wilkie. 1958. An analysis of the mechanical components in frog's striated muscle. J. Physiol. (Lond.) 143: 515–540.

Wilkie, D. R. 1950. The relation between force and velocity in human muscle. J. Physiol. (Lond.) 110: 249–280.

Author's address: *Dr. Arthur Chapman, Department of Kinesiology, Simon Fraser University, Burnaby 2, British Columbia (Canada)*

Effects of isometrical training on the force-time characteristics of muscle contractions

J. Sukop[1]
Charles University, Prague

R. C. Nelson
The Pennsylvania State University, University Park

Improvement in the diagnosis of the nervous and muscular systems in man is of primary interest in several scientific research laboratories. Some of the results have emphasized the importance of evaluating muscular contraction not only from the standpoint of the quantity of developed tension but also that of the time required for its development. Changes of both parameters have been investigated in subjects under the influences of fatigue (Natori, 1970; Royce, 1962), occluded blood circulation (Natori, 1970; Royce, 1958) and changes of extreme temperature (Clarke and Royce, 1962). Previous work has focused on the influences of age, sex, fatigue, and various sport activities (Potměšil and Sukop, 1969; Sukop and Reisenauer, 1967, 1968, 1969).

This approach also was applied by Willems (1971) during his evaluation of functional changes of muscle contraction as a result of a special weight-training program. He investigated the influence of 4 wks. of weight training on the muscle tension and time needed for developing 50 and 90 percent of maximal grip strength. The results revealed the irregularity and individuality of the improvement in functional fitness as a consequence of the training process. The pattern of change is characterized by a period of growth followed by a leveling-off, a so-called "plateau," and occasionally a decrease in performance levels is observed (Zaciorskij, 1969). Therefore, it is desirable to evaluate these force-time data not only after an extended training period, but also

[1]Dr. Sukop was a Distinguished Visiting Professor in the Biomechanics Laboratory at the Pennsylvania State University when this study was conducted.

during each individual training session. Results of such an experiment would add to our understanding of the process of strength development.

Unfortunately, the methods commonly used to record and measure the force-time parameters are very time-consuming, a factor which has limited the number of subjects and trials under investigation. A special on-line computer system was developed by the engineering staff of the Biomechanics Laboratory of the Pennsylvania State University to provide rapid recording and calculation of the data (Sukop, Nelson, and Petak, 1971). This system was utilized in this investigation of the changes in tension and time variables of an isometric elbow flexion contraction which resulted from an isometric training program.

PROCEDURE

Seven male physical education students served as subjects. All of them undertook 10 training sessions, executing 10 isometric contractions of the right elbow flexors in which the maximal force was developed in the shortest possible time. Changes in the course of tension development were evaluated by means of the "force-time curves" recorded for each trial during each session. This resulted in a total of 700 contractions for each subject. To minimize the effects of trial-to-trial variability, average curves were calculated for each set of 10. These curves were used to represent the performance of the subjects for each training session. The test protocol, which was the same for each session, began with the subject's performing two submaximal contractions (50–75 percent) followed by 10 maximal efforts spaced 1 min apart. The 10 training sessions were conducted on alternate days of the week over a 22-day period.

Muscle tension was evaluated in 33.3 msec intervals up to the moment when maximal force was developed. In this study, however, only the first eight values of tension in the interval 0–266.7 msec were used. This procedure, which provided tension values greater than 90 percent of maximum, produced the most important data concerning the force-time curves.

RESULTS

Figure 1 and Table 1 contain the force values for selected time intervals. In each case, an average group value or curve has been calculated from 10 repetitions for all subjects, a total of 70 contractions for each session. The curves on the left side of the figure depict those in which increasing values of tension in comparison with those of preceeding sessions were observed. A marked difference can be seen between the first and second sessions. This most likely is due to the factor of learning and adaptation to the test

Table 1. Average values of the tension of isometric muscle contraction (kiloponds) 7 men, 10 repetitions each

Session	Time intervals (msec)							
	33.3	66.7	100.0	133.3	166.7	200.0	233.3	266.7
1	2.1	7.1	13.1	17.3	20.3	22.6	24.2	25.3
2	2.0	9.0	15.3	19.8	23.0	25.3	26.9	27.8
3	2.3	9.8	16.7	21.1	24.2	26.2	27.7	28.5
4	2.4	9.9	16.8	21.2	24.3	26.5	28.0	29.0
5	2.5	9.9	16.8	21.2	24.3	26.4	27.9	28.9
6	2.2	10.0	17.6	22.3	25.5	27.7	29.4	30.4
7	2.4	9.9	17.2	21.5	24.7	27.0	28.8	30.0
8	2.2	9.9	17.7	22.5	25.5	27.6	29.1	30.3
9	2.7	11.1	19.2	23.9	27.2	29.4	31.1	32.2
10	2.2	10.3	18.4	23.5	26.7	29.0	30.7	31.9

apparatus and procedures. The systematic increase in force and rate of its application, which occurred between the second and fourth sessions, was due primarily to the training stimulus, although learning also may have been a factor.

The first plateau was observed between Sessions 4 and 5, as noted in the curves on the right side of Figure 1. Improvement again was seen between Sessions 5 and 6, followed by a second plateau (Sessions 6, 7, and 8). An increase in performance for Session 9 occurred; this increase also initiated the

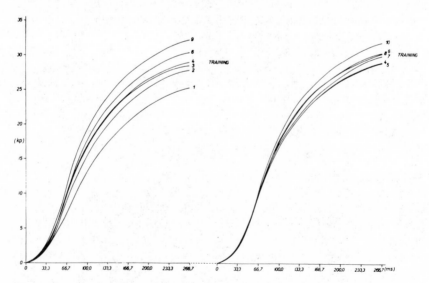

Figure 1. Group mean isometric force-time curves for 10 training sessions (N = 7). *ms*, msec; *kp*, kiloponds.

third plateau when compared with the final session. These results indicate that the group of subjects demonstrated an irregular pattern of improvement in which three plateaus emerged.

These data were analyzed further by comparing the mean forces at 33.3-msec intervals for successive training sessions (Figure 2). The absolute changes were generally the least during the early phase of tension development. As noted previously, the greatest increases occurred between Sessions 1 and 2, 5 and 6, and 8 and 9. The sessions connected by arrows demonstrate the fact that improvement was followed by a leveling-off phase (6–5 versus 7–6, and 9–8 versus 10–9). The only continuous increase in performance over successive sessions occurred between 1 and 4.

It is clear that, as a result of the training program, the subjects were able to develop a specific tension value in less time and, likewise, greater tension in a specified time period. This result is in agreement with that reached by Willems (1971), who reported that, after 4 wks. of training, his subjects produced the first 50 percent of maximal tension at a greater rate of tension development. Is this training effect caused solely by an increase in maximal tension or has the rate of tension development also been affected?

In an attempt to answer this question, relative values of tension for the average curves in each training session were calculated. The tension at the final interval (266.7 msec) for each session was considered as 100 percent, inasmuch as the absolute value for this interval changed during the training period. Each tension value for a given curve was converted to a percentage of the tension at 266.7 msec. These relative tension values are presented in Table 2.

The results indicate that, for Sessions 3–10, the relative tension values are

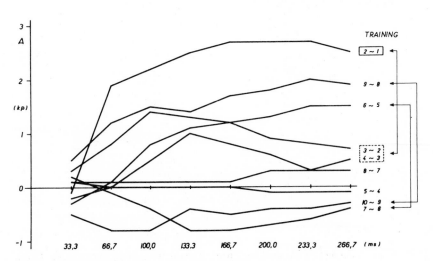

Figure 2. Group changes in muscular force for successive training sessions.

Table 2. Relative values of tension (%) of average contraction curves in the single training sessions (tension in 266. 7 msec = 100%)

Session	Time intervals (msec)							
	33.3	66.7	100.0	133.3	166.7	100.0	233.3	266.7
1	8.3	28.1	51.8	68.4	80.2	89.3	95.7	100.0
2	7.2	32.4	55.0	71.2	82.7	91.0	96.8	100.0
3	8.1	34.4	58.6	74.0	84.9	91.9	97.2	100.0
4	8.3	34.1	57.9	73.1	83.8	91.4	96.6	100.0
5	8.7	34.3	58.1	73.4	84.1	91.3	96.5	100.0
6	7.2	32.9	57.9	73.4	83.9	91.1	96.7	100.0
7	8.0	33.0	57.3	71.7	82.3	90.0	96.0	100.0
8	7.3	32.7	58.4	74.3	84.2	91.1	96.0	100.0
9	8.4	34.5	59.6	74.2	84.5	91.3	96.6	100.0
10	6.9	32.3	57.7	73.7	83.7	90.9	96.2	100.0

very similar. Only the values for Sessions 1 and 2 appear to differ to any extent. This most likely was due to the influence of learning. Additional support for this contention was derived from calculations of average relative curves and their standard deviations for all 10 sessions, Sessions 2–10 (9), 3–10 (8), 4–10 (7), and 5–10 (6). These results (Table 3) show that, by

Table 3. Statistical characteristics (mean, S.D.) of relative values of tension (%) calculated from average curves of 7, 8, 9, and 10 sessions. (tension in 266.7 msec = 100%)

Total number of sessions		Time intervals (msec)							
		33.3	66.7	100.0	133.3	166.7	200.0	233.3	266.7
10	$\bar{X}$	7.84	32.87	57.28	72.74	83.43	90.93	96.43	100.0
	s	0.60	1.79	2.08	1.74	1.30	0.71	0.43	
9 (2–10)	$\bar{X}$	7.79	33.40	57.89	73.22	83.79	91.11	96.51	100.0
	s	0.61	0.86	1.04	1.03	0.78	0.48	0.37	
8 (3–10)	$\bar{X}$	7.86	33.52	58.19	73.48	83.92	91.12	96.48	100.0
	s	0.61	0.83	0.65	0.78	0.72	0.51	0.38	
7 (4–10)	$\bar{X}$	7.83	33.40	58.13	73.40	83.79	91.01	96.37	100.0
	s	0.64	0.81	0.68	0.80	0.66	0.44	0.29	
6 (5–10)	$\bar{X}$	7.75	33.28	58.17	73.45	83.78	90.95	96.33	100.0
	s	0.60	0.82	0.72	0.86	0.71	0.45	0.29	

eliminating Sessions 1 and 2, the variability diminishes for most time intervals, indicating the marked similarity of the average relative curves for the last eight sessions.

Figure 3 graphically depicts the average relative curve for Sessions 3–10. The very small standard deviations at each time interval indicate that the group of subjects tested demonstrated little change in the general form of their force-time curves as a result of training. This occurred in spite of the fact that the mean tension value at 266.7 msec increased from 25.3 to 31.9 kiloponds, an increase of 6.6 kiloponds or 20.7 percent. This result suggests that little or no change occurred in the activation process which produces muscular tension. Further support for this assumption resulted from calculation for each curve the constant, k_a, of the exponential equation

$$Y_t = a_2 (1 - e^{-k_2 t}) - a_1(1 - e^{-kt})$$

(Clarke and Royce, 1962). The value for k_2 for Session 1 was 0.009, whereas in all others it was −0.012.

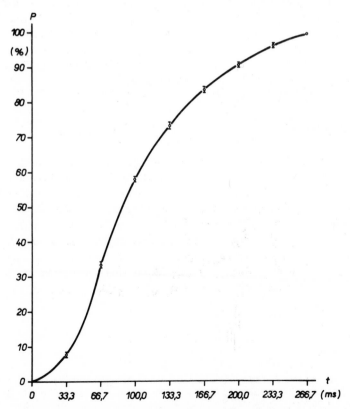

Figure 3. Average relative contraction curve with ± 1-S.D. intervals.

DISCUSSION

Beginning with the third session, the rate of tension development remained virtually unchanged. Of course, it should be emphasized that this was true for the average curves of our group. In this situation, the factor of repetition and individuality of the subjects is suppressed. These results are in agreement with those of Clarke (1964), who found that the tension capacity level of a muscle is independent of the rate of activation of the motor units. These are apparently independent functional aspects of the neuromuscular system that react differently to the stimulus of training. The changes in the capacity of the muscle to develop tension were not accompanied by the changes in the rate of activation. Relative curves used as a criterion for the coordination process of this muscle activity reflected little or no change during this experimental training program. The mechanism of gaining a higher quality of muscular contraction as a result of isometric training was limited to the increase in maximal tension capacity of the trained muscle group.

Future experimental research should attempt to answer the following basic questions. (a) Is isometric training really suitable for evoking the changes in the rate of activation process? (b) What duration of such a training program is necessary for these purposes? (c) To what extent is it possible to alter the capacity of the muscle in such a way as to change significantly the characteristics of the isometric force-time curve?

CONCLUSIONS

Evaluation of each strength training session permits the determination of periods of growth, stagnation, or decrease in functional capacity. Increases in tension values occur between the first and fourth sessions and at the sixth and ninth sessions. In addition, isometric training leads to an increase in the maximal force capacity of the elbow flexors. Furthermore, a specific tension level can be reached in less time and, likewise, greater tension can be reached in a specified time. Also, relative curves of muscle contraction are unaffected by strength training. Finally, it was found that changes in the tension capacity of the elbow flexors are not accompanied by changes in their rate of activation.

REFERENCES

Clarke, D. H. 1964. The correlation between strength and the rate of tension development of a static muscular contraction. Physiol. Einschl. Arbeitsphysiol. 20: 202–206.

Clarke, D. H., and J. Royce. 1962. Rate of muscle tension development and release under extreme temperatures. Physiol. Einschl. Arbeitsphysiol. 19: 330–337.

Hill, A. V. 1970. First and Last Experiments in Muscle Mechanics. Cambridge University Press, New York.

Natori, R. 1970. On some indicators of the capability to exert muscle force of human body from the standpoint of the time course of tension development of isometric tetanus. Jap. J. Phys. Fit. 19: 75–85.

Potmešil, J., J. Sukop. 1969. Observation of muscular contraction and relaxation of ski-runners of different efficiency. In: Proceedings of the Fifth Scientific Council, Prague, pp. 163–189.

Royce, J. 1958. Isometric fatigue curves in human muscle with normal and occluded circulation. Res. Quart. 29: 204–212.

Royce, J. 1962. Force-time characteristics of the exertion and release of hand grip strength under normal and fatigued conditions. Res. Quart. 33: 444–450.

Sukop, J., R. Nelson, and K. Petak. 1971. An on-line computer system for recording biomechanical data. Res. Quart. 42: 101–102.

Sukop, J. and R. Reisenauer. 1967. The influence of age and sex on the dynamics of development of muscular contraction and relaxation in the age of 13–15 years. III. In: Proceedings of the Scientific Council, Prague, 1967, pp. 159–175.

Sukop, J. and R. Reisenaur. 1968. The changes in muscular contraction and relaxation after the static load in 16-year boys and girls. Cs. Hyg. 13: 458–466.

Sukop, J., and R. Reisensuer. 1969. Changes in the course of muscle contraction during one day. Activ. Nerv. Super. 11: 309–311.

Willems, E. J. 1971. The relationship between the rate of tension development and the strength of a voluntary isometric muscular contraction in men. Scientific Report of the University of Leuven, Institute of Physical Education. Leuven, Belgium.

Zaciorskij, V. M. 1969. Cybernetics, mathematics, athletics. Izd. Fyz. Sport. (Moskva) 199.

Author's address: Dr. Jiri Sukop, Research Institute of Physical Culture, Ujezd 450, Prague 1 (Czechoslovakia)

The effect of training on maximal isometric back-lift strength and mean peak voltage of the erector spinae

T. E. J. Ashton
University of Windsor, Windsor

M. Singh
University of Alberta, Edmonton

Berger (1962) tested the static strength of the lower back muscles before and after 12 wks. of training. The isometric group improved their static strength mean by 11.13 lbs, whereas the static increase for the isotonic group was 7.3 lbs. The author concluded that static (isometric) strength was improved more by static training. Rasch and Morehouse (1957) had reported similar results on elbow flexion and isometric arm pressing. Although all subjects gained in maximal isometric back-pull strength after submaximal isometric training at 10, 20, and 30 percent of maximal pulls twice daily for 4 days and maximal training twice daily for 1 wk, Chapman and Troup (1969) noted no change in the relationship between the integrated electromyograms and external forces produced as a result of strength gains (mean gain being 14.5 kg). Thus, no change in the "efficiency," as defined by DeVries (1968), could be shown for the lumbar musculature. Similar results were obtained by Chapman and Troup (1970) over a period of 33 days. No changes in the integrated electromyogram pattern accompanied mean isometric strength increases of 13.3 kg. It might be inferred that the increase in strength was due solely to increases in motor unit activity and that no hypertrophy occurred. Corresponding results for concentric and eccentric back-lift strength training and their effect on lifting strength and electromyograms are lacking.

PROCEDURES

Subjects

Forty male student volunteers (average age, 23 yrs.) at the University of Alberta were ranked at pretest on maximal isometric back-lift score at a hip angle of 160° (20° from vertical) and randomly assigned to one of three training groups or to a control group (10 subjects per group). Each training subject performed three 3-sec maximal contractions three times per wk for 5 wk and then was retested on maximal isometric back-lift strength. Subjects in Group 1 underwent maximal isometric back-lift strength training at a hip angle of 160°. Group 2 trained maximally and concentrically, back-lifting through a hip angle range of 150–170°. Group 3 participated in maximal eccentric back-lift training through a hig angle range of 170–150 degrees. The fourth group acted as a control and was involved only in the pre- and posttest situations.

Apparatus

Strength measures involved the use of the back-lift technique as described by Singh and Ashton (1970) and a hydraulic back-lift dynamometer, a multipurpose device which can be used for isometric, concentric, and eccentric strength testing and training. It consisted of an electric motor connected to a double-acting hydraulic cylinder from which a cable passed over two ball-bearing pulleys and emerged at a point directly between the subject's feet and in front of him. The cable was connected to a load cell (Model UG31, BLH Electronics), which in turn was hooked to the handlebar. During isometrical testing and training, the subjects employed a mixed grip and were asked to lift straight up, extending their backs maximally while keeping their legs straight. The situation was similar for concentric training as the cable released at 2.25 in per sec. With eccentric training, subjects were asked to resist continually and maximally as the cable lowered at the same speed. All contractions lasted approximately 3 sec.

Muscle action potentials of the lumbar erector spinae were taken simultaneously at 1 and 2 in from the spine on the right side at a level between L4 and L5 by use of surface electrodes, 8 mm in diameter, as recommended by O'Connell and Gardner (1963). The distance between the two electrodes, which were held in place by a thin rubber belt, was 1.5 cm. A third electrode (ground) was fastened to the right ankle. All mean peak voltage readings were recorded in mv.

The electrogoniometer was attached to the lateral side of the left hip with adhesive tape so that the rheostat rested on the midpoint of the greater trochanter of the femur, with shafts connecting with the iliac crest and the midpoint of the lateral epicondyle.

The load cell, electrogoniometer, and electrodes were connected to a Honeywell biomedical recorder (Model 1912 Visicorder), with electromyographic spikes simultaneously rectified by an Accudata 136 physiological integrator.

All measurements were taken when the angle of the back was $20°$ from vertical (i.e., hip angle of $160°$).

RESULTS

Coefficients of reliability for maximal back-lift strength and electromyographical mean peak voltages of the lumbar erector spinae were 0.83 and 0.92, respectively. Corresponding standard errors of measurement were $\pm$ 29.22 lbs and $\pm$ 54.10 μV. The pre- and posttest means for each group on each of the two variables, as well as accompanying t tests and probabilities, are found in Table 1.

As can be seen from Table 1, all groups increased in maximal isometric back-lift strength (the three training groups at the 0.01 level of probability). Electromyographic mean peak voltage readings paralleled the strength increases at posttest except for the control group, the mean of which had dropped slightly.

In order to analyze differences in treatment effects, the significance of the observed differences among posttest group means on maximal isometric back-lift strength from Table 1 was tested by analysis of covariance (see Table 2).

The adjusted F value for groups, using pretest results as the covariate, was significant at the 0.05 level. Neuman-Keuls' test on ordered and adjusted posttest group means showed that each of the three training groups had

Table 1. Group means at pre- and posttest and t-tests

Variable	Groups (N = 10)	Pretest	Posttest	t	p
Strength	1	312.00	396.00	−4.260*	0.002
(lbs)	2	321.00	376.95	−3.652	0.005
	3	292.75	380.00	−3.937	0.003
	4	293.00	323.50	−2.274	0.05
Electromyogram	1	353.78	434.42	−2.307	0.05
(μV)	2	328.69	425.26	−3.105	0.01
	3	292.13	413.95	−1.644	0.14
	4	434.24	425.44	0.284	0.78

*Negative sign indicates that pretest mean minus posttest mean was negative.

Table 2. Analysis of covariance and summary of maximal isometric back-lift strength

Source	df	Mean square	Adjusted F	p
Between groups	3	8155.207	4.221	0.012
Within groups	35	1932.071		
RSQ = 0.308				

significantly higher scores after training than did the control group. A similar analysis of covariance was carried out on mean peak voltage of erector spinae, but the adjusted F value between groups was not significant (p = 0.44).

DISCUSSION

The eccentric and isometric groups made the greatest gains in mean maximal isometric strength (87.25 and 84 lbs, respectively). The mean strength of the concentric group also increased significantly, by 55.95 lbs. This would suggest that static strength of back muscles can be improved significantly by both static and dynamic training. In this case, dynamic training, namely eccentric, was as successful in eliciting isometric strength gains as was static training. Significant gains in mean maximal isometric strength by the three training groups were paralleled by significant increases in erector spinae mean peak voltages by the isometric and concentric groups. The probability that the increase of the eccentric group was significant was 0.14. For the most part, this would suggest that the strength increases were due mainly to an increase in motor unit activity, as found by Chapman and Troup (1969), with a possibility of an increase in efficiency for the eccentric group.

REFERENCES

Berger, R. A. 1962. Comparison of static and dynamic strength increases. Res. Quart. 33: 329–333.

Chapman, A. E., and J. D. G. Troup. 1969. The effect of increased maximal strength on the integrated electrical activity of the lumbar erectores spinae. Electromyography 9: 263–280.

Chapman, A. E., and J. D. G. Troup. 1970. Prolonged activity of lumbar erectores spinae. An electromyographic and dynamometric study of the effect of training. Ann. Phys. Med. 10: 262–269.

DeVries, H. A. 1968. Efficiency of electrical activity as a physiological measure of the functional state of muscle tissue. Amer. J. Phys. Med. 47: 10–22.

O'Connell, A. L., and E. B. Gardner. 1963. The use of electromyography in kinesiological research. Res. Quart. 34: 166–184.

Rasch, P. J., and L. E. Morehouse. 1957. The effect of static and dynamic exercises on muscular strength and hypertrophy. J. Appl. Physiol. 11: 29–34.

Singh, M., and T. E. J. Ashton. A study of back-lift strength with electro-goniometric analysis of hip angle. Res. Quart. 41: 562–568.

Author's address: *Dr. T. Ashton, Faculty of Physical and Health Education, University of Windsor, Windsor 11, Ontario (Canada)*

Muscular work and athletic records

H. R. Catchpole and N. R. Joseph
University of Illinois at the Medical Center, Chicago

The idea that work or energy is involved in the action of muscles seems self-evident and is embedded in the language we use to describe work itself, e.g., horsepower and foot-pounds of lift. We associate the equivalence of heat with motion, the interconvertibility of all forms of energy, and the first law of thermodynamics with the names of Benjamin Thompson (Count Romford), Joule, Helmholtz, Mayer, and Clausius from an era beginning about 1800.

A considerable task of 19th-century biology was to determine whether living organisms obey the laws of energetics or possess special and peculiar vital sources of energy. The great calorimetrists, Rubner, Atwater, and Benedict, proved that foods metabolized in the body yield the same amount of heat as when they are combusted in a calorimeter and thus validated the first law of thermodynamics for organisms. We take this work for granted when we count calories and estimate their probable effects on our weights and circumferences.

It is the distinctive function of skeletal muscles to perform external work or locomotion. Feats of running, lifting, throwing and jumping are an expression of this, and, when carried to the limits of capacity and skill, create Olympic or world records. Some 50 years ago, A. V. Hill (1922, 1933) estimated the actual amount of work delivered by the arm muscles of subjects, including three fast bowlers playing cricket. The latter were able to lift a 3/4-lb weight the equivalent of 100 ft, giving a maximal energy output of 9 or 10 kilogram-meters (kg m) per contraction. This value agreed with a calculation made from a record cricket ball throw of 150 yd. The energy of projection is imparted to the ball almost instantaneously, since it is certain that at the instant the ball leaves the fingers it has received all the available muscular thrust. What is the source of this energy?

All recent physiological texts assert, in more or less equivalent terms, that the energy comes from adenosine triphosphate (ATP) (Brown and Brengelman, 1965; Keele and Neil, 1971; Starling and Evans, 1968; Wilkie, 1968). Indeed, it appears to be assumed almost universally that a direct link exists between the chemical bond energy of adenosine triphosphate and the work of muscular contraction. Keul, Doll, and Keepler (1972) and Needham (1971) accept this role of ATP as a major premise, and it is implicit in recent models that are based on muscle ultrastructure. According to these views, muscle is a chemomechanical machine which converts (transduces) chemical bond energy to mechanical work.

We propose that the work of muscular contraction is not so linked to the chemical bond energy of ATP or to any related compound and suggest an alternative energy source. While the literature is confusing to a degree, our own calculations indicate that the existing ATP in muscle is insufficient to yield the energy found in the experiment of Hill or anything like it. ATP is a typical intermediary compound, normally present in small—even in minute—amounts, and the textbooks assure us that many other functions demand ATP. One could even question the wisdom of nature in saddling it with the instant burden of "fight or flight."

Attention is focused on the resting muscle, a tissue which in muscle physiology is described in terms that are remarkably soporific, including its name. It is said to be soft and supple and easily extensible. A child, when asked to show his muscles, promptly throws the biceps into violent contraction to demonstrate the exact opposite of a resting muscle. This attitude extends to physiologists who contrast the "rubberlike" resting muscle with the "steel-hard" contracted one, not entirely poetically. "Relaxation" to the resting state continues the metaphor.

These soothing and negative adjectives tend to conceal the fact that resting muscle is the only state of muscle that is primed, ready to contract, and potentially energy yielding. On the contrary, the impressively knotted and contracted muscle is devoid of work potential.

It is proposed that the resting muscle, composed of contractile proteins, water and ions, is in a highly ordered, low-entropy configuration and that it passes, on contraction, into a state of low order and high entropy (Joseph, 1971a, b). *This is the energy-yeilding process.* An alternative statement would be that water in resting muscle has a low dielectric constant and that that in contracted muscle has a high dielectric constant. The energy of contraction therefore can be described as dielectric configurational, or entropic energy. It is not chemical bond energy, and muscle is not a chemomechanical machine. Skeletal muscle is designed, par excellence, to convert dielectric energy into work at close to 100 percent efficiency.

It is shown that the dielectric energy of muscle accounts for maximal work. Two independent methods were used to estimate dielectric energy (Joseph, 1971a, b). In the first, dialectric energy is taken as the product of RT and the difference in total ionic concentrations between cell and blood, where R is the gas constant and T the absolute temperature. Electrolyte concentrations, determined analytically, are available for many tissues, including skeletal muscle. In the second method, dielectric energy is taken as the product of intracellular sodium concentration and the change of standard chemical potential of sodium ion between cell and blood, expressed in the proper units. The change in standard chemical potential of sodium ion is derived from the action potential of the cell.

Using the second method, for human skeletal muscle the change in standard chemical potential of the sodium ion is 2.8 kcal, corresponding to an action potential of +120 mV. Intracellular sodium concentration is 0.028 mol per kg of water. Then, dielectric energy per kg of muscle water is:

$$0.028 \times 2.8 \times 10^3 = 78.4 \text{ cal} = 330 \text{ J}$$

Hill estimated the combined water content of the adult human biceps and brachialis, the dominant muscles involved in throwing, to be about 250 g. These muscles delivered 9 kgm of work, equivalent to 36 kgm per kg of tissue water, which converts to 353 J per kg of water. Thus, the maximal work of the arm in ball throwing is accounted for by the estimated dielectrical energy of the responsible muscles.

On the basis of the same value for the dielectric energy of human skeletal muscle, an estimate is made of the maximal high jump. The water content of the flexors of the thigh is taken to be approximately 2.5 kg. The available dielectric energy, 900 J, equals 90 kgm. A man weighing 76.7 kg, which is the Olympic average, according to Tanner (1964) would be able to lift his center of gravity 90 cm per 76.7 kg or 1.17 m above its vertical standing position. Assuming this to be 1 m (according to Tanner, the average height of Olympic jumpers was 191.5 cm), the highest position attainable would be 2.17 m or 7 ft. Since Pat Matzdorf of the University of Wisconsin cleared a record 7 ft, 6.5 in, this approximate calculation seems satisfactory.

If the source of muscular energy is dielectric or configurational, what is the role and mode of operation of the metabolic aspect of muscle action? In the concept presented, muscle operates essentially between two thermodynamic states: well-ordered, low-entropy and low-dielectric constant (resting) and disordered, high-entropy and high-dielectrical constant (contracted). This gives a clue to the pattern of metabolic events.

Since intramuscular water can exist in two or more labile states, it can change its solvent properties at the same time that it changes its dielectric constant and liberates energy. In the resting state, it acts as a nonpolar solvent, whereas in the stimulated, contracted state, it is the (expected) polar solvent. Muscle metabolism follows these states of water.

In resting muscle, reactivity of substrates (including ATP) is prevented by the hydrophobic medium, and overall metabolism is low. As the dielectric constant suddenly rises to its value in free liquid water, all glycolytic enzyme systems are activated, metabolism rises rapidly and the characteristic muscle heat is liberated, continuing for many milliseconds. The function of muscle metabolism is to restore the resting organization of low-entropy and low-dielectric constant that was lost on contraction. As the resting state is achieved, glycolytic activity falls, and the system again is primed to contract. Because one of the contractile proteins is an enzyme, ATPase, configurational change is linked closely to the initiation of metabolic activity.

In summary, muscle is a dielectric, entropic, or configurational machine working at constant temperature between two states, resting and contracted, with 1° of freedom. This limitation places restraints on possible metabolic processes, inasmuch as metabolism becomes a dependent variable. It should be noted that calculations of cellular dielectric energy depend on ionic distributions predicated on the heterogeneous equilibrium of Gibbs (1927). All notions of energy-related ionic transport (ionic "pumps") are rejected (Joseph 1966, 1971a, b; Joseph, Engel. and Catchpole, 1961).

The present treatment avoids the esthetic anomaly of built-in "muscular inefficiency," for which observations, for example, of the flight of a hummingbird, provide minimal support. The enthalpic consequences of the second law of thermodynamics appear as part of the recovery process, not of contraction itself.

Regarding mechanical models of muscle that are based on morphology, which are currently in vogue, it always has been recognized that muscle is highly structured at various levels. Grossly, each muscle has a characteristic form which alters on contraction in an invariant way. In the 17th century, Leeuwenhoeck already had discerned cross-striations, which have been revealed more clearly by stains and modern microscopes. The electron microscope has revealed a wealth of further detail and finer ordering, while X-ray analysis adds its quota of substructure. These macro- to ultramicroscopic structures constitute the "moving parts" of the machine. But these are not the "origin" of the motion any more than are the busy pistons, rings, wrist pins, connecting rods, and crankshaft of a gasoline or diesel engine. In muscle, a transient change in the dielectric constant of water initiates disassembly and energy liberation, followed by self-assembly and energy absorption, in which every order of structure participates.

REFERENCES

Brown, A. C., and G. Brengelman. 1965. *In:* T. C. Ruch and H. Patton (eds.), Physiology & Biophysics, p. 1032. W. B. Saunders, Philadelphia.

Starling, E. H., and L. Evans. 1968. Principles of Human Physiology, pp. 786 and 835. 14th Ed. H. Davson and M. G. Eggleton (eds.). Lea & Febiger, Philadelphia.

Gibbs, J. W. 1928. On the equilibrium of heterogeneous substances. *In:* Collected Works. Vol. 1. Longmans Green, New York.

Hill, A. V. 1922. J. Physiol. 56: 19–41.

Hill, A. V. 1933. Living Machinery. Ball & Sons, London.

Joseph, N. R. 1966. Nature 209: 398–399.

Joseph, N. R. 1971a. *In:* H. R. Elden (ed.), Biophysical Properties of the Skin, pp. 551–596. John Wiley & Sons, New York.

Joseph, N. R. 1971b. Physical chemistry of aging. *In:* N. R. Joseph (ed.), Interdisciplinary Topics in Gerontology. Vol. 8. S. Karger, Basel.

Joseph, N. R., M. B. Engel, and H. R. Catchpole. 1961. Nature 191: 1175–1178.

Keele, C. A. and E. Neil. 1971. Samson Wright's Applied Physiology, p. 213. Oxford University Press, London.

Keul, J., E. Doll, and D. Keppler (eds.). 1972. Energy Metabolism of Human Muscle. University Park Press, Baltimore.

Needham, D. M. 1971. Machina Carnis: The Biochemistry of Muscular Contraction in Its Historical Development. Cambridge University Press, New York.

Tanner, J. M. 1964. The Physique of the Olympic Athlete. Allen & Unwin, Ltd., London.

Wilkie, D. R. 1968. Muscle, p. 1. St. Martin's Press, New York.

Author's address: *Dr. H. R. Catchpole, Department of Pathology, College of Medicine, University of Illinois at the Medical Center, 1853 West Polk Street, Chicago, Illinois 60612 (USA)*

Advances in instrumentation and techniques

Photographic methods

New chronophotographic methods for three-dimensional movement analysis

W. Baumann

German Sport School, Cologne

The analysis of the kinematics of human body motions is usually carried out in accordance with two essential assumptions: (a) the plane photographic picture is a sufficient, true model of the spatial motion and, (b) the spatial orientation of the fictitious plane of motion is known exactly. Very often, these assumptions do not hold true and are not proven experimentally. This applies especially to recordings of performance under actual competitive conditions. In this field, the cinematographic method still seems to be exclusive, despite its obvious shortcomings in space and time resolution (Baumann, 1968; Gutewort, 1968). However, because most biomechanical investigations are conducted in a laboratory situation, there is no need to favor conventional cinematography over other photographic methods. In applications requiring a high degree of accuracy of the primary kinematic parameters, chronophotographic techniques mostly outperform cinematography. The sophisticated pulsed-light photogrammetry (Gutewort, 1968, 1971) showed the potential use of this measuring method. However, the restrictions concerning the reduced ambient light and the considerable technical expenditure seem to hamper a more extensive application. We developed further chronophotographic methods, emphasizing an improvement in performance and the application of constructional units that are both inexpensive and for the most part available commercially.

CHRONOCYCLOGRAPH WITH DIGITALLY CONTROLLED PICTURE FREQUENCY AND SYNCHRONIZER

This method is based on marking the measuring points by continuously operating incandescent lamps of high luminous density. Periodical interruption of the beam path by a rotating shutter disk provides the time information on the stillphotograph. Unlike conventional devices, the present chronocyclograph (see Figure 1) offers a number of constructional features that extend accuracy and range of application.

The *lens* is a nearly symmetrical Gaussian double lens with minimal error of distortion. Moreover, the lens is stopped down to f/22, thus assuring adequate suppression of ambient light, greatest possible depth of field, and reduction of residual image defects.

The *shutter* disk is located between the lens elements, where the diameter of the light bundle transmitted by the lens is minimal, thus allowing small disk movement, i.e., high shutter speeds. Since the shutter operates in the plane of the exit pupil, the entire image field is exposed simultaneously. As shown in Figure 1, the disk is provided with eight equidistant boreholes serving as shutter apertures. One of them is partially masked, producing an identification marker once per revolution; every eighth exposure of the different object points is interrupted additionally and thus especially marked out. By this means, a major disadvantage of chronocyclograph techniques is avoided completely, namely, the unsatisfactory identification of simultaneously exposed object points that arise chiefly from temporary motionless points (see Figure 2).

An electrooptic *position sensing assembly* detects the position of the disk at that time when the marker hole passes the optical axis of the lens. The output pulse is used to control the shutter operation and serves as a synchronized pulse for external data recorders.

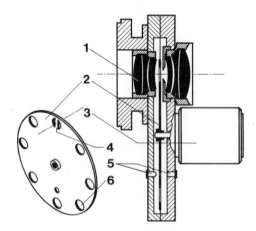

Figure 1. Chronocyclograph and shutter disk. *1*, lens, type Symmar f. 5.6/100 mm, stopped down to f/22; *2*, shutter disk, cut from 0.3-mm aluminum sheet; *3*, stepping motor, 7.5° per step, starting frequency range, 90–120 steps per sec, maximal stepping rate, 1,200 steps per sec; *4*, marker strap; *5*, electrooptic position sensing assembly; *6*, shutter apertures, equidistance accurate to the nearest 1/100 mm. The casing is made of aluminum and can be matched to most single-lens reflex cameras up to maximal size of 6.5 x 9 cm.

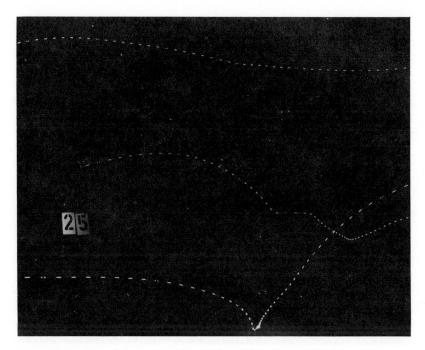

Figure 2. Photogram. Motion of a leg taken at 125 frames per sec. Dashes allow identification and timely coordination by simply counting off.

Significant improvement in performance related to time resolution was obtained by using a stepping motor as the *driving unit* for the shutter disk. Because stepping motors are a convenient source of digital motion, they ensure synchronous running with the driving pulses within the operation frequency range. Within a certain starting frequency range, the motor may be started or stopped without loss of step. High stepping rates can be achieved by programmed acceleration. The versatile controllability of the disk and motor assembly is utilized fully by an *electronic control unit,* which allows synchronization of two chronocyclographs in the case of stereophotographic recording. An accurate crystal oscillator provides the initial reference frequency, which is fed into a programmable divider. When the oscillator is switched on, an output frequency matched to the starting frequency is supplied to the driver circuits of both stepping motors. The motors start and will be stopped, respectively, when the marker holes are centered in the optical axes. The motors then are set in synchronous motion and division ratio is increased step by step, causing an acceleration of the stepping motors. When a frequency corresponding to the preselected frame rate is attained, the divider will be locked and the motors run synchronously with the desired stepping rate. This "ready for shot" state is checked permanently by means of a coincidence circuit, which receives the respective synchronized pulses. If one motor falls out of step, both motors will be stopped automatically and synchronized again.

The measuring time intervals are adjustable in the range from 6 to 30 msec, being equivalent to a picture frequency range of 166.6–33.3 cycles per sec. The maximal overall time error is ±20 μs, or 0.2 percent of 100 cycles per sec. In connection with the enhanced space resolution of large picture formats, this method offers a wide range of application to the stereophotogrammetric recording of motion.

The identification marking allows the use of plane table photogrammetric techniques because synchronous image points in a pair of stereoscopic pictures can be identified properly. This implies maximal adaptability of the camera positions to the object to be analyzed. If uncalibrated cameras are used, accurate adjustment of the outer orientation and a suitable system of reference points are necessary to calculate the spatial object coordinates X, Y, and Z.

The high-intensity miniature filament bulbs that are employed make it possible to take indoor photographs by normal artificial light.

The system has been tested with Hasselblad EL 70/M camera bodies and proved to be excellent in a number of investigations. With a picture format of 6 x 6 cm, image scale of 1:100, and base length of 1,800 mm, and evaluation effected at magnification factor 20, maximal errors of the object coordinates are ±1 mm for X and Y and ±3 mm for Z. Accuracy of the derivable instantaneous velocities and accelerations depends on the formulas applied for balancing, interpolation, and differentiation as well as on the actual motion itself.

STEREOSCOPIC CHRONOPHOTOGRAPHY USING PULSE-OPERATED INFRARED EMITTERS

An obvious disadvantage of chronophotographical techniques is the independence on ambient light. As long as the unalienable reduction of the object to certain well-defined object points is achieved by reflecting or self-luminous markings, which contrast only in the visible spectrum, the application is restricted to areas where the ambient level of illumination is low. Although the method described makes it possible to take indoor photographs in artificial light, it is extremely difficult to take daylight photographs successfully. Apparently, the use of radiation sources in the invisible region of the spectrum with spectrally matched filters and adequate sensitized emulsions would offer a solution to the problem. High-power GaAs emitters meet the requirements of radiant intensity, output efficiency, and frequency response in the near infrared.

A crystal controlled oscillator (see Figure 3) with a modulo-n divider assures pulse periods accurate to the nearest 1 msec. The desired "frame rate"

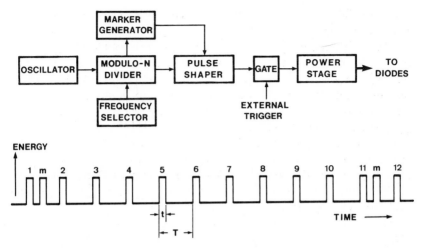

Figure 3. Block diagram of control circuitry and radiated infrared pulse sequence. The circuitry is battery-powered and, except for the power transistors, completely designed with low-power TTL integrated circuits. The emitters used are diodes of type TIXL 16 with a peak wavelength of 0.93 μm. _1,2,3,_ . . . = successive pulses; _m_ = marking pulses; _T_ = pulse period, 22 settings from 1 to 30 msec corresponding to a picture frequency range from 1 kilocycle per sec to 33.3 cycles per sec.

is set manually by a selector switch. A marker generator provides the identification markings after each 10th measuring pulse. The power stages drive the infrared-emitting diodes with square-wave forward currents of 10–20 A, thus producing maximal optical output power. Because the power stage is switched on only when the gate is triggered, proper synchronization between the event and the beginning of exposure and pulse sequence can be obtained. An incorporated photodiode initiates the diode emission by an event-triggered electronic flash. Other techniques employing ultrasound or electromyogram signals may be a convenient supplement.

The measuring time intervals are adjustable in the range from 1 to 30 msec. The time error is less than ±1 μs. The reduction of point exposure time to 50 μs not only eliminates any image blur, but provides considerable improvement in time resolution. By means of spectrally matched filters e.g., Schott RG 830 or Kodak Wratten No. 87 C (Eastman Kodak, Rochester, N.Y.), and infrared-sensitized plates, e.g., ORWO J 950, sufficient suppression of visible ambient light can be achieved. Because infrared reflections from a background containing chlorophyll may cause contrast reduction, bright sunlight on such a background should be avoided. In all other cases, daylight exposures are possible without restriction.

These object markings, together with the control circuitry described, allow the application of both stereophotogrammetric and plane table photogrammetric techniques for the recording of spatial motions. The test results using two uncalibrated Hasselblad cameras EL 70/M with Zeiss Planar lenses f/3.5 100 mm, that is, stereophotogrammetry with amateur equipment, are better than those achieved with the chronocyclographs. However, these results seem to indicate the limits of accuracy, which should not be exceeded when human body motions are concerned.

REFERENCES

Baumann, W. 1968. Über die kinematografische Bewegungsanalyse. Med. Welt 19: 2168–2174.

Gutewort, W. 1968. Die digitale Erfassung kinematischer Parameter der menschlichen Bewegung. *In:* J. Wartenweiler, E. Jokl, and M. Hebbelinck (eds.), Biomechanics, pp. 53–60. S. Karger, Basel, and University Park Press, Baltimore.

Gutewort, W. 1971. The numerical presentation of the kinematics of human body motions. *In:* J. Vredenbregt and J. Wartenweiler (eds.), Biomechanics II, pp. 290–298. S. Karger, Basel, and University Park Press, Baltimore.

Author's address: *Dr. Wolfgang Baumann, Institute für Biomechanik der Deutschen Sporthochschule, 500 Köln 41, Carl-Diem-Weg 5 (West Germany)*

Validation of a mathematical model for correction of photographic perspective error

T. P. Martin
State University of New York, Brockport

M. B. Pongratz
E G and G, Los Alamos

Photographic perspective error long has plagued the researcher attempting to obtain accurate three-dimensional coordinates from two-dimensional film. Accurate velocity and acceleration values are dependent on accurate positional information, and photographic perspective error is responsible for the concealment of true three-dimensional position coordinates. The purpose of this investigation was the validation of mathematical formulas derived to correct for photographic perspective error. These corrective formulas were derived from trigonometric relationships between two two-dimensional, perpendicular views of any th te-dimensional position in space and were designed to transform measured coordinate information (from film) to true three-dimensional coordinate data. The corrective formula for any point P is:

$$x = \frac{XYx\left(1 - \frac{ZYz}{Dz}\right)}{\left[1 - \frac{(XYx)\,(ZYz)}{(Dz)\,(Dx)}\right]}$$

$$y = ZYy\left(1 - \frac{x}{Dx}\right)$$

$$z = ZYz\left(\frac{Dx - x}{Dx}\right)$$

where x = true x value; y = true y value; z = true z value; XYx = x coordinate measured in XY film plane; Zyy = y coordinate measured in ZY film plane;

Zyz = z coordinate measured in ZY film plane; Dx = distance from Film Plane 2 to origin along x axis; and Dz = distance from Film Plane 1 to origin along z axis. A review of related literature on the topic of perspective error and the derivation of the corrective formula utilized in this investigation can be found in Martin and Pongratz (1973) or obtained from the senior author.

METHOD

Two cameras, set in a horizontal plane, with optical axes perpendicular and intersecting, established the three-dimensional reference system (see Figure 1). The model was constructed by attaching three horizontal yardsticks at varying heights on a vertical range pole. As a result of this placement, the yardsticks were labeled "nose" (+10 in vertical), "left arm" (−10 in vertical), and "right arm" (−15 in vertical) (see Figure 2). Seven points were determined on each yardstick by designating Point 1 as the intersection of the edge of each yardstick with the vertical axis and placing the other six points at 5-in intervals from the vertical axis. Thus, there existed 21 designated points on the model. By knowing the perpendicular distance of each point from the vertical axis and the angle made by each yardstick in the XZ plane, the true coordinates of each point were determined easily (Table 1). Two Nikon 35-mm cameras, 1.4 lens, were set 10 ft from the origin and utilized to obtain biaxial film records of the model (see Figure 2). Kodak Tri-X film (ASA 400) (Eastman Kodak, Rochester, N.Y.) was employed and the film was analyzed

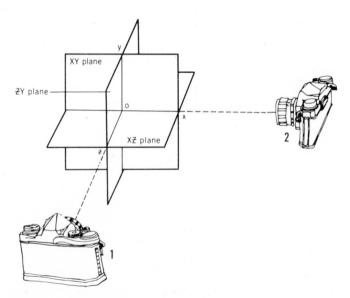

Figure 1. Three-dimensional reference system.

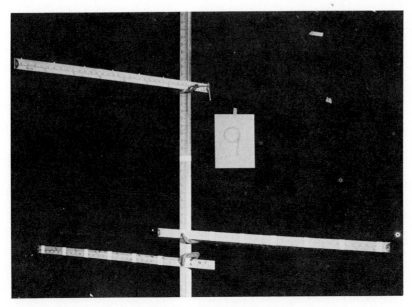

Figure 2. Model as photographed along x axis (camera 2).

on an X-Y coordinate digital converter, "Coordicon" (Wayne George Corp.). In this way, two sets of two-dimensional coordinates (x,y and z,y), accurate to 1/1000 of an inch, were obtained for each of the 21 points on the model. The mean of the two y values was used to represent the y coordinate, resulting in x,y,z film coordinates for each point. The film coordinates for each point then were corrected for scale (factor of 1.525), yielding one set of data, and corrected for scale and perspective error, yielding a second set of data (Table 1). The measured coordinates (true) of the 21 points on the model were utilized as the criterion for comparison with these sets of data.

ANALYSIS

Errors in position determination were calculated as the distance between each true point and its respective uncorrected and corrected points ascertained by film analysis.

$$\text{Position Error} = \sqrt{(x_{ti} - x_{mi})^2 + (y_{ti} - y_{mi})^2 + (z_{ti} - z_{mi})^2}$$

Uncorrected and corrected displacements between adjacent points on each yardstick were determined and subtracted from the true displacement of 5 in to indicate absolute errors in displacement determination.

Table 1. Point coordinates (inches)

Code		True			Film coordinates corrected for magnification			Film coordinates corrected for magnification and perspective		
		x	y	z	x	y	z	x	y	z
Nose	11	0.000	10.000	0.000	-0.020	10.239	-0.066	-0.020	10.242	-0.066
	12	3.919	10.000	3.077	4.044	10.725	3.149	3.942	10.413	3.046
	13	7.838	10.000	6.154	8.278	11.173	6.577	7.854	10.521	6.147
	14	11.757	10.000	9.231	12.738	11.609	10.271	11.755	10.591	9.265
	15	15.676	10.000	12.308	17.414	12.080	14.265	15.613	10.668	12.409
	16	19.595	10.000	15.385	22.372	12.561	18.514	19.481	10.726	15.508
	17	23.514	10.000	18.462	27.593	13.046	23.116	23.313	10.762	18.625
Left arm	21	0.000	-10.000	0.000	-0.014	-10.183	-0.026	-0.014	-10.185	-0.026
	22	3.183	-10.000	-3.867	3.143	-10.184	-4.024	3.264	-10.203	-3.916
	23	6.366	-10.000	-7.733	6.028	-10.177	-8.209	6.419	-10.199	-7.770
	24	9.549	-10.000	-11.600	8.662	-10.164	-12.656	9.503	-10.177	-11.654
	25	12.732	-10.000	-15.466	11.273	-10.166	-17.350	12.730	-10.146	-15.509
	26	15.915	-10.000	-19.333	13.611	-10.188	-22.254	15.802	-10.127	-19.324
	27	19.098	-10.000	-23.199	15.848	-10.201	-27.494	18.907	-10.084	-23.162
Right arm	31	0.000	-15.000	0.000	-0.014	-15.333	-0.027	-0.014	-15.336	-0.027
	32	-3.162	-15.000	3.891	-3.140	-15.373	3.895	-3.035	-15.299	3.993
	33	-6.323	-15.000	7.781	-6.593	-15.463	7.515	-6.159	-15.301	7.901
	34	-9.485	-15.000	11.672	-10.312	-15.581	10.949	-9.298	-15.304	11.798
	35	-12.646	-15.000	15.562	-14.300	-15.717	14.224	-12.429	-15.294	15.697
	36	-15.808	-15.000	19.453	-18.599	-15.891	17.310	-15.568	-15.294	19.556
	37	-18.969	-15.000	23.343	-23.120	-16.128	20.278	-18.670	-15.321	23.433

$$\text{Displacement Error} =$$

$$\left| 5.00 - \sqrt{(x_{mi} - x_{mi+1})^2 + (y_{mi} - y_{mi+1})^2 + (z_{mi} - z_{mi+1})^2} \right|$$

Velocity was calculated as displacement divided by time, with time set hypothetically at 1 sec for each displacement. Therefore, the comparative velocity errors are identifical to the displacement errors. Theoretically, there existed a constant velocity of 5 in per sec for all displacements. This resulted in criterion acceleration values of zero. Thus, the absolute value of the uncorrected and corrected changes in velocity, divided by time (1 sec), indicates the comparative acceleration errors.

RESULTS

The data very clearly demonstrate the effect of perspective error on the determination of position, displacement, velocity, and acceleration values. The farther that a point was located either in front of or behind the focus plane, the greater was the perspective error. In addition, it is interesting to note that if a point was located in front of the focus plane of one camera and behind the focus plane of the second camera (e.g., left and right arms), compensating errors existed which operated to increase the accuracy of the y coordinate and thus the accuracy of position determination. This was not the case for the nose, which was in front of the focus planes of both cameras.

It is clear from the uncorrected and corrected point coordinates and from the comparison of position errors that the mathematical formulas operate to increase the accuracy of the measurements obtained by film analysis. It appears that the corrective formulas are precise and valid. The errors noted with the use of the correction can be attributed to inaccuracies in the construction of the model, establishment of perpendicular-intersecting optical axes, placement of the model within the reference system, and the determination of the true point coordinates—all errors related to the artificial testing situation.

In order to concentrate on the accuracy of spatial determinations of position, the temporal variable, inherent in kinematic analysis, was controlled by the use of still photography. However, theoretical time intervals of 1 sec between adjacent points on each yardstick were utilized to reflect the effect of errors in position determination on velocity and acceleration values. The velocity and acceleration values presented in Table 2 indicate the effect of perspective error on the accuracy of these values. In addition, they reflect the importance of a correction for perspective error on the determination of true values.

Table 2. Comparative errors (inches)

Code	Position M*	Position M & P†	Displacement M*	Displacement M & P†	Velocity M*	Velocity M & P†	Acceleration M*	Acceleration M & P†
Nose								
11	0.248	0.252						
			0.205	0.040	0.205	0.040		
12	0.740	0.415					0.261	0.047
			0.466	0.007	0.466	0.007		
13	1.322	0.521					0.342	0.001
			0.808	0.006	0.808	0.006		
14	2.153	0.592					0.360	0.017
			1.167	0.023	1.167	0.023		
15	3.343	0.678					0.380	0.022
			1.547	0.044	1.547	0.044		
16	4.905	0.745					0.433	0.016
			1.980	0.060	1.980	0.060		
17	6.900	0.804						
Left arm								
21	0.186	0.187						
			0.094	0.075	0.094	0.075		
22	0.245	0.218					0.012	0.082
			0.083	0.008	0.083	0.008		
23	0.610	0.209					0.085	0.033
			0.168	0.041	0.168	0.041		
24	1.389	0.191					0.203	0.068
			0.371	0.028	0.371	0.028		
25	2.389	0.152					0.062	0.130
			0.433	0.103	0.433	0.103		
26	3.726	0.170					0.264	0.039
			0.698	0.064	0.698	0.064		
27	5.390	0.212						
Right arm								
31	0.335	0.337						
			0.016	0.030	0.016	0.030		
32	0.373	0.341					0.012	0.028
			0.004	0.002	0.004	0.002		
33	0.598	0.363					0.060	0.002
			0.064	0.004	0.064	0.004		
34	1.242	0.378					0.098	0.004
			0.162	0.001	0.162	0.001		
35	2.245	0.389					0.134	0.026
			0.295	0.026	0.295	0.026		
36	3.630	0.393					0.185	0.009
			0.480	0.035	0.480	0.035		
37	5.345	0.448						

*Corrected for magnification.
†Corrected for magnification and perspective.

SUMMARY

Mathematical formulas derived for the correction of photographic perspective error were tested by constructing and photographing a model that was designed to illustrate perspective error. Photographic perspective was shown to be an important source of error in the determination of position, displacement, velocity, and acceleration values. The application of the derived formula to the film data resulted in a substantial increase in the accuracy of the calculated values. It appears that the mathematical correction for photographic perspective error is valid and can be of definite value in cinematographic research in biomechanics. The small errors noted with the use of the correction can be attributed to measurement error.

REFERENCES

Martin, T. P., and M. B. Pongratz. 1973. A mathematical correction for photographic perspective error. Res. Quart., Submitted for publication.

Author's address: *Dr. Thomas P. Martin, Cinematography Laboratory, Sport Science Department, State University-Brockport, Brockport, New York 14420 (USA)*

A new three-dimensional filming technique involving simplified alignment and measurement procedures

B. Van Gheluwe

Vrije Universiteit Brussel, Brussels

From the beginning, cinematographic techniques always have been popular with researchers investigating a variety of human movements. Recently, cinematographic equipment has been simplified considerably, and film material has become less expensive. The result is that the student or expert will use the camera for research in most motion studies. This may be the reason for the increased interest in the use of analytical film techniques in sports during the last few years.

Recording techniques have become increasingly complicated and sophisticated, inasmuch as all nonplanar movements necessitate the use of three-dimensional methods involving two or more cameras. But, even by employing the most modern and advanced equipment, the positioning and alignment of the cameras always will be necessary, as well as the measuring of the mathematical parameters defining them. This is a tedious and time-consuming job that cannot be evaded in the conventional cinematographic recording techniques.

This chapter outlines a general method that requires only a minimum of camera alignment and avoids any direct measuring procedure.

METHOD

In earlier studies, most parameters defining the position of the cameras in space within an arbitrary chosen reference frame had to be measured on the field. This involved painstaking and time-consuming work utilizing plumb bobs, spirit levels, and other alignment tools. Also, in order to avoid complex mathematical formulas and calculations, only simple geometric camera configurations were feasible.

With the proposed method, the camera positioning is completely "free," except that the optical axes of the cameras must intersect with one another.

The parameters defining the exact camera positions in space are back-calculated from the image coordinates for a set of reference points compared with their known life-size coordinates. Because each camera films these points, their positions in space can be reconstructed mathematically. From the resulting parameters, the solution of a set of selected simultaneous equations provides the life-size coordinates for any spatial point of interest.

Theoretical Basis of the Method

The method is based on the analytical properties of coordinate transformations from one orthogonal reference frame to another. First, a general orthogonal frame (O, X, Y, Z) is defined in space (Figure 1), and then the subsequently analyzed motion can be described analytically through the X, Y, and Z coordinates. The number of cameras used is not limited; however, for the sake of simplicity, only two cameras are considered in the following example.

In addition, an orthogonal reference frame is assigned to each camera, (O_1, X_1, Y_1, Z_1) and $O_2, X_2, Y_2, Z_2)$, respectively (Figure 1). The respective optical centers of each lens are O_1 and O_2, and $O_1 Z_1$ and $O_2 Z_2$, respectively, represent the optical axes which intersect at the origin O of the reference frame.

Theoretically, there is an analytical relationship between the image coordinates (X_1, Y_1) or (X_2, Y_2) of any arbitrary point in space and its spatial coordinates (X_1, Y_1) or (X_2, Y_2) in relation to the reference frame of each camera (Figure 1). Providing that the focal length of each lens is known, the following mathematical relations exist (Figure 2).

$$\frac{X_1}{x_1} = -\frac{Z_1}{f_1} \text{ and } \frac{X_2}{x_2} = -\frac{Z_2}{f_2}$$

$$\frac{Y_1}{y_1} = -\frac{Z_1}{f_1} \text{ and } \frac{Y_2}{y_2} = -\frac{Z_2}{f_2}$$

(1)

where f_1 and f_2 represent the focal lengths of the lenses for the respective cameras. The focal length is the only geometrical parameter not back-calculated by this method because it can be measured in the laboratory easily and accurately at any time before or after filming.

Applying the laws of orthogonal coordinate transformations from the

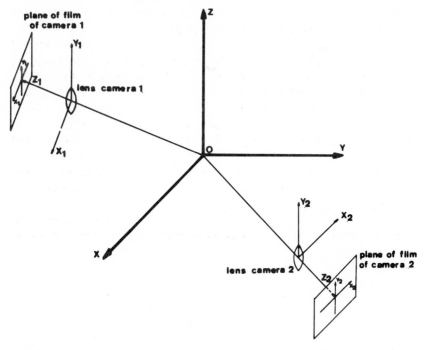

Figure 1. Geometric position of both cameras with respect to the reference frame (O, X, Y, Z).

general reference frame (O, X, Y, Z) to the reference frame fixed to each camera results in:

$$X = F_x^1 (X_1, Y_1, Z_1) = F_x^2 (X_2, Y_2, Z_2)$$

$$Y = F_y^1 (X_1, Y_1, Z_1) = F_y^2 (X_2, Y_2, Z_2) \qquad (2)$$

$$Z = F_z^1 (X_1, Y_1, Z_1) = F_z^2 (X_2, Y_2, Z_2)$$

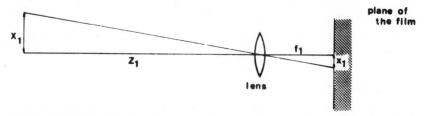

Figure 2. Geometric relation between one of the image coordinates x_1 and the respective coordinate x_1, defined within the reference frame assigned to Camera 1. This is also valid for Camera 2.

where F_x^1, F_y^1, F_z^1, and F_x^2, F_y^2, F_z^2 represent the classical transformation functions between the given reference frames.

Using the image coordinates (x_1, y_1) and (x_2, y_2) from specially selected refrence points with known spatial X, Y, Z coordinates, the transformation function F_x^1, F_y^1, F_z^1, and F_x^2, F_z^2, F_z^2 can be reconstructed mathematically solving the set of Equations 1 and 2.

Theoretically, the X, Y, and Z coordinates of any arbitrary point now can be calculated from its image coordinates using Equations 1 and 2, but faulty solutions do result from straightforward calculation of the mathematical equations giving the X, Y, Z coordinates. This problem is due to the mathematical instability of the equations, i.e., small variations in the equation coefficients, originating from measurement errors, cause completely different solutions to these equations.

In order to avoid these mathematical instabilities, highly sophisticated "least square" methods (as used in high-energy physics) are applied to the equations.

Reliability of the Method

To test the reliability of the method, several trials were performed. First, a reference frame was built. Three straight steel tubes were set in a configuration so that each exactly intersected the other perpendicularly, the point of intersection being the origin of the general reference frame in Figure 1. On each tube, black and white strips (10 cm wide) were painted (Figure 3). This reference frame then was placed in the background of the photographic field of each camera, the optical axis intersecting in the origin. Several landmarks were measured carefully and their known coordinates were used as control on the coordinates calculated by the proposed method.

The results of this method proved that an accuracy rate of better than 1 cm can be achieved by measuring carefully the image coordinates, especially of the reference points used to define the spatial position of each camera. For this reason, measuring these image coordinates with a relative accuracy of 1/100 seemed to be sufficient. In addition, several controls were incorporated into the method in order to warn the investigatior of possible inaccuracy during data collection.

A comparison is made between this system of three-dimensional recording on film and conventional techniques.

Advantages

First, no measuring procedure on the field is required. Therefore, this method becomes suitable in situations in which it is not always possible to measure needed distances and angles because of lack of time, obstacles in the line of

480   Van Gheluwe

Figure 3. The general reference frame: three straight steel tubes are set in a configuration so that each exactly intersects the others perpendicularly, the point of intersection being the origin.

measurement, and intrusions imposed on athletes and sports event organizers by camera set-ups. Another advantage of three-dimensional recording on film is that it allows quick and easy camera positioning, according to the following procedure.

Step 1. Put the reference frame in the middle of the action area.

Step 2. Set the cameras at those positions that will produce the desired pictures.

Step 3. Align the cameras with the origin of the reference frame; the mark in the viewer indicating its geometric center must coincide with the origin of the reference frame. Use sturdy tripods so that the cameras do not move during the filming.

Step 4. Film the reference frame with both cameras.

Step 5. Remove the reference frame, if necessary.

Step 6. Filming now can begin.

Disadvantage

In order to handle the necessary and complex calculations, a "fast" computer is indispensable. With a C.D.C. 6400, calculation of the life-size coordinates of an arbitrary point in space requires an average of 5 sec for central processing.

CONCLUSION

The proposed three-dimensional filming technique has provided reliable and valid results and is recommended for use in research in biomechanics where for some reason conventional techniques are not applicable.

Author's address: *Bart Van Gheluwe, Vrije Universiteit Brussel, H.I.L.O., A. Buyllaan, 105, Brussels (Belgium)*

Photographic method of analyzing the pressure distribution of the foot against the ground

M. Miura, M. Miyashita, H. Matsui, and H. Sodeyama
University of Nagoya, Nagoya

Most human activities are based on the movement of both legs. Therefore, in order to analyze the mechanics of human activities, it is essential to record the forces exerted against the ground through the feet.

Many investigators have devised various kinds of force platforms for the analysis of the human movements (Elftman, 1939; Fenn, 1930; Murray, Seireg, and Scholz, 1967; Payne, Slater, and Telford, 1968). Most of these methods permitted only the direct measurement of the force exterted on the platform. In recent years, in addition to measuring the force for complete information on the human activity, performances have been filmed on the platform.

In the present study, a simple photographic apparatus was made to identify the force motion relationship during execution of human activities. With this method, the pressure distribution of the foot and the posture were filmed simultaneously.

APPARATUS AND METHODS

A wooden platform was constructed that was 20 m long, 1 m wide, and 1 m high. A section 1 m long and 0.7 m wide was cut from the middle of the platform, and a section of plate glass 10 mm thick was inserted (see Figure 1).

A special rubber mat with an even surface and many small projections on the underside was developed. Each projection was a pyramid 5 mm high and with a base measuring 5 x 5 mm. The mat was placed on the plate glass, under which a mirror was leaned against the wall at an angle of 45°. The plate glass was lighted from the side by a fluorescent light.

When the foot of the subject came in contact with the rubber, the projection was pressed against the plate glass, causing the contacting surface

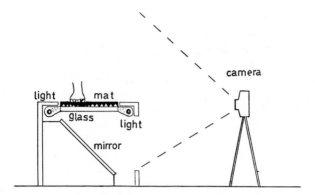

Figure 1. Platform, rubber mat, and mirror.

of the projection with the plate glass to become enlarged. This contact surface was proportional to the magnitude of the applied pressure, which was represented by the gray scale on the surface of a black and white photograph. Therefore, when a man stood on the mat, both his posture and the pressure distribution of his soles could be photographed from the side by a 35-mm motion camera. In the present study, the speed of the camera was 50 frames per sec.

The gray scaled pressure distribution was analyzed through a color densitometry system. By this method, the density values of the photograph of the sole were displayed in colors (blue, light blue, yellowish brown, yellow, red, and pink) on the screen. This picture on the screen then was photographed in color with a 35-mm still camera.

APPLICATIONS

This technique has been used to study the pressure distribution on the soles during a vertical landing (Figure 2), the support phase in running (Figure 3), and the standing long jump (Figure 4). By comparing the pressure distribution patterns with body positions, it is possible to examine the relationship between the ground reaction forces and the motion of the body. For example, during the support phase in running, the pressure increases as the body moves directly over the supporting foot (Figure 3). During the vertical landing (Figure 2), the pressure pattern changes as first the toes and then the ball of the foot make contact, followed by the heel.

CONCLUSION

It is concluded that this apparatus is of considerable potential value in the study of biomechanics even though at present its primary use is in qualitative analysis of ground reaction forces.

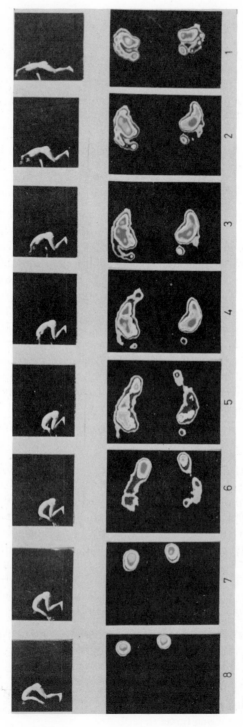

Figure 2. Vertical landing. The subject lands vertically from a chair 50 cm high.

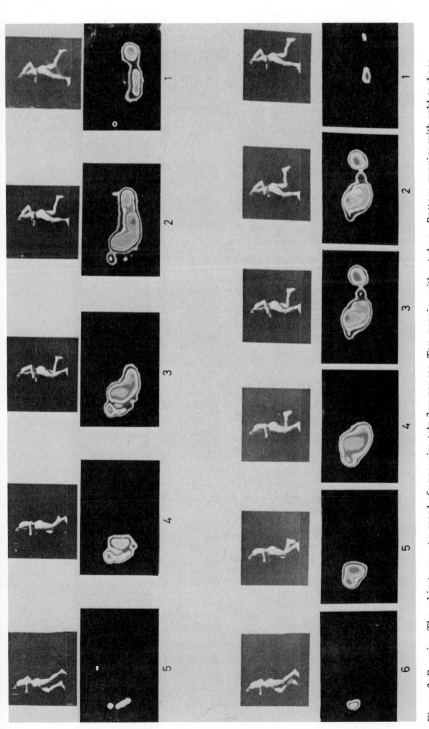

Figure 3. Running. The subject runs at a speed of approximately 3 m per sec. *Top*, running without shoes. *Bottom*, running with rubber shoes.

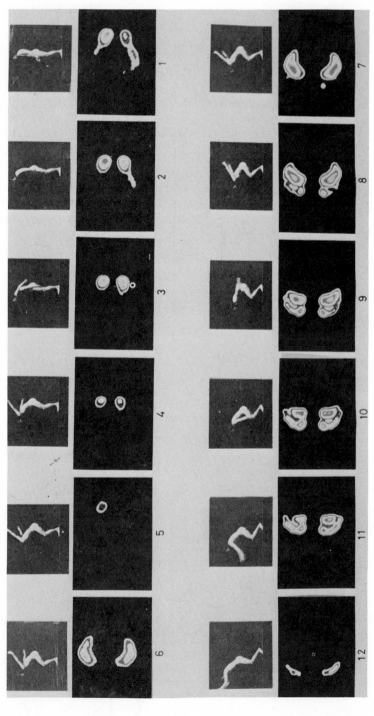

Figure 4. Forward jumping. The subject jumps forward without shoes from the upright position.

REFERENCES

Elftman, H. 1939. Force and energy changes in the leg during walking. Amer. J. Physiol. 138.

Fenn, W. O. 1930. Work against gravity and due to velocity changes in running: Movement of the center of gravity within the body and foot pressure on the ground. Amer. J. Physiol. 93: 433.

Murray, M. P., A. Seireg, and R. C. Scholz. 1967. Center of gravity, center of pressure, and supportive force during human activities. J. Appl. Physiol. 23: 6.

Payne, A. H., W. J. Slater, and T. Telford. 1968. The use of a force platform in the study of athletic activities. A preliminary investigation. Ergonomics 11: 2.

Author's address: *Dr. Mochiyoshi Miura, Department of Physical Education, University of Nagoya, Nagoya (Japan)*

Other techniques

Versatile uses of a computerized graphic tablet system

M. G. Owen and M. Adrian
Washington State University, Pullman

The computerized graphic tablet is an example of hardware, developed for computer graphics systems, that has been utilized in a variety of biomechanics studies to increase the efficiency and accuracy of data collection. Although initially used exclusively for data collection from films, the graphic tablet and its supporting components now are being used to collect data from electrogoniograms, photographs, and drawings or tracings. A description of the components of the system and the techniques for accomplishing these applications are discussed.

The components of the system and their relationship to each other are illustrated in Figure 1. In this particular system, the Graf/Pen (Science Accessories Corp.), the tablet consists of a sheet of glass measuring 14 sq in, the frame of which has strip sensors embedded along one edge and across the top. Touching the stylus to the tablet surface causes the control unit to generate a spark that, on jumping the spark gap in the tip of the stylus, causes fast rise-time soundwaves to be propagated through the air to the sensors. Reception by the sensors stops counters that were initiated with the spark. These counters contain binary numbers proportional to the X and Y distances from the stylus spark to the sensors. The binary numbers are transferred from the control unit to an Interdata miniature computer and converted to decimal numbers that represent the coordinates of the data points touched with the stylus.

The Interdata is controlled by a machine language program that is stored on one track of the cartrifile tape and invoked by typing instructions on the teletype. The program routes the data point coordinates to either a CRT display or the teletype to provide immediate feedback of results to the investigator. The coordinates then are stored on another track of the cartrifile tape.

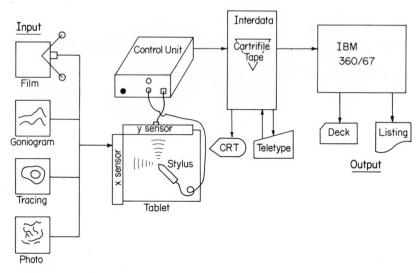

Figure 1. Data collection system.

At the convenience of the investigator, the Interdata is interfaced with the IBM 360/67 computer, and the stored contents of the cartrifile tape are transmitted and concatenated with a program that produces both punched cards and a standard output listing containing the coordinates of the data points collected. The cards are in a form appropriate for submission with any FORTRAN program desired by the investigator. Examples of such programs include those that calculate linear and angular displacement; velocity and acceleration; joint angle; area moment of inertia; etc.

The reliability and validity of the system were established during installation and initial debugging by collecting and then plotting data points for a known function. Each investigator who uses the tablet establishes his own reliability by performing a procedure of repeated measures. Reliability coefficients of 0.99 (Barthels, 1973) and 0.98 (Burkett, 1973) were obtained by two investigators previous to the collection of data points for their respective studies.

When used to collect data for typical biomechanics studies, the graphic tablet is placed in one of two basic orientations, flat on a table or supported in a vertical position by a specially constructed framework. The latter position is used when data are collected from films, as illustrated in Figure 2. The projector and the graphic tablet are mounted on the same framework so that the distance between them remains constant and the angle of projection onto the tablet is always 90°. A stop-action motion analyzer projector is used to project single film frames onto the rear surface of the tablet. Points such as the knee joint, marked on the subject at the time of filming, are the referents for collection of data. The investigator, seated in front of the tablet, touches

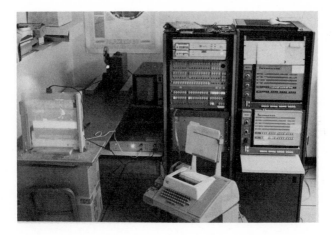

Figure 2. Film data collection.

these points with the stylus at the place where they are projected onto the tablet (Barthels, 1973; Burkett, 1973). This procedure greatly reduces the time required to analyze films.

When the graphic table is placed flat on a supporting surface, point coordinates can be obtained from any kind of paper records. Figures 3–5 illustrate examples from three studies in which this technique was used.

In a study involving the physical properties of rat femurs (Fry, 1973), photographs of cross-sections of the bone were magnified by a factor of 12, and then the outline of the section and the boundary of the cortical area were traced on 10-mm-per-cm graph paper. The tracings were placed on the graphic

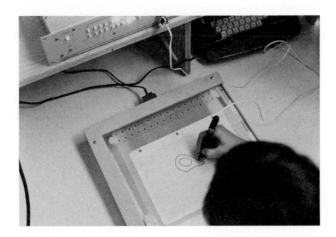

Figure 3. Tracing.

Figure 4. Photograph.

tablet, and the investigator collected the data point coordinates at the intersection of each bone boundary-grid line. In this way, the areas of many small sections of the irregularly shaped bone could be determined accurately. From these small sections, total area and the area moment of inertia for each bone section were calculated by an appropriate computer program.

Currently in progress are studies using the remaining two techniques illustrated in Figures 4–5. Data for a study comparing the segmental method for obtaining center of gravity with the reaction board method are being collected from photographs placed on the graphic tablet. Marked points on the subject then are touched with the stylus in the same manner as that for collection of data from film.

Figure 5. Goniogram.
Figures 3–5. Data collection from paper records.

In another study in progress, the jumping patterns of intercollegiate female basketball and volleyball players are being investigated by placement of electrogoniometers on the hip, knee, ankle, and metatarsophalangeal joints. Output is obtained through the Minneapolis Honeywell Visicorder, Model 1508, with time-line intervals set at 0.01 sec. The goniograms thus produced are placed on the graphic tablet, and each intersection of a time line and the line representing angular displacement of the joint is touched with the stylus. The coordinates collected in this manner then are submitted to the computer with an appropriate program to obtain angular velocity and acceleration from electrogoniometer output.

In conclusion, it can be seen that the advantages of this system are the relative speed with which data can be collected, the accuracy of the collection method, and the variety of data records that can be handled.

REFERENCES

Barthels, K. 1973. Three dimensional kinematic analysis of the hand and hip in the butterfly swimming stroke. Ph.D. dissertation, Washington State Univ., Pullman, Wash.

Burkett, L. N. 1973. Three dimensional film analysis of overarm throwing patterns of Down's syndrome children. Ph.D. dissertation, Washington State Univ., Pullman, Wash.

Fry, R. A. 1973. Effect of exercise on the properties of bone. Ph.D. dissertation, Washington State Univ., Pullman, Wash.

Author's Address: *Dr. Marjorie G. Owen, Department of Physical Education for Women, Washington State University, Pullman, Washington 99163 (USA)*

Analysis by hybrid computer of ground reactions in walking

A. Cappozzo, M. Maini, M. Marchetti, and A. Pedotti
University of Rome, Rome

The aim of the present study was to develop an efficient technique for characterizing human gaits for use in clinical rehabilitation and prosthetic practice. Because impairment of gait occurs frequently, it is of fundamental importance that improved methods and devices for the functional recovery of disabled persons be developed. In this context, the possibility of a workable description of the gait is of primary importance.

Photographic techniques that are used commonly provide extensive quantities of data. Within the scope of this study, however, these techniques are too time-consuming and the resulting data too difficult to interpret. The measurement of the ground reaction (i.e., the resultant of forces attributable to the interaction between feet and ground during locomotion) by means of a force platform may offer a simpler solution to this problem. The ground reaction depends on the dynamics of the whole body and represents a highly significant synthesis of it. Thus, it has all of the properties of a sensitive determinant of the locomotor act and can be measured easily with a high degree of accuracy.

On the basis of these considerations, this study deals with a characterization of human gait through the information provided by the measurement of the ground reaction. For this purpose, it is of great importance that the accumulated information assume a form that is easy to interpret. The quantities directly obtainable as functions of time by the force platform are the vertical and horizontal components of the ground reaction acting on one foot.

In a recent study (Jacobs, Storecki, and Charnley, 1972), an analysis of the vertical component in normal and pathological gaits was performed. Further and more significant information can be gained if such an analysis also takes into account the horizontal component and the displacement of the point of application. As described, the displacement of the point of application of the ground reaction acting on one foot can be evaluated by simple computation of the output of the force platform.

The more efficient representation of all of these quantities is a vector representation (Contini, Gage, and Drillis, 1964; Elftman, 1939). To each stride, two vector diagrams, one for each foot, are matched. In order to evaluate the information obtained and the practical efficiency of this representation, a number of typical and atypical gaits have been examined.

METHODS

Steady-state level walking with constant mean direction of progression was selected for analysis. The present investigation deals with the components of the ground reaction force lying in the plane of progression. These components are considered to be the most significant ones (Bresler and Frankel, 1950) and permit their representation in a single plane.

The subjects walked at a constant speed on a wooden track 20 m long, with the force platform located two-thirds of the distance along the track. The force platform (Kistler, Type 9261A) contained four piezoelectric transducers, each of which measured the three orthogonal components of the force acting on it.

Only the vertical and horizontal components lying on the plane of progression were taken into account. The electronic apparatus connected to the platform (Kistler charge amplifier, Type 5001) had three output voltage signals referring to:

$$F_h = F_{h1} + F_{h2} + F_{h3} + F_{h4}$$
$$F_{va} = F_{v1} + F_{v2}$$
$$F_{vp} = F_{v3} + F_{v4}$$

where the subscripts *1, 2, 3,* and *4* refer to the four transducers, and *a* and *p* refer to the anterior and posterior pair of transducers respectively, with reference to the direction of progression. These signals were recorded on a magnetic tape and then analyzed by a hybrid computer (EAI 8945 of the CRA, ENEL). The relative outputs were:

1. The horizontal component F_h.
2. The vertical component F_v, obtained by

$$F_v = F_{va} + F_{vp}$$

3. The point of application of the ground reaction x, obtained by

$$x = \left[\frac{F_{vp}}{F_{va} + F_{vp}} \right] 1$$

where 1 is the distance between the anterior and posterior pair of transducers. 4. The vector, the components of which were F_h and F_v, was displayed on a persistent-image oscilloscope screen for sampled instants of time (where the intervals of time were constant), representing the actual geometric relationship with the platform (i.e., angle of slope and point of application). This vector diagram was obtained by a special program implemented on the computer. The horizontal straight line that appears in the photographs represents the actual distance between the anterior and posterior pair of transducers with respect to the force vectors. The three functions of time, F_h, F_v, and x, were recorded by a plotter, and the vector diagram was photographed (Figure 1). The whole computer program was designed in such a way that it eventually could be implemented using the hardware of one compact, relatively inexpensive unit.

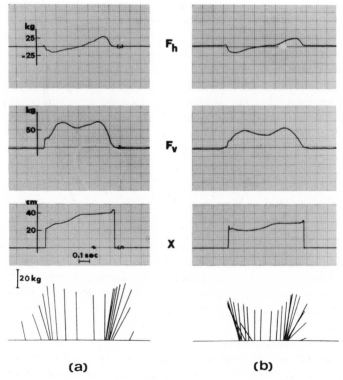

Figure 1. Plots of horizontal (F_h) and vertical (F_v) components, and of the abscissa of the point of application of the ground reaction (x), vector diagram (one vector every 20 msec). (a) Normal female barefooted subject (C. S.); (b) handicapped (drop-foot) female barefooted subject (L. P.).

The subjects, 29 normal and 2 handicapped persons (22 males and 9 females), performed a total of 223 steps. All normal subjects were tested with and without shoes while walking at normal speed. Six women wore "platform shoes" (types of shoes currently in fashion), whereas all the other subjects wore typically light shoes. Five normal male subjects also were examined at speeds ranging from 84 to 140 steps per min, and one subject (an athlete) was examined while running. Also tested were an elderly poliomyelitic woman wearing orthopedic shoes and a girl affected by peripheral nerve damage caused by intossication (drop foot) who wore shoes with an orthopedic Codivilla spring. The latter subject also was examined without shoes.

RESULTS

The first problem studied was the choice of the most meaningful ground reaction data and the most effective means of presenting these results to meet the following objectives: (a) the possibility of identifying characteristic intra-individual patterns, and (b) the possibility of identifying a characteristic interindividual pattern for a specific population.

The populations considered in this work are characterized by the following factors: sex, efficiency of the locomotor apparatus, type of shoes (or barefooted), and speed of gait. In Figure 1, the plots of the vertical (F_v) and horizontal (F_h) components and of the abscissa of the point of application (x) versus time, one belonging to a normal female subject (Figure 1, a) and one to a handicapped female subject (Figure 1, b), both barefooted, are compared. It is clear that the x-versus-t factor is the only one showing the obvious difference between the two gaits. The plot belonging to the disabled individual, in fact, shows that the point of application "comes back" (something that never happens for barefooted, normal subjects).

It must be emphasized that a further elaboration of the functions $F_v(t)$ and $F_h(t)$ (i.e., a harmonic analysis) still does not permit a clear distinction between the two gaits. The information required in these functions is better emphasized in the more compact vector diagram, and to this extent the vector diagram is much more expressive.

In Figure 1, two vector diagrams belonging to the two subjects cited previously are shown. In these diagrams, the density of the vectors is immediately readable as the speed with which the point of application moves, and the eventual "coming back" of the point of application also is easily readable. The slope and magnitude of the vectors also contain useful information. In order to emphasize the efficiency of the vector diagram from the point of view of our aims, tests within the same population and tests belonging to different populations were compared. The conclusions given can be understood as general and having been derived from a number of homogeneous tests.

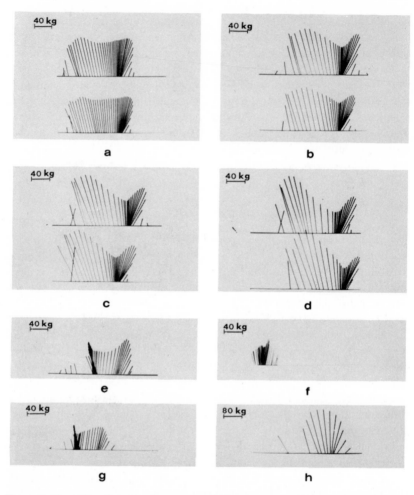

Figure 2. Examples of vector diagrams. *a–d*, Normal male subject (A. P.) walking at different speeds (84, 108, 120, and 140 steps per min, respectively). *Top*, wearing light shoes; *bottom*, barefooted. *e*, Normal female subject (C. S.) wearing "platform shoes." *f*, Elderly polioymelitic female (Z. B.) wearing orthopedic shoes. *g*, Handicapped (drop-foot) female (L. D.) wearing orthopedic shoes. *h*, Running stride of an athlete. *a–g*, One vector every 20 msec; *h*, one vector every 15 msec.

In Figure 2, *a, b, c,* and *d,* patterns belonging to a normal male subject wearing light shoes or barefooted are shown. Light shoes do not affect the pattern except at the very beginning and the very end of the support phase. Modifications of the pattern with respect to the speed of progression are evident. The increasing of the inertial forces with speed results in increasingly larger variations of the vector magnitudes.

The differences between normal male and female patterns can be determined from Figure 1, *a* and Figure 2, *a, b, c,* and *d* (both groups of subjects were barefooted). The female pattern, unlike the male one, shows a character-istic closeness of the vectors and sometimes the point of application at the beginning of the support is stationary. The patterns shown in Figure 1,*a* and Figure 2,*b* are typical patterns of normal barefooted adults walking at a natural speed.

The smoothness of the progression of the vectors should be noted; these patterns are the image of a movement perfectly carried out despite its great complexity. The first vector from the left appears in most normal patterns facing the direction of motion; it is due to the heel strike. In Figure 2, *e,* a typical pattern belonging to a woman wearing platform shoes is shown. At the beginning of the support, there is a striking coming back of the point of application of the vectors. The patterns belonging to pathological conditions are easily recognizable when compared with the normal patterns described previously. The elderly poliomyelitic subject produced the pattern shown in Figure 2, *f*. Patterns of Figure 2, *g* and Figure 1, *a* are for the subject with drop foot, with and without shoes, respectively. Even if a slight esthetic improvement in the appearance of the gait was observable, it can be seen (by inspection of the relative patterns) that while the subject wore flat shoes with the Codivilla spring instead of being barefoot, a normal pattern was not reached. The pattern shown in Figure 2, *h* is for a running stride.

CONCLUSIONS

The vector diagram representation of the ground reaction force provides an efficient description of human gait. It can be obtained in real time at a relatively low cost for instrumentation. In this study, a comparison is made between a normal pattern and other patterns modified by various factors. On the basis of these encouraging results, future testing will include different populations, especially those with pathological gaits, to establish typical patterns for each group.

REFERENCES

Bresler, B., and J. P. Frankel. 1950. The forces and moments in the leg during level walking. Trans. ASME.

Contini, R., H. Gage, and R. Drillis. 1964. Human gait characteristics. *In:* R. M. Kenedi (ed.), Biomechanics and Related Bio-Engineering Topics. Perga-mon Publishing, Elmsford, N.Y.

Elftman, H. 1939. Forces and energy changes in the leg during walking. Amer. J. Physiol. 125: 339–356.

Jacobs, N. A., J. Storecki, and J. Charnley. 1972. Analysis of the vertical component of force in normal and pathological gait. J. Biomech. 5: 11–34.

Author's address: *Dr. A. Cappozzo, Instituto di Fisiologia Umana, Università di Roma, 00100 Roma (Italy)*

A force platform system for biomechanics research in sport

A. H. Payne
University of Birmingham, Birmingham

The author has reported previously on the requirements for a suitable force platform for use in biomechanics research in sports (Payne, 1968). Briefly, it is necessary that the platform be rigid, light, and very stiff and have a natural frequency that is not excited by the body forces under investigation. There should not be any significant "cross-talk" between axes of measurement, and recording must ensure that results are not modified by, for example, high-inertia recording pens. Of equal importance is the requirement that experiments be carried out, as much as possible, under natural conditions for the particular movement under study.

In recent years, the technology of force platform design has improved to the point that complete units can be purchased commercially. These satisfy the requirements of rigidity, lightness, natural frequency, and electronic reliability, but the models presently available are rather small (40 x 60 sm) for use in sports and measure only three orthogonal components of force. Many researchers continue to carry out experiments in small rooms or restricted spaces with platforms positioned above the floors, thus necessitating built-up runways.

FACILITY AT THE UNIVERSITY OF BIRMINGHAM

In order to obtain conditions as nearly ideal as possible, a double force platform was constructed in an outdoor sports field ("Redgra" composition) at the University of Birmingham (see Figure 1). This platform consists of a concrete-lined pit capable of containing either one or two square force platforms measuring 2 ft, 6 in (76 cm) and positioned side-by-side in such a

This work was made possible with a grant from the Science Research Council.

way that their top surfaces are at ground level. These top surfaces are detachable so that metal, wood, or rubber compound surfaces may be used. A run-up strip of Tartan rubber compound (3M Co., St. Paul, Minn.) is available for use by athletes accustomed to this type of surface. Each force platform can be moved quickly in and out of the pit, by one operator, via a ramp, on a small hand-hydraulic lift truck.

Figure 1. The force platform system at the University of Birmingham. *Top:* the platforms are bolted onto concrete plinths in the pit. *Bottom left:* the pit is covered completely, and the tops of the platforms are at ground level. *Bottom right:* continuous-motion clock synchronizes film and trace.

504   Payne

The identical platforms are of a modified Cunningham and Brown strain-gauged cylindrical post construction (Cunningham and Brown, 1952), i.e., a 6-in (15-cm) thick aluminum honeycomb platform in each case is supported on a massive cast-iron frame by instrumented Dural posts, one at each corner (see Figure 2). The strain gauges are connected in a 3-KHz AC Wheatstone Bridge network, and the out-of-balance signals are registered by

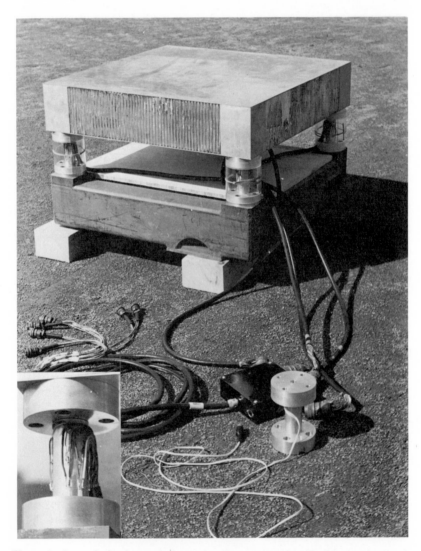

Figure 2. One of the force platforms standing on wooden blocks, showing in the foreground the "dummy" post which contains part of the Z-bridge. *Inset:* close-up of a supporting post. Note the gap at the top of the central cylinder. Contact with the top flange is confined to a 0.4-in diameter surface (see Figure 3).

means of a mirror galvanometer ultraviolet recorder onto special photographic paper.

The strain gauges are attached so that each platform is capable of
measuring forces in all six modes, namely, vertical (Z), two horizontals (X
and Y), and moments of force (M_x, M_y, and M_z) about these axes. Cross-talk
between vertical and horizontal axes is less than 3 percent, although a
correction is necessary when reading M_x and M_y values in the presence of
horizontal forces, inasmuch as the appropriate strain gauges are about 21 cm
below the top of the platform. In early tests, it was found that bending of the
aluminium honeycomb platform, although very small (0.025 mm at the
center when loaded with a subject weighing 220 lbs or 100 kg) in vertical
loading, produced unacceptable cross-talk in the horizontal modes. This
problem was overcome by the use of semiflexible connections between the
posts and the platform (see Figures 2 and 3).

The natural frequencies of the platforms were measured by connecting
them in turn to a mechanical vibrator driven by a variable-frequency oscillator. Comparison of the phases of input forces and the output from an
accelerometer attached to the platform top showed that the lowest structural

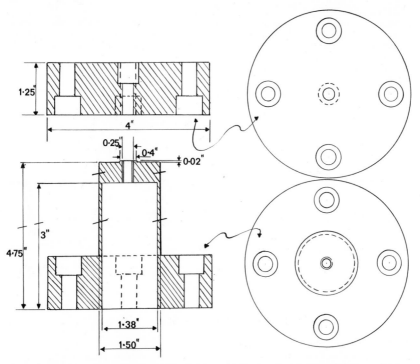

Figure 3. Sketch of one support post (not to scale) showing how semiflexible connection is
achieved with top flange by means of a 0.4-in diameter projection. The top is bolted to the
central cylinder through this projection (design was in Imperial measurements).

frequency was in the horizontal bending mode. Values were 164 and 168 Hz, with harmonics recorded at over 400 Hz.

Certain movements by some subjects, e.g., high-speed running over the platforms, give rise to troublesome oscillations when these lower frequencies are excited; and, although these are recognized easily, they do represent a definite limitation at present. This is the price that must be paid for the advantage of having all the strain gauges on just four supporting posts in each platform. Damping systems and electronic filters to alleviate the problem are being considered.

Calibrations have been made by static loading, with dynamic checking of Z, X, and Y calibrations from cine-analysis of standing broad jumping.

USE OF THE FORCE PLATFORMS

In order to relate body movements with the force-time traces they produce, it has been necessary, at least in the first few recordings of a particular

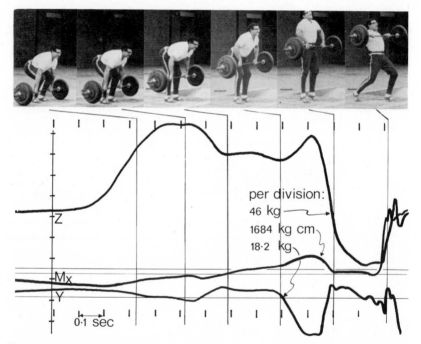

Figure 4. The weightlifting clean on one platform only: 125 kg on bar; body weight, 105 kg. Forces in kg, and moment of force in kg-cm, are shown on vertical scale against time in 0.1-sec divisions along the horizontal.

A moment of force downward at the front of the platform shows positive. Y is the horizontal force at right angles to the bar and shows negative when the force is directed to the front of the weightlifter.

movement, to record the movement on cine film. A clock in the field of view was wired to produce pulses on the trace for later synchronization of film and trace.

The facility at Birmingham University has been used so far to determine the feasibility of experiments with various movements in sports. For instance, a study of weightlifting (see Figure 4) appears to have excellent possibilities as far as mechanical measurements are concerned and requires only one platform. On the other hand, the complex shifts of body weight into the tennis serve and the lack of any significant transmission of the actual ball-to-racket impact through the body to the ground seem at present to limit the use of the force platform results to that of force-trace pattern comparisons (see Figure 5). However, simple observations of trace patterns can be very rewarding; for example, the force trace of the rear foot thrust in shot putting (see Figure 6) shows the sliding of the foot before and after the drive by that leg, and the front foot is seen to drive in a direction opposite to the direction of the throw (see Figure 7).

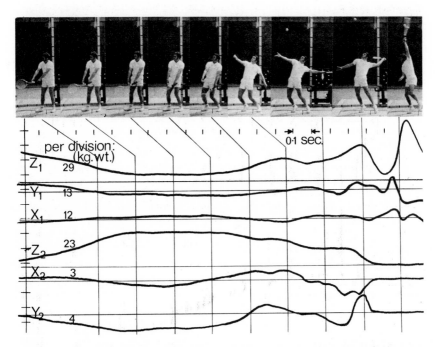

Figure 5. The tennis serve: front foot on Platform 1 and rear foot on Platform 2 (number suffixed to letter mode indicates platform). Y_1 and Y_2 refer to horizontal forces in the line of ball flight and are positive when directed along that line. X_1 and X_2 are horizontal forces at right angles to the line of ball flight and are positive when directed toward the camera. The player is of British county standard.

508   Payne

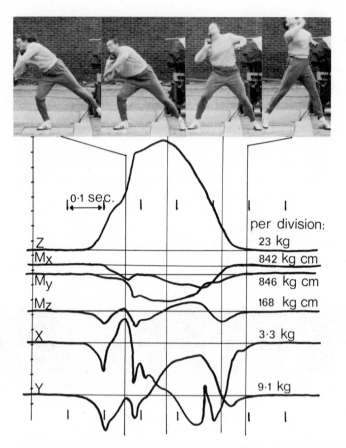

Figure 6. Rear foot action in shot putting. M_x shows positive when moment of force is downward to front of platform. M_y shows positive downward on side nearest camera. M_z is positive anticlockwise when looking down on platform from above. X is positive when a force is directed toward the camera. Y is positive when a force is applied opposite the direction of throw.

ACKNOWLEDGMENTS

The author wishes thank Dr. J. Dunn, Dr. D. Anderson, and Mr. S. K. Joshi for their help with this project.

REFERENCES

Cunningham, D. M., and G. W. Brown. 1952. Two devices for measuring the forces acting on the human body during walking. Proc. Soc. Exp. Stress Anal. 9: 75–90.
Payne, A. H. 1968. The use of force platforms for the study of physical

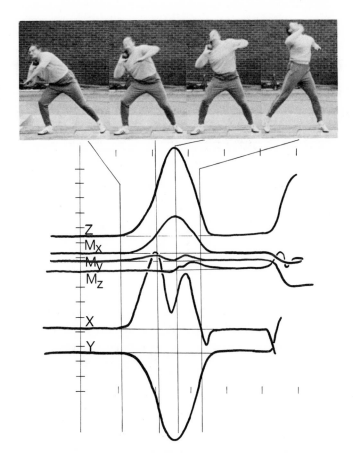

Figure 7. Front foot action in shot putting. For simplicity in showing the traces, this is a different throw from that in Figure 6. Both throws were by a female athlete and in the region of 44 ft or 13.40 m. Calibrations are the same as in Figure 6.

activity. *In:* J. Wartenweiler, E. Jokl, and M. Hebbelinck (eds.), Biomechanics, pp. 83–86. S. Karger, Basel, and University Park Press, Baltimore.

Author's address: *Dr. A. H. Payne, Physical Education Department, University of Birmingham, Birmingham B15 2TT (England)*

An optoelectronic instrument for remote on-line movement monitoring

L. E. Lindholm
Chalmers University of Technology, Göteborg

The usual method of recording body movements is high-speed filming. A disadvantage of this method is that the results from the filming are not immediately available for analysis. The recorded film first must be developed, and then a series of measurements must be performed on each individual film frame. Filming for 10 sec at a speed of 1,000 frames per sec requires that 10,000 frames be processed!

METHODS

At the Department of Applied Electronics of Chalmers University of Technology in Gothenberg, Sweden, a new instrument for on-line recording of movements has been developed. The instrument has been named SELSPOT from *Sel*ective *Spot* Recognition.

In this system, a special position-sensitive photodetector is used. The detector provides an electrical output signal specifying the x and y positions of an input light-spot signal that is relative to fixed internal coordinates. This detector was developed in 1957 by Professor T. Wallmark, now at the Department of Electron Physics III, Chalmers University of Technology. By using this detector together with time- or frequency-multiplex technique, the position of several light sources can be determined simultaneously. The frequency response of the detector is about 1 MHz, which theoretically permits the simultaneous use of, for example, 1,000 light sources, with each light source having a light flash frequency of 1,000 Hz.

As light sources, small light-emitting diodes (LED's) are used (see Figure 1). These diodes provide extremely short pulses of infrared light; they are

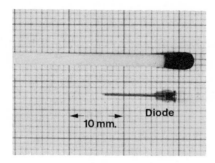

Figure 1. Light-emitting diode.

affixed to the part of the body that is being measured, for example, a moving limb. The picture of the LED's is projected on the detector by means of an optical lens system. The output signal from the detector is fed to a special analog signal processor. This processor contains noise-suppressing and linear-izing circuits. By this means, the noise is suppressed to about 60 dB, and the deviations from linearity are decreased from about 60 to 0.5 percent.

The outputs from the system, specifying the x and y positions of each diode, are available as analog signals or as binary numbers. Observe that each SELSPOT channel contains two signals, one x and one y. Recording of 10 SELSPOT channels thus requires a 20-channel recorder. For a large number of SELSPOT channels, it is advantageous to feed the signals directly into a computer instead of first recording them.

In Figure 2, there is a picture of the detector housing with associated optics (Canon lens f/50; 1:0.95). It is not necessary to focus sharply the picture of the LED's on the detector; a blurred picture will do as well. The complete SELSPOT system is not sensitive to stray light from the surround-ings. "Light noise" is suppressed in the processing unit.

Figure 2. Detector housing and optics.

The resolution of the instrument is one part in 10,000 (the limit is set by the signal-to-noise ratio of the signal processor and not by the detector). Deviation from linearity is about 0.5 percent. The range of the instrument (the maximal distance between the detector and the LED's) is about 10 m with the Canon lens but it can be increased easily by using more powerful LED's.

The instrument now in use at Chalmers contains 32 channels but can be expanded easily to 128 channels by using a plug-in system. Each channel is equivalent to high-speed filming at 1,000 frames per sec.

APPLICATIONS

The SELSPOT can be used in a variety of applications. In the medical field, for example, movements of all kinds can be recorded. At Chalmers, recordings have been made of movements of limbs, prostheses, eyeball, eyelid, vocal cords, respiration, and head positions of automobile drivers.

One interesting application is the early detection and classification of disturbances of movements that can occur in early stages of some diseases, for example, Parkinson's disease. Such disturbances might be too small to be detected by the human eye but can be revealed by the SELSPOT in cooperation with a pen recorder or computer. Also, the progress of the disease might be followed and the results of therapy measured in absolute values.

One possible application for the SELSPOT might be in aids for the handicapped. For example, SELSPOT can provide an "eye" function. Also, in industry, SELSPOT can be used for precision measurements and control of machinery. It also might be used as an "eye" for an industrial robot.

Author's address: *Dr. L. E. Lindholm, Department of Applied Electronics, Chalmers University of Technology, S-402 20 Göteborg (Sweden)*

Analysis and synthesis of body movements utilizing the simple n-link system

A. Dainis
Duke University, Durham

Analysis procedures and computer programs have been developed for certain mechanical systems to determine the forces responsible for recorded motions, e.g., by Plagenhoef (1966, 1968). However, the investigation of variations of an action essential in optimization calculations would be difficult to carry out using this approach. A mathematical model based on the equations of motion containing complete information regarding the dynamics of movements allows mathematical procedures for movement synthesis as well as analysis. It also presents other useful possibilities.

One purpose of this study was to investigate the usefulness of a computer program that can synthesize movements by calculating the motion resulting from an input of forces representative of muscular actions. A related purpose was to look for a method that would yield complete information regarding the dynamics of movements and yet not require the high-quality equipment and the considerable labor involved in a conventional film analysis.

METHODS

A basic mechanical model of the human body consists of a set of rigid segments or links of fixed length and mass distribution connected by hinges that allow movements in the appropriate directions. In this study, the model was restricted to movements in the two-dimensional sagittal plane, and further simplification was obtained by considering only symmetrical move-

This work was supported by and partially carried out in the Department of Physical Education, University of North Carolina, Chapel Hill, N. C., and served as basis of an M.A. thesis in that department.

ments about that plane, in which case both upper and lower expremities could be treated as single segments. When this system of links is supported at a fixed point by one extremity, it is capable of describing a considerable number of gymnastics movements on the horizontal bar, parallel bars, and rings, as well as some movements carried out with the support on the feet.

Figure 1 shows a four-link system and indicates the notation that will be used. Starting from Lagrange's equations of motion and using the intersegment angles ϕ as the independent coordinates, for a system of n links the following n coupled differential equations result:

$$\sum_{i=1}^{n} m_i \sum_{s=1}^{i} \sum_{m=1}^{i} \frac{\partial^2 \bar{r}_i}{\partial \phi_s \partial \phi_m} \cdot \frac{\partial \bar{r}_i}{\partial \phi_k} \phi_s \dot{\phi}_m + \sum_{s=1}^{n} \sum_{i=k}^{n} (m_i \frac{\partial \bar{r}_i}{\partial \phi_s} \cdot \frac{\partial \bar{r}_i}{\partial \phi_k} + I_i) \ddot{\phi}_s$$

$$= -T_k - g \sum_{i=1}^{n} m_i \sum_{m=k}^{i}{}' s_m \cos \theta_m \quad (k = 1, \ldots, n) \qquad (1)$$

The quantities m_i, I_i, and α_i denote the mass, moment of inertia about its center of mass, and the ratio of the distance of the center of mass to the total length of the ith segment, respectively. The torque T_k exerted by the $(k-1)^{th}$ segment on the kth segment is measured clockwise as positive, $\bar{r}_i$ is the vector from the point of support to the center of mass of the kth segment, g is the gravitational acceleration, and the prime on the summation sign indicates that the last term of the sum is to be multiplied by α_i. These equations express at any instant the relationship between the positions, (ϕ), angular velocities $(\dot{\phi})$, and accelerations $(\ddot{\phi})$ of the segments and the torques acting between them.

Two computer programs were developed utilizing APL/370 programming language, and the computations were carried out on an IBM/370 computer accessed by telephone line from an IBM 2741 terminal. This form of direct communication with the computer has the advantage of allowing data to be entered as the computation proceeds, with any desired output being obtained before the next step of the calculation is initiated.

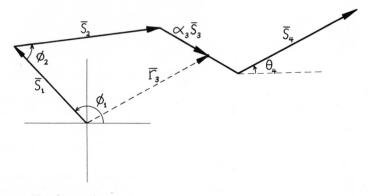

Figure 1. The simple link model.

The synthesis program accepts as input the torques acting between the segments and the time increment during which the torques are applied. The resultant motion is obtained by integrating Equation 1 over the time increment using a three-point predictor-corrector method (Milne, 1953) modified to use unequally spaced abscissas. After each entry, the new state of the system described by the angles (ϕ) their velocities ($\dot{\phi}$), and accelerations ($\ddot{\phi}$) is printed out. By adjusting the torques and observing their effects at sufficiently small time intervals, it should be possible to reproduce specific actions and thus to investigate their dynamics. Testing the practicability of this method was one of the aims of this study. One of the options offered by this program is the input of the accelerations ($\ddot{\phi}$) in place of the torques. This results in the output of torques required to produce the accelerations.

The analysis program accepts as input the configuration of the system described by the angles (ϕ) and outputs required by the torques to bring the system to this configuration from the preceding one. The time intervals separating successive configurations either can be entered or calculated by the program for *time-dependent movements* in which the entire system is swinging under the action of gravity. The torques are very sensitive to errors in the input angles, and some method must be used to reduce this effect. In the method used by Plagenhoef (1968), the data points for each joint were fitted simultaneously over the entire movement by a polynomial curve to smooth out irregularities caused by experimental errors. This required the input of all data initially, and the programs could be used only with equal time increments.

The analysis program described here calculates the torques producing the next-to-last input body configuration by approximating the motion between the last three entered sets of angles by quadratic functions of time. This constant acceleration approximation averages the motion over three consecutive body positions, reducing the effects of individual errors, and results in acceptable torques from data that have been smoothed graphically by hand. This technique allows the use of unequal time increments and the analysis of time-dependent movements without a prior knowledge of the time increments between successive body configurations. The time scale of a time-dependent movement is calculated from the pendular motion of the system by integrating along the path of the center of mass of the system and requires the knowledge of the torque exerted by the point of support on the first segment.

RESULTS

Initial use of the synthesis program has indicated that the reproduction of specific movements using torques as input is difficult, and perhaps not practical as an unsupported technique. This is caused by the strong interdependence of the action of the torques and the inability to predict the effects of velocities and inertias of the segments. However, the somewhat trivial example of a swing with a relaxed body is treated easily because the torques are known beforehand. The input of accelerations in place of torques offers the advantage that the input quantities are not interdependent in their actions, but, even so, difficulty was encountered. It is quite possible that, with practice and familiarity with force patterns, this method could become practical.

The analysis program with the calculation of time intervals provides an easy method for examining the forces responsible for many movements and readily allows experimentation with different techniques of execution. Using a three-link system, the program was applied to two different giant swings done on the horizontal bar. Figure 2, *a* shows a swing done with excellent technique taken from a diagram published by George (1968), and Figure 2, *b* shows the segment positions of a poorly executed giant swing obtained from film. The intersegment angles from these diagrams were graphed and visually fitted with smooth curves. These graphs enabled interpolation for angles occurring between the body positions given by the diagrams, and the readings were used directly as input.

Figure 3 shows the torques responsible for producing the two giant swings. Frictional forces at the point of support were neglected, i.e., $T_1 = 0$. This should be a good approximation but would not be necessary if the frictional forces were known. The same physical parameters for the model were used in both analyses, and both systems started from the vertical straight-line configuration with a velocity of $-75°$ per sec about the support. Giant swing (*a*) required 1.87 sec for completion and finished with a center of

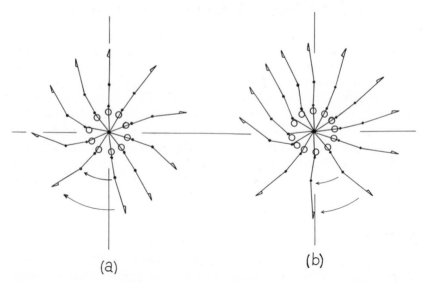

(a)   (b)

Figure 2. Giant swing with (*a*) good technique and (*b*) poor technique.

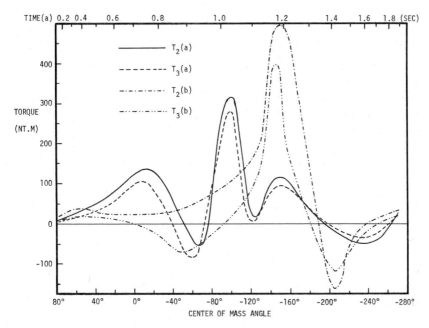

Figure 3. Torques at the shoulder and hip joints responsible for the giant swings plotted against the angular positions of the center of mass of the system. The time scale refers only to giant swing (*a*).

mass velocity of −101° per sec, whereas giant swing (*b*) took 1.75 sec to complete and arrived with an angular velocity of −151° per sec. Much interesting information can be inferred from the graphs in Figure 3. Any other variations of the swing could be similarly analyzed and compared.

CONCLUSIONS

Solutions of the equations of motion of a biomechanical system representing the human body can be very useful in investigating the dynamics of movements and allow the examination of motions under precisely controlled conditions that are impossible to achieve in real performance situations. Because the simple link model is not very demanding computationally and can be used readily to investigate a considerable number of movements, its results should serve as a valuable guide for the treatment of more complex models.

REFERENCES

George, G. S. 1968. A second look at swing. Mod. Gymnast 10: 36.
Milne, W. E. 1953. Numerical Solutions of Differential Equations. John Wiley & Sons, New York.

Plagennoef, S. C. 1966. Methods of obtaining kinetic data to analyze human motions. Res. Quart. 37: 103–112.

Plagenhoef, S. C. 1968. Computer programs for obtaining kinetic data on human movement. Biomechanics, pp. 221–234.

Author's address: *Dr. Andrew Dainis, Department of Physical Education, University of Maryland, College Park, Maryland 20742 (USA)*

Globographic analysis of human posture

J. H. Prost
University of Illinois, Chicago

For this study, "posture" was defined as a completely static state. Two samples of data were analyzed. One was a collection of photographs showing people taking relaxed poses under a variety of environmental circumstances. The other was a collection of photographs of two subjects taking expressive poses under three specified conditions: standing, posing on a chair, and posing on the floor. The postures were analyzed by use of the globographic method to quantify and describe the spatial configurations of the limbs. The purpose of the study was to discover the structuring principles that people use to compose their postural displays.

METHODS

The spatial configurations of the limbs were analyzed by the globographic method (Albert, 1876; Dempster, 1955; Strasser and Gassmann, 1893). With this method, the body is divided into a series of links or segments. The spatial relations between the links are described by angular measurements. The links used here were: right and left links for the (a) upper arm, (b) lower arm, (c) hand, (d) upper leg, (e) lower leg, (f) foot, and (g) a single, undivided link for the torso. The angular measurements for these links indicated the bony relations at the compound shoulder, elbow, and wrist joints, hip, knee, and ankle joints, and rotational positions of the upper arm, lower arm, and upper leg (Prost, 1974 a).

For the expressive postures, two subjects were asked to perform various emotional poses, e.g., happy, sad, angry, afraid, surprised, under three conditions: standing, sitting on a chair, and posing on the floor. The floor poses took three generalized forms: kneeling, sitting, and lying. The chair for both subjects was the same, i.e., a straight-backed dowel chair. The performances of the subjects were photographed so that the slides could be used to determine the globographic values for each pose.

The sample of environmental varieties was collected to illustrate postures taken under a variety of environmental conditions, i.e., with numerous kinds of chairs, sofas, etc. The photographs were selected from a number of commercially available magazines. Since the photographs had not been shot under controlled conditions, some of the globographic values could not be retrieved, e.g., rotational values.

The globographic values for the two samples were displayed in the "postural" field produced by the conjoint Euclidean arrangement of the variables. The field was multidimensional and continuous throughout. Each pose, specified by its globographic values, was a point locale within this field. The arrangement of the points was analyzed to discover the principles that organized the distributions (Prost, 1974a, b).

RESULTS

Environmental Varieties

The sample showed some of the postures that people take under a variety of environmental conditions. It was organized in the field as a continuous display. There were no "natural," or mathematically separable, subgroups. There were two areas of high density, representing poses of "standing" and "sitting," but the poses from these areas intergraded in such a way as to form a continuum between the two clusters. The continuous nature of the display showed that words such as "standing" and "sitting" can have only statistical definitions; they are not referents for mutually exclusive phenomena.

The variety of postures clearly were biomechanical responses to the variability of the environments. The postures could be shown to be stable and relaxed limb configurations adapted to the objects used for support.

Expressive Postures

Two subjects assumed, as best they could, expressive poses, e.g., happy, sad, afraid, angry, surprised. The poses were taken under three conditions: standing, sitting on a chair, and posing on the floor. The floor poses resulted in several variants, i.e., kneeling, floor sitting, and lying.

Figure 1 shows the display of the poses. They clustered around five centers: standing, chair sitting, floor sitting, kneeling, and lying. The standing and chair-sitting centers were coincident with the high-density areas of the environmental varieties sample. Point locales for each of the centers were determined by statistical modes. The modal values for all standing poses specified the "standing mode;" modal values for all chair-sitting poses specified the "chair-sitting mode," etc. In Figure 1, the poses and the "modal" centers for floor sitting, kneeling, and lying are plotted in terms of their linear distance from the standing mode (ordinate) and chair-sitting mode (abscissa). In the multidimensional field, the poses actually surround each of the "modes" like spokes around the hub of a wheel.

The expressive poses are grouped into extended, or elongated, forms radiating from the neighborhoods of each of the modes. Each mode has its own group for each expressive type, i.e., happy standing, happy chair sitting, happy floor sitting, sad standing, sad chair sitting, sad floor sitting, afraid standing, etc. Figure 2 is the same as Figure 1, with lines drawn to connect

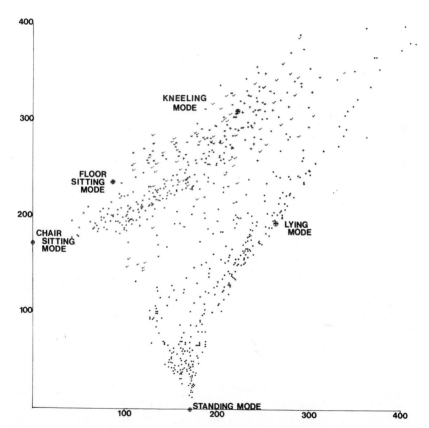

Figure 1. Two-dimensional graph showing the general distribution of 722 cases of expressive postures for two subjects. Five "modal" attitudes (see text) are indicated by *large circled crosses.* Each case is plotted with different symbols for the five generalized postural attitudes: *dots* for "standing," *plus signs* for "lying," *circles* for "chair sitting," *alphas* for "floor sitting," and *checks* for "kneeling." The cases are plotted using the linear distance from the chair-sitting mode (abscissa) and the linear distance from the standing mode (ordinate). Sixteen cases, falling beyond the righthand corner, have been eliminated for convenience.

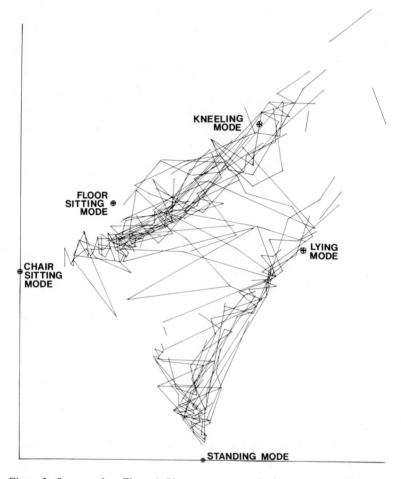

Figure 2. Same graph as Figure 1. Lines connect cases in the same expressive group (e.g., cases of happy standing, sad standing, happy chair sitting, sad chair sitting, etc.). Ungrouped cases have been eliminated (about 40 percent of the sample). The graph shows the general orientation of the groups. Some cases enter into two or more groups; these are the cases that are connected by more than one or two lines.

cases belonging to the same emotional group. Note that the groups extend away from the modes and that some groups cross between modes to create interlocking groups. Some groups around each mode also interlock with one another through what might be called "ambiguous" postures, e.g., a pose that is both "surprised" and "happy," etc. The extended nature of the groups is correlated with intended emotional intensity, i.e., "slightly happy" is closer to a mode than is "very happy". The modes appear to be formalized expressions of the generalized types; for example, the standing mode is a very formal, military form of standing; the chair-sitting mode is a straight-backed, legs-together, "proper" form of sitting.

CONCLUSIONS

The postural field is organized by two factors, the biomechanics of positioning the body in particular environmental situations and expressive adjustments within the biomechanical limitations. It was impossible to determine what basic rules were used to compose the various emotional patterns, i.e., instinct, culture, etc. The expressive groups for both subjects were in some cases similar and in others unexplainably different (i.e., contrast in habits).

The use of the globographic method and the concept of the postural field showed that postural varieties are related through a continuum and that expressive modalities are organized within the continuum around formalized centers.

REFERENCES

Albert, E. 1876. Zur Mechanik des Huftgelenkes. Med. Jahrb. Gesellsch. Artzte. Wien. 107–129.
Dempster, W. T. 1955. The anthropometry of body action. Ann. N. Y. Acad. Sci. 63: 559–585.
Prost, J. H. 1974a. Varieties of human posture. Hum. Biol., In press.
Prost, J. H. 1974b. Filming body behaviour. *In:* P. Hockings (ed.), Visual Anthropology. H. Mouton and Co., The Hague, In press.
Strasser, H., and A. Gassmann. 1893. Hulfsmittel und Normen zur Bestimmung und Veranschaulichung der Stellungen- Bewegungen und Kraftwirkungen am Kugelgelenk, insbesondere am Huft- und Schultergelenke des Menschen. Anat. Hefte. 2: 389–434.

Author's address: *J. H. Prost, Department of Biological Sciences, University of Illinois at Chicago Circle, Chicago, Illinois 60680 (USA)*

The quick-release method for estimating the moment of inertia of the shank and foot

P. R. Cavanagh and R. J. Gregor

The Pennsylvania State University, University Park

The accurate determination of body segment parameters is a problem of crucial importance to biomechanics research. Within the limitations of a rigid-body anthropometric model, the calculation of forces and torques from movement data can be only as valid as the data concerning the mass, length, and inertial parameters used in the calculations. It is essential, therefore, that systematic investigations be made of any techniques that enable body segment parameters to be estimated in the same living subjects used in experiments from which movement data were collected.

The quick-release method of determining the moment of inertia of a limb segment about its proximal joint was first suggested by Fenn, Brody, and Petrilli (1931) and was used subsequently by Drillis and Contini (1966) and by Bouisset and Pertuzon (1968). The latter investigators reported a coefficient of variation for the moment of inertia of the forearm and hand of 19.69 percent in 11 subjects. However, this figure gives no indication of the variation encountered in a single subject under the same conditions or variation caused by different initial torques and positions of force application before release.

The purpose of this study was to investigate the quick-release method for the determination of the moment of inertia of the shank and foot about the knee joint in the sagittal plane. Specifically, the effects of different torques and positions of force application on the resulting estimate were examined.

METHODS

Four male subjects were used in these experiments. Their ages, heights, and weights are shown with other data in Table 1. For the quick-release experi-

Table 1. Characteristics of the subjects used in the experiments and estimates of the moment of inertia

Subject	Age (Yr)	Ht (in)	Weight (N)	Segment length (in)	Segment mass (kg)	I_{k1} *	$\bar{I}_{k2}$†	$\bar{I}_{k3}$†	$\bar{I}_{k4}$†
							(ft-lbs. per sec^2)		
1	41	69	176	16.45	11.50	0.1950	0.1029	0.1402	0.1908
2	25	71	152	17.35	10.41	0.1963	0.1357	0.1423	0.1579
3	25	72	160	17.63	10.54	0.2051	0.1176	0.1468	0.1894
4	34	70	213	17.40	13.19	0.2506	0.1353	0.1425	0.2381

*I_{k1} is estimate of I_k from water displacement and anthropometric data.
†I_{k2}, I_{k3}, and I_{k4} are mean estimates from the quick-release method at the near, mid, and far positions, respectively.

ments, the subjects were seated on a table so that the shank and foot hung freely with the thigh strapped to the table surface. In this position, a linear accelerometer with its sensitive axis in both sagittal and transverse planes was attached to the left limb at approximately 70 percent of distance between the center of rotation of the knee joint and the lateral malleolus (shank segment length). A loop of aircraft cable resting on a plastic pad over the anterior border of the tibia transmitted tension horizontally from the subject's limb to an LVDT mounted on an adjustable bracket. This arrangement is shown in Figure 1. The subjects were asked to build up tension slowly to a given level and then to maintain it. A manually operated plunger was impacted at a time unknown to the subject in order to facilitate release. Surface electromyograms were recorded from the flexors and extensors of the knee. The accelerometer, force transducer, and electromyogram outputs were recorded on an ultraviolet light galvanometer recorder, and measurements were taken subsequently from the chart record with dividers.

The experimental design employed is shown in Figure 2. The effect of two variables on the resulting moment of inertia was examined. First, to obtain a given torque, three different points of application of the restraining force on the limb were used. Each position involved a different force that was calculated for each subject before the start of the experiment. The three positions chosen were 65, 75, and 90 percent of the distance from the knee

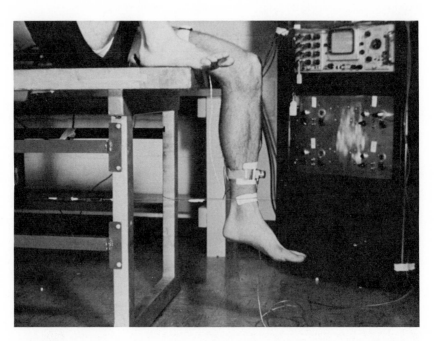

Figure 1. A subject in position for quick release (quadriceps electrodes not in place).

EXPERIMENTAL DESIGN

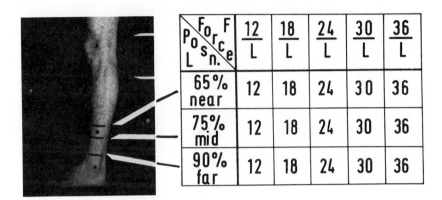

Force Posn.	12 L	18 L	24 L	30 L	36 L
65% near	12	18	24	30	36
75% mid	12	18	24	30	36
90% far	12	18	24	30	36

Table of Torques, T (lbf.ft)

Figure 2. Torques maintained before release. Near, mid, and far positions are indicated on the limb of one subject. *lbf. ft*, ft-lb.

joint center to lateral malleolus; these were designated as near, mid, and far positions, respectively. The second effect studied was that of different torques at the instant of release. A threefold range of torques from 12 to 36 ft-lbs. was used in each of the three positions for all subjects.

The position at which the restraining force was applied was randomized, as were the torques at a given position on the limb. Any trial that showed activity in the hamstring muscles prior to release was discarded.

An additional estimate of the moment of inertia was calculated by a mass estimate from water displacement and values of 0.303 and 0.446 l for the radius of gyration about the center of gravity and position of the center of gravity respectively, where l was the shank segment length (Dempster 1955; Drillis and Contini, 1966). These estimates are shown in Table 1.

RESULTS

The retouched raw record from a typical trial is shown in Figure 3. After release of the steady force, the acceleration peaked within 10 msec, and a reflex response was seen in the hamstrings after approximately 60 msec. A

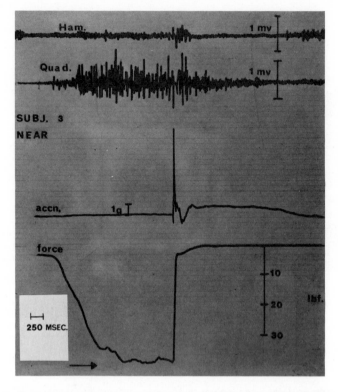

Figure 3. Typical galvanometer record (retouched) of a quick release showing electromyograms, accelerometer, and force transducer outputs. *Ham.*, hamstrings; *Quad.*, quadriceps.

silent period in the quadriceps electromyogram shortly after release was also apparent in many records.

One Subject, 50 Trials

With the same force applied at the same position on the limb of a single subject, data on 50 trials were collected during one testing session. A range of 0.0492 about a mean of 0.1776 ft-lbs. per sec^2 was encountered, with a coefficient of variation of 6.5 percent.

Four Subjects, Position by Torque Design

The results obtained from the design shown in Figure 2 were analyzed by a two-way analysis of variance. Each cell contained four entries for the estimate of I_k, each entry being the mean of two trials for that subject.

The results indicated that the magnitude of the torque exerted at the moment of release had no effect on the estimate of I_k obtained. Unexpectedly, however, the position of force application to produce the same torque had an effect on the resulting estimate that was statistically significant ($p <$ 0.01). The trend, at all torques, was for a greater value of I_k when the force was applied more distally on the limb. The individual means at each position are shown in Table 1.

DISCUSSION

Since all the trials were conducted on a single day, the variability reported for 50 trials of a single subject does not include such effects as a relocation of the accelerometer and remeasurement of the distance on the limb. This presumably would increase the coefficient of variation from the calculated value of 6.5 percent measured under identical conditions of torque and position of force application.

The greater values of I_k obtained when the restraining force was applied at more distal positions on the limb were due to smaller values for the initial acceleration, despite the fact that the net torque about the knee joint apparently was identical. In an attempt to account for these different estimates, the following possible sources of experimental error were identified: (a) joint center location, which affects the calculation of both torque and angular acceleration; (b) validity of the anthropometric model, in which a two-link system is treated as a one-segment rigid body; (c) misalignment of the accelerometer axis, which renders cosine errors possible in both sagittal and transverse planes; (d) muscle activity at release, which results in a different mechanical system at moment of release; (e) uneven release, causing several peaks on the accelerometer trace, were noted for some trials; (f) hip flexion at release, which may have occurred despite attempts to restrain the thigh; and (g) inaccurate chart measurement, which is complicated by the large dynamic range of accelerometer and force records.

With the exception of the first error mentioned, all sources identified would influence the calculated estimate of I_k independent of the torque exerted or the position at which the restraint was applied. However, error in the location of the joint center would affect the value of I_k to an extent dependent on where the force was applied on the limb. Calculations indicated that a 2-in error in joint center location only would introduce a relative error of less than 10 percent between the estimates from the two extreme positions used in the experiments. We thus are unable to account fully for the differences observed. It is also apparent that the estimates of I_k from the anthropometric data (I_{k1} in Table 1) are all higher than the largest estimates obtained from the quick-release experiments. This disparity is not wholly unexpected in view of the problems associated with the use of data from small numbers of aged cadavers, and the result cannot be taken as evidence that the larger values from the quick-release method are necessarily the correct ones.

RECOMMENDATIONS AND CONCLUSIONS

Several modifications would improve the accuracy of the results obtained with the quick-release method as used in the present experiment.

1. The ankle joint could be immobilized with a brace or cast to improve the validity of the anthropometric model.

2. The center of rotation of the knee joint for the small range of motion used could be located more accurately using a device similar to that described by Finley, Wirta, and Taylor (1969).

3. Any quick-release apparatus should be validated by a simple mechanical system before it is applied to the living human subject.

4. A procedure should be devised in which the relaxed limb is accelerated by externally applied forces rather than by internal muscular forces.

In conclusion, it would appear that the quick-release method for the determination of the moment of inertia is neither as simple nor as reliable as some investigators have suggested.

REFERENCES

Bouisset, S., and E. Pertuzon. 1968. Experimental determination of the moment of inertia of limb segments. *In:* J. Wartenweiler, E. Jokl, and M. Hebbelinck (eds.), Biomechanics, pp. 106–109. S. Karger, Basel, and University Park Press, Baltimore.

Dempster, W. T. 1955. Space requirements of the seated operator. WADC Technical Report 55–159. Office of Technical Services, U. S. Dept. of Commerce, Washington, D. C.

Drillis, R., and R. Contini. 1966. Body segment parameters. Technical Report 1166.03. N. Y. U. School of Science and Engineering, New York.

Fenn, W. O., H. Brody, and A. Petrilli. 1931. The tension developed by human muscles at different velocities of shortening. Amer. J. Physiol. 97: 1–14.

Finley, F. R., R. W. Wirta, and D. R. Taylor. 1969. Rehabilitation biomedical engineering center and program. Final Report. Moss Rehabilitation Hospital, Philadelphia.

Author's address: *Dr. Peter R. Cavanagh, Biomechanics Laboratory, The Pennsylvania State University, University Park, Pa. 16802 (USA)*

Computer techniques in the biomechanics of sport

P. Sušanka

Charles University, Prague

Modern biomechanics research depends heavily on the use of computer technology. Digital computer processing is an essential aspect of research involving cinematographic procedures where large volumes of data must be handled. In order to perform a kinematic or kinetic analysis of human movement, the investigator typically simplifies the human body into a system of linked masses, the ends of each link being defined by anatomical landmarks. Once the masses, positions of the centers of gravity, and vertical parameters of the system have been defined, a solution to various mechanical problems is possible (Plagenhoef, 1971). This chapter outlines a methodology of computer analysis of human motion and its application to the analysis of sports movements.

METHODS

Subjects were filmed at speeds between 96 and 600 frames per sec with ZL 16, Beaulieu R16, and Eclair GV 16 cameras. The data obtained from the analysis of these films were input to a FORTRAN program that enabled certain features of the movement to be analyzed. The basic model used in the analysis involves the use of matrix methods to the two- or three-dimensional equations of motion for the main kinematic pairs identified (Brát, 1965; and Sušanka, 1970). The initial stages of development of this model, described in this chapter, involved an examination of the linear displacement of individual points and the angular displacement of body segments. Subsequently, algorithms were created that would enable open kinematic chains of two or three links to be studied both kinetically and kinematically.

To permit analysis to be made in both real time and cycle time, separate interpolation routines were written for both equally spaced data and nonequidistant data. Approximations by the method of least squares were used in the calculation of derivatives, and the programs allowed a free choice of start and endpoints and as well as the detailed study of important phases of the movement. The output from the program consisted of displacement, velocity, and acceleration as functions of time, as well as horizontal and vertical components of force and their resultants together with certain joint moments. The displacement of the segmental and whole body centers of gravity also were computed.

Two track and field event movements were studied in an attempt to produce results with practical application. These were the pole vault and the shot put. In addition to the output parameters mentioned previously another variable was found useful in the analysis of the pole vault, i.e., functional length of the fiberglass pole and its role of change. This length was defined as the linear distance between the highest hand grip and the point of pole contact with the ground.

RESULTS

In Figure 1, the solid line shows the time course of the rate of change of functional length of the pole during a vault of 4.60 m by Subject O. J., whose

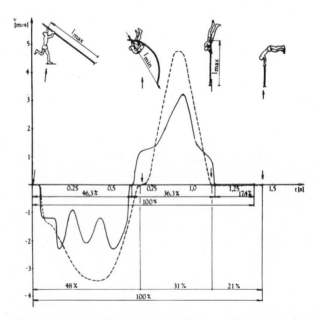

Figure 1. Rate of change of functional pole length for a skilled vaulter (*dashed line*) and a less-skilled vaulter (*solid line*).

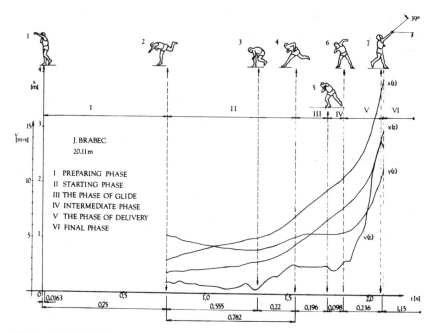

Figure 2. Displacements and velocity of the shot as functions of time during a high-level performance.

personal best is 4.90 m. The dashed line represents the theoretical time course derived from the analysis of a 5.45-m jump of a different subject.

The path of the center of gravity of the shot during a throw of 20.11 m by Subject J. B. is shown in Figure 2, where certain phases of the throw have been identified from the film.

DISCUSSION

A comparison of the two vaulting patterns shown in Figure 1 indicated certain irregularities in the results from a poorer vaulter that may be due to incorrect technique. It is possible that the use of these results in conjunction with a video-tape recorder during training would enable the coach to make specific recommendations for the modification of the athlete's technique.

From the kinematic analysis of the shot put shown in Figure 2, together with other results from world class performers, it was observed that the velocity of the shot falls almost to zero at a stage represented approximately by Position 3 in Figure 2. This suggests that the movements preceding this posture may be simplified, inasmuch as they contribute nothing to the forward propulsion of the shot. When the throw was performed without the balance position (Position 2 in Figure 2); however, with both feet in contact with the ground during the lowering of the shot, the movement became more fluent, and a higher velocity at release was obtained. Certain phases of this analysis were comparable with the analog simulation of an oblique throw by Tutěvic (1969).

CONCLUSIONS

The application of computer techniques to the biomechanical analysis of human movement enables the techniques of sports movements to be examined critically. Once a model has been developed, it may have wider application in the general analysis of human body movement in medicine, kinesiology, and other related areas.

REFERENCES

Brát, V. 1965. Maticová metoda kinematického řešení prostorových mechanismů s nižšími kinematickými dvojicemi. Rozpravy Csav 75: 2.

Plagenhoef, S. 1971. Patterns of human motion: a cinematographic analysis. Prentice-Hall, Englewood Cliffs, N. J.

Sušanka, P. 1970. Některé aspekty kinematické analýzy pohybu. FTVS UK, rigorosní práce.

Tutěvic, V. N. 1969. Těorija sportivnych metanij. Fizkult. sport.

Author's address: *Dr. Petr Sušanka, Department of Biomechanics of FTVS-UK, Charles University, Ujezd 450, Praha 1 (Czechoslovakia)*

Miscellaneous

Kinanthropometry and biomechanics

M. Hebbelinck
Vrije Universiteit Brussel, Brussels

W. D. Ross
Simon Fraser University, Burnaby

The study of man and movement is a complex task. As Alexis Carrel once remarked, "We do not apprehend man as a whole. We know him as composed of distinct parts. And even these are created by our methods."

Perhaps we might conceive of the study of human movement as an edifice completely obscured by canvas except for a few rents and pinholes that afford various glimpses of reality. At the previous biomechanics seminar in Rome, our current seminar chairman discussed the evolution of biomechanics as a scientific specialization. In our analogy, he pointed to a pinhole that, by the diligence and ingenuity of scientists such as those we have assembled here, is being enlarged.

We see biomechanics as one of a family of evolving subdisciplines in the study of human movement. Because of our academic focus, we have presented these in an organizational schema for a university kinesiology program, as shown in Figure 1. Of course, our terminology and identification of subdisciplines is largely speculative. However, in the spectrum of the study of human movement, we appreciate the international vitality involved in teaching and research in the areas of anatomy, kinanthropometry, biomechanics, physiology, learning, psychology, sociology, and cultural aspects.

KINANTHROPOMETRY

Our particular focus at this time is kinanthropometry, or that area in the study of human movement having to do with the measurement of size, shape,

This study was supported by a President's Research Grant from Simon Fraser University, Burnaby.

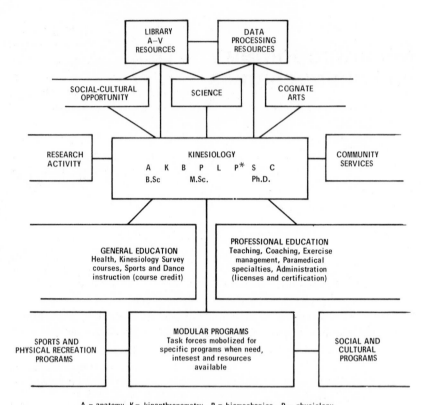

A = anatomy, K = kinanthropometry, B = biomechanics, P = physiology,
L = learning, P*= psychology,   S = sociology, C = cultural aspects.
These are sub-disciplines related to Kinesiology (Kinanthropologie, French)
or the study of human movement in all its biological and sociological ramification.

Figure 1. Organizational schema for a university kinesiology program.

proportion, composition, maturation, and gross function as each is related to such concerns as growth, exercise, performance, and nutrition.

Our examples of the relationship between biomechanics and kinanthropometry in this chapter are drawn from sports largely because it is possible to evaluate performance in this area. However, it is a disservice to both sub-disciplines to view this application as an exclusive preoccupation.

THE LITERATURE IN KINANTHROPOMETRY

The wealth of information available in the area of kinanthropometry can be appreciated only from systematic searching through the literature. Since 1970, the Canadian Selective Dissemination of Information Project, or CAN/ SDI, has provided current, routine, computerized searching of two major indexing publications: ISI Source and Citations Indexes, and Biological Abstracts and Bioresearch Index.[1] Since 1970, these searches have generated about 3,750 citations for the authors, with roughly 10 percent judged pertinent to the area of kinanthropometry and suitable for collecting.

Perhaps the simplest retrieval system for biomechanics or kinanthropometry is based on a continuous citation search of key authors. Although reflecting the special interests of two investigators, the lists[2] show a substantial return.

In both East and West Germany, two indexing and abstracting services[3] provide a rich informational source of materials in the sports sciences and all the subbranches, including biomechanics and kinanthropometry as we have defined it.

KINANTHROPOMETRY AND BIOMECHANICS OF SPORT

At the championship level, contributory factors in performance, such as skill, opportunity, intelligence, motivation, and training, may be assumed to be close to optimal for the particular sport.

There is ample evidence to demonstrate that, in some sports, the biomechanical demands are such that there are specific prerequisites for physique.

---

[1]CAN/SDI now searches 10 different magnetic tapes that correspond to seven different commercial indexes and abstracts in the U. S. Library of Congress MARC II book cataloging records, the technical reports made available by the U. S. National Technical Informational Services, and MEDLARS from the U. S. National Library of Medicine. Since the inception of the service, a continuous search program in kinanthropometry using less than three dozen key words makes use of the BA-PREVIEWS—Biological Abstracts and Bioresearch Index (Bioscience Information Service) and ISI—Source and Citation Indexes (Institute for Scientific Information). These two data bases were selected because each offers specific desirable features: ISI (equals all sections from Current Contents) indexes include about 2,500 journals in the natural and applied sciences, medicine, engineering technology, and social and behavioral sciences. It is one of the fastest indexing services, providing references to papers before the papers actually appear in the periodical literature. BA-PREVIEWS, on the other hand, is much slower at indexing, but much more comprehensive in the biosciences than ISI, routinely searching some 7,700 sources. The coverage in ISI is restricted to journals; that of BA-PREVIEWS is applied not only to journals but to proceedings of congresses, symposia, conferences, etc., technical reports, government serials, annual institutional reports, books in series, and dissertations. BA-PREVIEWS also provides excellent coverage on non-English-language publications.

[2]For kinanthropometry: Aastrand, P. O., or Astrand, P. O., or Behnke, A. R., or Carter, J. E., or Cheek, D. B., or Greulich, W. W., or Ross, W. D., or Sheldon, W. H., or Tanner, J. M., or Thompson, D. W., or Weiner, J. S.

For biomechanics: Bahler, A. S., or Basmajian, J. V., or Chaffin, D. B., or Chapman, A. E., or Dempster, W. T., or Fung, Y. C., or Grieve, D. W., or Hill, A. V., or Hirsch, C., or Miller, D. I., or Nelson, R. C., or Whitney, R. J.

[3]In the German Democratic Republic: Dokumentationsstelle, Deutsche Hochschule fur Körperkultur, Leipzig.

In the German Federal Republic: Sportdokumentation, Bundesinstitut fur Sportwissenschaft, Cologne.

Size is often a prime selective factor. Mean values for selected size parameters of the participants in the 1968 Olympics in Mexico in relation to a unisex reference human (Ross and Wilson, 1973) are presented in Figures 2–4. It is evident that there are often unique size characteristics associated with various events. The statistical significance of these differences in the Mexico data are reported elsewhere by Carter and Hebbelinck (1974) and have been reported previously for the Rome data by Tanner (1964).

In exploring some of the implications for kinanthropometry in the study of sports performance, we have selected 11 events from the 1968 Olympics data. Height, weight, and range values for some of these events are shown in Table 1. There are pronounced individual differences among sports, as demonstrated further in the analysis of variance on height, weight, arm length, leg length, biacromical breadth, and biiliocristal breadth, shown in Table 2. Even within the given sport classification, absolute size often is associated with

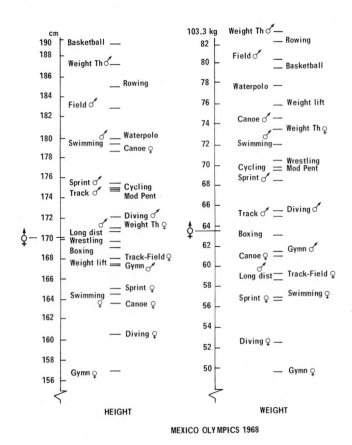

Figure 2. Mean height and weight values for the participants in the 1968 Mexico Olympics. Data by courtesy of L. de Garay, L. Levine, and J. E. L. Carter.

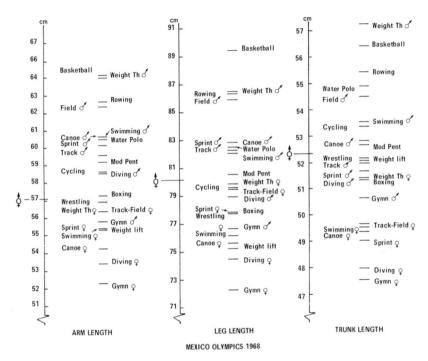

Figure 3. Mean arm length, leg length, and trunk length of the participants in the 1968 Mexico Olympics. Data by courtesy of L. de Garay, L. Levine, and J. E. L. Carter.

success. It is interesting to note that even in the highly select company of Olympic competitors, leading eight, leading three, and winners in 100-m, 200-m, 400-m, 100-m, and 400-m hurdles, as well as steeplechase, jumpers, and throwers tend to be systematically taller and heavier than their peers, as reviewed by Owen (1970). Tittel and Wutscherk (1972) show a similar tendency for swimmers and rowers.

In interpreting sports performance from the point of view of biomechanics, it should be recognized that there may be secular trends. These may reflect differing genetic pools of talent, changing technique, and differing opportunity for participation. In Table 3, we have illustrated these trends by mean size parameters gleaned from reports by Kohlrausch (1930), compared with data assembled by Tanner (1964), Correnti and Zauli (1964), Hirata (1966), Carter and Hebbelinck (1974), and Hirata and Teipel (1973). In general, compared with the situation in 1928, the athletes of the modern era are "taller" in those events in which height appears to have a biomechanical advantage, such as basketball, rowing, and weight-throwing events. Also, the sprint and middle-distance runners since 1960 appear to be taller and heavier than their counterparts in 1928. This trend is not evident in those sports in which increased height has no biomechanical advantage, such as gymnastics.

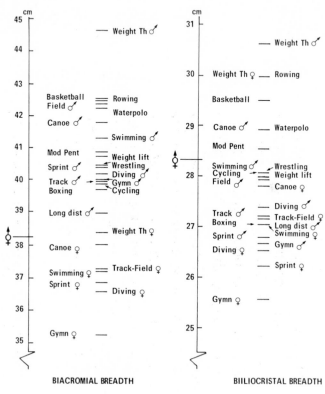

BIACROMIAL BREADTH   BIILIOCRISTAL BREADTH

MEXICO OLYMPICS 1968

Figure 4. Mean biacromial breadth and biilicristal breadth of the participants in the 1968 Mexico Olympics. Data by courtesy of L. de Garay, L. Levine, and J. E. L. Carter.

Table 1. Mean height, weight, and range values for selected Olympic events, Mexico, 1968

Event	Height (cm)			Weight (kg)		
	$\bar{X}$	Maximum	Minimum	$\bar{X}$	Maximum	Minimum
100,200,4 × 100	175.4	188.7	163.0	68.4	89.7	54.7
800–1,500	177.3	191.0	163.8	65.0	79.4	51.4
5,000–10,000	171.9	186.0	155.8	59.8	71.7	48.7
Marathon	168.7	176.6	159.0	56.5	66.1	51.7
Long, high, triple	182.8	204.3	170.2	73.2	89.8	57.7
Shot, discus, hammer	186.1	195.9	177.4	102.3	126.2	86.0
Javelin	179.5	186.5	174.5	76.6	84.6	63.4
Basketball	189.1	212.0	171.3	79.7	109.1	60.9
Gymnastics	167.4	178.6	157.8	61.5	69.7	50.8

Table 2. Analysis of differences between selected sports (men). Mexico Olympics 1968

Variable	F-Ratio	Between-sport differences
Height	47.43†	GYM‡ LDMT DIV PT CYC SPR SWM WTP CAN ROW BKB
Weight	56.56†	LDMT GYM DIV SPR CYC PT SWM CAN WTP BKB ROW
Endomorphy	21.38†	GYM LDMT SPR CYC CAN DIV PT BKB ROW SWM WTP
(Endomorphy)§	(8.55†)	(GYM LDMT SPR CYC CAN DIV PT BKB ROW SWM)
Mesomorphy	14.70†	LDMT BKB CYC SWM SPR WTP PT ROW DIV CAN GYM
Ectomorphy	14.57†	WTP ROW PT GYM CAN CYC DIV SPR SWM BKB LDMT
Biacromiale	24.40†	LDMT CYC GYM DIV SPR PT SWM CAN WTP ROW BKB
Billiocristale	25.86†	GYM SPR LDMT DIV SWM CYC PT WTP CAN BKB ROW
Arm Length	31.79†	GYM LDMT DIV CYC PT SPR WTP SWM CAN ROW BKB
Trunk Length	14.20†	GYM LDMT DIV SPR PT CAN CYC SWM WTP ROW BKB
Leg Length	27.84†	GYM DIV LDMT CYC PT SWM WTD SPR CAN ROW BKB

After Carter and Hebbelinck, 1974

*F .05(10,601) = 1.85.
†F .01(10,601) = 2.36.
‡GYM, gymnastics; LDMT, long distance and marathon; PT, pentathlon; CYC, cycling; SPR, sprinting; SWM, swimming; WTP, water polo; CAN, canoeing; ROW, rowing; BKB, basketball.
§Endomorphy (skinfold) analysis run without water polo group.

Table 3. Height (cm), weight (kg), and ponderal index ($10^3 \cdot \sqrt[3]{W}$ kg / Hcm) for Olympic competitors in 1928, 1960, 1964, 1968, and 1972

Event	Amsterdam 1928	Rome 1960	Tokyo 1964	Mexico 1968	Munich 1972
100 m, 200 m, 4 × 100 m					
Height	172.7	177.7	176.0	175.4	176.0
Weight	64.7	70.8	70.8	68.4	69.8
Ponderal Index	23.25	23.28	23.51	23.32	23.39
800 m and 1,500 m					
Height	174.4	180.5	177.0	177.3	177.2
Weight	66.7	68.9	65.8	65.0	67.2
Ponderal Index	23.25	22.71	22.81	22.68	22.94
5,000 m and 10,000 m					
Height	167.7	172.2	173.6	171.9	172.4
Weight	60.3	60.7	60.3	59.8	61.3
Ponderal Index	23.11	22.82	22.84	22.75	22.87
Marathon					
Height	166.1	168.3	170.3	168.7	170.0
Weight	59.6	60.2	60.8	56.6	60.2
Ponderal Index	23.52	23.29	23.09	22.76	23.05
Jumpers					
Height	178.9	179.9	181.4	182.8	181.6
Weight	69.1	71.9	73.2	73.2	73.8
Ponderal Index	22.94	23.06	23.06	22.88	23.10
Throwers					
Height	180.9	191.3	187.3	186.1	187.4
Weight	88.6	101.7	101.4	102.3	107.3
Ponderal Index	24.64	24.40	24.90	25.13	25.36
Javelin					
Height	180.9	182.2	183.0	179.5	181.4
Weight	88.6	87.3	83.4	76.5	87.5
Ponderal Index	24.64	24.35	23.87	23.65	24.47
Basketball					
Height		189.7	189.4	189.1	192.6
Weight		79.5	84.3	79.7	85.5
Ponderal Index		22.67	23.15	22.76	22.87
Rowers					
Height	181.1		186.0	185.1	185.5
Weight	76.9		82.2	82.2	84.5
Ponderal Index	23.48		23.38	23.49	23.66
Gymnasts					
Height	169.6		167.2	167.4	168.0
Weight	61.8		63.3	61.6	64.1
Ponderal Index	23.31		23.84	23.59	23.82

In viewing any of the traditional height-weight ratios, it is noted that the following male modern competitors are "stouter" than their 1928 counterparts: sprinters and 400-m runners, long jumpers, high jumpers, triple jumpers, discus throwers, shot putters, hammer throwers, and rowers. The modern runners in distances greater than 400 m and javelin throwers, except in 1972, have become "leaner." Recently, Hirata and Teipel (1973) have presented excellent comparative data on height, weight, and ponderal indices of participants in all Olympic sports for men and women for the years 1964 (Tokyo) and 1972 (Munich). In males, the trend to increasing stoutness appears to be continuing in the decathlon and in all the throwing events. In both male and female swimmers, the trend is toward leanness. These changes in size have some implication for sport biomechanics.

Shape, or the outer morphological configuration of sport participants, appears to be a selective factor, as implied from the various techniques employed by Carter (1970), Carter and Hebbelinck (1974), Correnti and Zauli (1964), Kohlrausch (1930), Tanner (1964), and Tittel and Wutscherk (1972). Each method of body typing has its advantages. However, our personal preference is for a phenotypical anthropometric system for both men and women such as that presented in detail by Carter (1972) and discussed by the authors in other publications (Hebbelinck, Duquet, and Ross, 1973; Hebbelinck and Ross, 1974). We are also intrigued with the possibilities of projecting somatotype data on an X, Y coordinate axis and of quantifying dispersion distances, as proposed by Ross and Wilson (1974).

As stated by Carter (1970) "champion performers at various levels of a particular sport exhibit similar patterns of body size and somatotype, with patterns tending to become narrower as levels of performance increase." This concept is illustrated in Figure 5, which shows the Heath-Carter somatotype distribution of the mean somatotypes for 11 selected events at the 1968 Olympics. Among the biggest competitors are weight-throwers, rowers, and basketball players. However, the unique mechanical demands of these events require different conformation of physique, as shown by the divergent somatoplots. Javelin throwers and gymnasts have practically identical somatotypes, although the javelin throwers, at 179.5 cm and 76.7 kg, are much bigger than the gymnasts, at 167.4 cm and 67.1 kg. Thus, kinanthropometric data on size *and* shape may be of prime importance in biomechanical interpretation of various sport performances.

Proportionality assessment has been a human concern since man first scratched his likeness on the walls of caves. As a mathematical concept or esthetic appreciation, it has acquired great importance in arts, having supplied anthropometric canons elaborated by the Babylonians and then by the Egyptians, the Greeks (Polykleitos), and the Romans (Vitruvius), and in the Renaissance there were notable studies by Leon Battista-Alberti (1404–1472), Leonardo da Vinci (1452–1519), and Albrecht Dürer (1471–1528). Galileo commented on size relationships in animal forms 300 years ago, and, in 1729, Swift immortalized his perceptions in *Gulliver's Travels.* Brody (1945), Dempster (1955), von Döbeln (1956), Huxley (1932), Martin and Saller (1959), and Thompson (1917, 1966) have related kinanthropometry and biomechanics in discussions of human proportions.

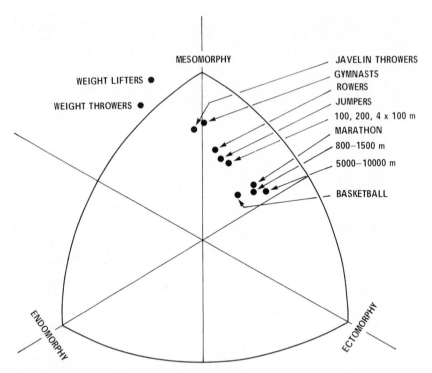

Figure 5. The Heath-Carter somatotype distribution of the mean somatotypes for 11 selected events at the 1968 Mexico Olympics.

Hardly in the traditions of esthetic anthropometry, as shown in Figure 6, we have prepared a series of caricatures of humans with proportions designed to give mechanical advantage in sports. The normal man is contrasted with the spindley, long-legged cross-country skier, the short-armed and short-legged weightlifter, the chesty and powerfully developed legs and feet of the soccer player, the small, muscular, perfectly symmetrical gymnast; and the large hands and feet of the swimmer. Do we indeed find the counterparts of these sports-machines in real life?

There are often proportional differences that conceivably may give a mechanical advantage in sports performance. Tanner (1964) and Tittel and Wutscherk (1972) have described proportional differences and speculated on their biomechanical importance. However, some of these differences may be trivial, as noted in Table 4, which shows sizes adjusted to a standard reference height of 170.18 cm. We suspect that absolute size is far more important biomechanically than are proportions. Eiben (1972), in an excellent kinanthropometric study on female track and field athletes, also showed that relative measures of lengths and widths did not supply much information. Direct body measures rather than relative measures were important biomechanically.

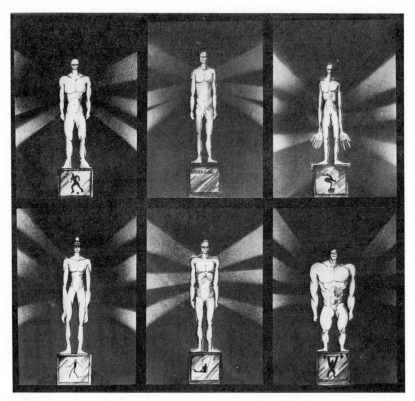

Figure 6. Caricatures of sports types.

Other studies, such as those by Krakower (1935) and Della (1950) on jumpers, by Kopf (1960) on discus throwers, and by Williams and Scott (1967) and Lewis (1970) on rowers, have explored the biomechanical implications of size and proportional relationships to performance. We feel that anatomical evidence, particularly in viewing individual differences in origins and insertions of muscles, could help biomechanical interpretation of performance variance.

Maturation and growth also have implications in the biomechanics of sports. From a dimensional point of view, where height has the dimension L, strength (force) or cross-sectional area L^2, and volume and weight L^3, and from Newton's second law, where time has the dimension L, it has been noted by Astrand and Rodahl (1970) that running time is independent of height for 12-, 14-, and 18-yr-old girls and for 11-, 12-, and 18-yr-old boys. There was a relationship between height and running time for 14-yr-old boys reflecting not so much a mechanical advantage of size but the effect of the growth spurt in which taller, earlier maturing boys were experiencing the added benefits of maturation.

Table 4. Skeletal mean values adjusted to standard height (170.18 cm). Mexico Olympics, 1968

Event	Trunk length*	Leg length	Arm length	Shoulder breadth	Hip breadth
100,200, 4 × 100	50.2	80.3	58.4	39.1	26.0
800–1,500	50.4	80.6	57.7	38.5	26.5
5,000–10,000	50.7	79.5	57.8	38.7	26.9
Marathon	51.0	78.2	56.9	38.8	27.1
Long, high triple	49.9	81.2	58.1	38.5	25.8
Discus, shot, hammer	51.8	78.7	58.3	40.6	28.0
Javelin	51.0	78.3	57.5	39.8	28.2
Basketball	50.3	80.3	57.6	38.2	26.5
Rowers	50.9	79.5	57.6	39.1	27.6
Weightlifters	52.8	76.3	56.1	41.3	28.3
Gymnastics	51.5	78.1	56.7	40.6	27.1
X̄	51.0	79.2	57.5	39.4	27.1
Ref ♀	52.4	79.9	57.1	38.2	28.3

*Trunk length = suprasternale - symphysion; leg length = trochanterion - sphyrion; arm length = acromiale - stylion; shoulder breadth = biacromial diameter; hip breadth = biiliocristal diameter.

In skiers under 14 years of age, Ross and Day (1972) and Ross, McKin, and Wilson (1974) have shown that size is not related to performance; however, somatotype apparently was related, inasmuch as the ectomeso-morphic youngsters in the samples had a disproportionate number of very good and good ratings. As shown in Figure 5, the ectomesomorphy is also a prototype for a number of sports, including all running and jumping events, basketball, rowing, and others.

However, as pointed out by Dupertuis and Michael (1953) and by Borms, Hebbelinck, and Ross (1973), there is evidence to show that fatness is related to early maturation and that, conversely, leaness or ectomesomorphy may be associated with late maturation. If a premium is placed on winning in youth sport, and selection for advanced training is made too early, the ectomeso-morphic late maturer who may be mechanically ideal for a particular sport may be discouraged or not selected.

Changing proportions during growth are of interest. For example, during the adolescent growth spurt, hands and feet reach adult size first, followed by calf and forearm, then by hips and chest width, then by the shoulders, and lastly by trunk length and chest depth. This sequence of events, particularly when the hands and feet are relatively large with respect to the rest of the body, especially the shoulders, is a tempting kinanthropometric speculation of biomechanical advantage enjoyed by young swimmers with respect to propulsion and resistance in the water.

Composition of the human has been studied by a variety of techniques, outlined by Brozek and Henschel (1961), and by the physical performance implications, reviewed by Parizkowa (1968). In addition to the estimation of body fat and lean tissue masses, the concept of the DNA unit and intriguing evidence of increasing skeletal muscle cells with growth, as re-viewed by Cheek et al. (1971), presents the human design in a new light. Here again is an area of common interest for investigators in kinanthro-pometry and biomechanics who look to the cellular biologists for addi-tional information, particularly about the role of exercise and nutrition during critical growth stages.

Measurement of gross function by a variety of strength, suppleness, and stamina tests as well as by motor performance tests provides for fitness appraisal and guidance. Apart from a recognition of the importance of this aspect of kinanthropometry, it is beyond the scope of this study to draw further implications.

Our introduction to kinanthropometry hopefully is an invitation to investigators in all areas of biomechanics to examine more critically the physical characteristics that underly human performance. This invitation for a kinanthropometric focus is not limited to sports. There is individual vari-ance in characteristics of size, shape, proportion, composition, and matura-tion that is reflected in all human function and movement. Indeed, we might regard kinanthropometry and biomechanics as closely related members of a family of disciplines aimed at understanding human movement in all its biological and sociological ramifications.

ACKNOWLEDGMENTS

We thank Maurice Deutsch, Science Librarian, Simon Fraser University, for design of computer retrieval systems for kinanthropometry.

Data from 1968 Mexico Olympic Games were used by kind permission of A. L. deGaray, L. Levine, and J. E. L. Carter, senior editors, *Genetical and Anthropological Studies of 1265 Athletes of the XIX Olympic Games.*

REFERENCES

Astrand, P.O., and K. Rodahl. 1970. Body dimensions and muscular work. *In:* Textbook of Work Physiology. McGraw-Hill, New York.
Borms, J., M. Hebbelinck, and W. D. Ross. 1973. Somatotype and skeletal maturity in twelve year old boys. *In:* O. Bar-Or (ed.), Pediatric Work Physiology. Proceedings of the 4th International Symposium, pp. 85–91, Wingate Institute, Israel.
Brody, S. 1945. Bioenergetics and Growth. Hafner Publishing, New York.
Brozek, J., and A. Henschel. 1961. Techniques for Measuring Body Composition. National Academy of Sciences–National Research Council, Washington, D. C.
Carter, J. E. L. 1970. The somatotypes of athletes–A review. Hum. Biol. 42: 535–569.
Carter, J. E. L. 1972. The Health-Carter Somatotype Method. San Diego State College, San Diego.
Carter, J. E. L., and M. Hebbelinck. 1974. Physical anthropology of the athletes. *In:* A. L. deGaray, L. Levine, and J. E. L. Carter, (eds.), Genetical and Anthropological Studies of 1265 Athletes of the XIX Olympic Games, Mexico City, 1968. In press.
Cheek, D. B., A. B. Holt, D. E. Hill, and J. L. Talbert. 1971. Skeletal muscle cell mass and growth: the concept of the deoxyribonucleic acid unit, Pediat. Res. 5: 312–328.
Correnti, V., and B. Zauli. 1964. Olympionici 1960. Ricerche di Antropologia Morfologica sull, Athletica Leggera. Institute of Anthropology of the University of Palermo, Rome.
Della, D. G. 1950. Individual differences in foot leverage in relation to jumping performance. Res. Quart. 21: 11–14.
Dempster, W. T. 1955. Space requirements of the seated operator. Geometrical, kinematic and mechanical aspects of the body with special reference to the limbs. WADC Technical Report, pp. 55–159. Wright-Patterson Air Force Base, Ohio.
Döbeln, W. von. 1956. Human standard and maximal metabolic rate in relation to fat free body mass. Acta Physiol. Scand. 126 (Suppl.).
Dupertuis, W. D., and N. B. Michael. 1953. Comparison of growth in height and weight between ectomorphic and mesomorphic boys. Child Dev. 24: 203–214.

Eiben, O. G. 1972. The Physique of Women Athletes. Hungarian Scientific Council for Physical Education, Budapest.

Garay, A. L. de, L. Levine, and J. E. L. Carter. 1974. Genetical and anthropological studies of 1265 athletes of the XIX Olympic Games, Mexico City, 1968. In press.

Hebbelinck, M., W. Duquet, and W. D. Ross. 1973. A practical outline for the Heath-Carter somatotyping method applied to children. *In:* O. Bar-Or (ed.), Pediatric Work Physiology. Proceedings of the 4th International Symposium, pp. 71–84, Wingate Institute, Israel.

Hebbelinck, M., and W. D. Ross. 1974. Body type and performance. *In:* L. A. Larson (ed.), Fitness, Health and Work Capacity, pp. 82–93. Macmillan, New York.

Hill, A. V. 1950. The dimension of animals and their muscular dynamics. Proc. Roy. Inst. Gr. Brit. 34: 450.

Hirata, K. 1966. Physique and age of Tokyo Olympic champions. J. Sports Med. Phys. Fit. 6: 207–222.

Hirata, K., and M. Teipel. 1973. Ponderal index of Munchen Olympic champions. Paper presented at the International Meeting of Standardization of Physical Fitness Tests, Jyväskylä, Finland, July 22–25.

Huxley, J. 1932. Problems in Relative Growth. Dial Press, New York.

Kohlrausch, W. 1930. Zusammenhaege von Körper–form und Leistung. Ergebnisse der anthropometrischen Messungen ar den Athleten der Amsterdammer Olympiade Arbeitsphysiol. 2: 187–198.

Kopf, H. 1960. Konstitutionsforschung und Sportmedizin. Aerztl. Praxis. 12: 811–813.

Krakower, H. 1935. Skeletal characteristics of the high jumper. Res. Quart. 6: 75–79.

Lewis, A. A. 1970. Physique as a determinant of success in sport with particular reference to Olympic oarsmen. N. Zeal. J.H.P.E.R. 1970: 5–21.

Martin, R., and K. Saller. 1959. Lehrbuch der Anthropologie. Band I and II. Gustav Fischer Verlag, Stuttgart.

Owen, J. W. 1970. Heights and weights of athletes in the Olympic games in Mexico City. Brit. J. Sports Med. 4: 289–294.

Parizkowa, J. 1968. Body composition and physical fitness. Curr. Anthropol. 9: 273–288.

Ross, W. D., and J. A. P. Day. 1972. Physique and performance in young skiers. J. Sports Med. Phys. Fit. 12: 30–37.

Ross, W. D., D. R. McKim, and B. D. Wilson. 1974. Kinanthropometry and Young Skiers. Proceedings of Canadian Association for Sports Sciences, Vancouver, 1972. Charles C Thomas, Springfield, Ill. In press.

Ross, W. D., and B. D. Wilson. 1974. A somatotype dispersion index. Res. Quart. In press.

Ross, W. D., and N. C. Wilson. 1973. A phantom stratagem for proportional growth assessment. Vth International Symposium on Pediatric Work Physiology, Den Haan, Belgium, 1973.

Tanner, J. M. 1964. The Physique of the Olympic Athlete. A Study of 137 Track and Field Athletes at the XVIIth Olympic Games, Rome 1960. George Allen and Unwin Ltd., London.

Thompson, W. D. 1917. On Growth and Form. University Press, Cambridge, Mass.

Thompson, W. D. 1966. On Growth and Form. An Abridged Edition by J. T. Bonner. University Press, Cambridge, Mass.

Tittel, K., and H. Wutscherk. 1972. Sportanthropometrie. Joh. Ambrosius Barth, Leipzig.

Williams, J. P. G., and Scott (ed.). 1967. Rowing: A Scientific Approach. Kaye and Ward Ltd., London.

Author's address: *Dr. Marcel Hebbelinck, Laboratory of Human Biometry and Movement Analysis, Vrije Universiteit Brussel, A. Buyllaan, 105, Brussels 5 (Belgium)*

Cinematographic analysis of the intermittent modifications occurring during the acquisition of a novel throwing skill

J. R. Vorro and D. J. Hobart
University of Maryland School of Dentistry, Baltimore

Numerous investigators agree that, as a new task is practiced, the manipulation aspect tends to be the major contributor to success (Harris and Smith, 1954; Smader and Smith, 1953; Vorro, 1973; Wehrkamp and Smith, 1952). In addition, Wehrkamp and Smith (1952), Rubin, von Trebra, and Smith (1952), Smader and Smith (1953), and Hobart (1972) agree that travel time decreases as a new skill is acquired. These studies have not, however, sought to determine the point during practice at which the change occurred.

METHODS

It was the purpose of this study to analyze the intermittent kinematic changes that occurred as proficiency was acquired in a novel throwing skill. The task consisted of an underhand ball toss for accuracy. The arm of the subject was in a position of extension with the hand in full pronation. The motion was essentially uniplanar and consisted of simple shoulder flexion through a range of motion of approximately 60°.

All testing and practice throws were made with the subjects facing a target and seated in a specially constructed chair. Attached to the right side of the chair was a plywood frame that acted as the starting point for each throw. Attached to the anterior surface of the frame was a panel of Plexiglas, the starting panel. It was adjustable so that the right limb of each subject could be positioned perpendicular to the floor. A microswitch that controlled a circuit containing a neon lamp was attached between the frame and the starting panel (Figure 1). The circuit design was such that, as the switch was activated by the subject's pressing against the panel, the lamp was illuminated. The subjects were required to have the switch closed before the initiation of each throw. This was accomplished with no detectable muscular activity on the electromyogram record. With the beginning of movement, the time necessary for the light to darken was approximately 0.01 sec. Thus, the starting point for each throw could be located easily on the film as the first frame with the starting lamp darkened. Three adjustable straps were attached to the chair, and when fastened across the subject's shoulder, chest, and waist, comfortably prevented gross body movements during each throw. A black Plexiglass throwing cup was attached to the dorsal surface of the fully pronated forearm so that its distal edge was at the level of the ulnar styloid process (Figure 1). The cup was equipped with a photoelectrical cell and a light source at its base. When a ball was in the cup, another neon lamp located directly above the starting lamp was illuminated. As the ball left the base of the cup, it activated the photoelectrical cell, which then opened the circuit and turned off the neon lamp. Thus, the end of the throwing movement, or the instant the ball broke contact with the base of the cup, was located easily on the film.

A 16-mm Model H16 Bolex motion picture camera, operating at a film transport speed of 64 frames per sec, was used. The camera was equipped with a 17-mm f2.7 Wollensak lens. Correct illumination was achieved by using two photographic umbrellas and four floodlights (Hobart and Provenza, 1973).

Elapsed time and synchronization between the motion pictures and the electromyogram record were achieved by use of an electronic timer located in the field of view of the camera (Hobart et al., 1973) (Figure 1).

White adhesive tape was placed over the right shoulder joint and the right ulnar styloid process. Black dots were drawn on the tape to represent the center of rotation of the shoulder and the center of the ulnar styloid process, respectively. As an additional assistance in the film analysis, a coordinate system was used to plot the movement of the limb during each toss. The origin and a point on the ordinate axis of the coordinate system were two black dots located in the field of view of the camera, posterior to the subject and perpendicular to the floor (Figure 1). Thus, all movements occurred in a positive quadrant.

The data were reduced with the use of a Hewlett-Packard digitizer (Model 9864A) and calculator (Model 9810A). Individual frames were projected on the digitizing surface, and the position of the limb in relation to the coordinate system was recorded for each frame of every trial set. All angular measurements then were computed and recorded by the calculator.

The time factor necessary for the calculations of velocity and acceleration was determined by dividing the total movement time by the number of frames used to record the motion. Previous research established that the transport speed of the camera varied only ± 0.001 sec over any 1-sec period.

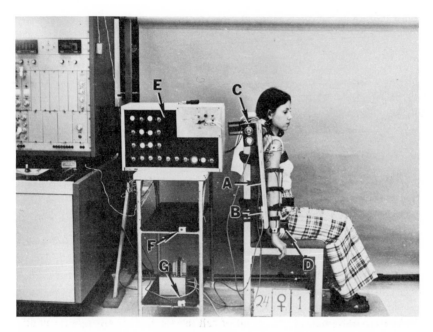

Figure 1. The testing arrangement including (A) the Starting Panel, (B) Microswitch, (C) neon lamps, (D) ball cup, (E) electronic timer, (F) Coordinate system ordinate, and (G) coordinate system origin.

Because of this consistency, the element of time was used as a common basis for trial comparison. The frame with the limb angle closest to 0° was used as the synchronization point for each trial.

A repeated measures analysis of variance was used for the data analysis. Posteriori comparisons were made using the Student-Newman-Keuls test. The 0.01-level was set for the statistical analyses.

RESULTS AND DISCUSSION

After the learning data had been analyzed and evidence of significant improvement in performance obtained, mean release velocity and mean release acceleration rates were calculated for each of the five trial sets (Hobart and Vorro, 1974). These measurements were taken at ball release and are presented in Table 1. Analysis of the data revealed that no significant changes occurred between either release velocity or release acceleration rates, respectively, as skill was acquired.

Mean release angles were calculated from the final frame for each of the throws and are presented in Table 1. Analysis of the data indicated significant differences among these mean release angles (Table 2). Posteriori comparisons

Table 1. Summary of release information and movement time

Test item	Trials				
	1–4	26–29	51–54	76–79	101–104
Relative velocity (rad/sec)	5.858	5.999	6.169	6.063	6.005
Relative acceleration (rad/sec/sec)	−88.013	−102.134	−88.725	−92.814	−113.207
Relative angle (rad)	1.063	0.915	0.910	0.873	0.888
Movement time (sec)	0.233	0.208	0.202	0.200	0.202

established the existence of significant differences between the initial release angle and each of the remaining test trials.

Table 1 also presents the mean total time of movement, from the first visible sign of limb movement, to the point of ball release. Statistically significant differences were found to exist among these mean times, and posteriori comparisons indicated that the significant differences were between the movement time of the initial trial set and each of the subsequent trial sets (Table 3).

Velocity means were calculated for each frame of the trial sets beginning with the second frame prior to the synchronization point and ending with the sixth frame (Table 4). Because of intra- and intersubject variations of movement time and angle of release, the sixth frame was the final frame for which each subject had complete data available. Although a general pattern of

Table 2. Analysis of variance of release angle

Source	df	mS	F	p
Trials	4	0.1161	9.516	< 0.01
Residual	95	0.0122		

Table 3. Analysis of variance of movement time

Source	df	mS	F	p
Trials	4	0.0039	4.875	< 0.01
Residual	95	0.0008		

Table 4. Velocity means (rad/sec)

Frames	1–4	26–29	51–54	76–79	101–104
			Trials		
−2	0.470	0.682	0.672	0.638	0.504
−1	1.193	1.721	1.384	1.454	2.204
0	1.718	3.575	3.133	2.691	3.142
1	4.847	6.746	7.337	4.064	4.409
2	4.162	5.986	7.254	4.961	5.906
3	4.822	7.154	8.960	6.432	7.116
4	6.406	7.883	9.647	7.069	7.756
5	6.846	8.282	7.912	7.828	8.058
6	7.354	8.411	7.957	7.761	7.566

velocity increase could be distinguished progressing from the initial frame, the differences between trials were not statistically significant.

Acceleration means were calculated similarly for each frame of the trial sets. They are presented in Table 5. Analysis indicated a statistically significant change in acceleration in the zero frame between Trials 1 and 4 and 26 and 29. Significant changes also occurred in Frame 1 between Trials 76 and 79 and all earlier trials (Table 6).

In order for success to be achieved at a motor task, proportionate changes must occur between correct and incorrect elements of the task. Intermittent data analysis throughout the practice sessions revealed that the total time taken to execute the task decreased between Trials 1 and 4 and 26 and 29. In addition, the angle of ball release decreased significantly between the same trials. The reduction of 0.148 rad ($8°$) during this interval represented a considerable adjustment to the task and effectively influenced the reduction in movement time. Considering the nonsignificant changes in release velocity,

Table 5. Acceleration means (rad/sec/sec)

Frames	1–4	26–29	51–54	76–79	101–104
			Trials		
−2	19.928	−9.750	17.993	28.739	41.911
−1	34.695	64.207	33.460	63.220	66.871
0	75.670	133.624	59.496	43.923	102.210
1	65.949	74.223	29.054	152.316	95.184
2	87.774	88.455	45.917	70.071	69.659
3	54.162	85.347	43.372	64.269	76.480
4	89.915	40.668	24.401	73.994	52.600
5	34.448	39.686	22.130	35.144	8.243
6	48.292	21.018	10.469	0.968	−10.509

Table 6. Analysis of variance of acceleration means

Source	df	mS	F	p
Frame (F)	8	69,918.063	8.458	< 0.01*
Trials (T)	4	24,933.473	4.718	< 0.01*
FXT	32	11,632.809	2.201	< 0.01
Subjects (S)	19	6,497.582		
S + F X T	171	8,266.833		
Residual + T X S	684	5,285.181		

*Not applicable to present experiment.

release acceleration, and velocity pattern, the release angle was a prime contributor to success at the task.

Within this experiment, the alterations necessary for success were performed to the extent that the early changes occurring between the initial two trial sets, in movement time, and in release angle resulted in improved performance at the task.

REFERENCES

Harris, S. J., and K. U. Smith. 1954. Dimensional analysis of motion. VII. Extent and direction of manipulative movements as factors in defining motions. J. Appl. Psychol. 38: 126–130.

Hobart, D. J. 1972. A cinematographic and electromyographic analysis of the modifications occurring during the acquisition of a novel throwing task. Unpublished Ph.D. dissertation, University of Maryland, College Park, Md.

Hobart, D. J., J. Diggs, D. L. Kelley, and D. V. Provenza. 1973. An electronic timer for synchronized electromyography and cinematography. Amer. J. Phys. Med. 52: 243–249.

Hobart, D. J., and D. V. Provenza. 1973. Practical solutions to problems in cinematographic analysis. JOHPER 44: 99–100.

Hobart, D. J., and J. R. Vorro. 1974. Electromyographical analysis of the intermittent modifications occurring during the acquisition of a novel throwing skill. In: R. C. Nelson and C. A. Morehouse (eds.), Biomechanics IV, University Park Press, Baltimore, and S. Karger, Basel.

Rubin, G., P. von Trebra, and K. U. Smith. 1952. Dimensional analysis of motion. III. Complexity of movement pattern. J. Appl. Psychol. 36: 272–276.

Smader, R., and K. U. Smith. 1953. Dimensional analysis of motion. VI. The component movements of assembly motions. J. Appl. Psychol. 37: 308–314.

Vorro, J. R. 1973. Stroboscopic study of motion changes that accompany modifications and improvements in a throwing performance. Res. Quart. 44: 216–226.

Wehrkamp, R., and K. U. Smith. 1952. Dimensional analysis of motion. II. Travel-distance effects. J. Appl. Psychol. 36: 201–206.

Author's address: *Dr. Joseph Vorro, Department of Anatomy, University of Maryland School of Dentistry, Baltimore, Maryland 21201 (USA)*

Electromyographic analysis of the intermittent modifications occurring during the acquisition of a novel throwing skill

D. J. Hobart and J. R. Vorro

University of Maryland School of Dentistry, Baltimore

Recently, several authors have published evidence that muscle action potentials are modified during the acquisition of a novel motor skill. Most agree that practice results in an improvement in the coordination or the timing of muscle response (Finley, Wirta, and Cody, 1968; Hobart, 1972; Kamon and Gormley, 1968; Payton and Kelley, 1972; Person, 1958). The changes in the integrated electrical activity attributable to practice have been investigated by several authors, with each producing different results. Brush (1966) found no changes attributable to practice; Finley et al. (1968) found an increase; and Payton and Kelley (1972) found a decline in the integrated electrical activity of one muscle and no change in another. Finley and Wirta (1967), supported by Hobart (1972), suggested that a shift in the electrical activity of the muscles may occur because of practice. Some muscles increase their activity, others decrease, each contributing to a smoother performance of the task.

It was the purpose of this investigation to examine the intermittent modifications in the total integrated electrical activity as well as in the timing of muscle response as a novel motor skill was acquired.

METHODS

The task consisted of an underhand toss with the arm extended and the hand fully pronated. A padded splint was attached to the cubital fossa surface of the limb, and a small cup was placed on the dorsal surface of the forearm at the level of the ulnar styloid process (Figure 1). A small sponge-rubber ball was propelled to the target in a pendulum-type throwing motion that consisted of shoulder flexion through approximately $60°$. This particular task was chosen because the movement occurred basically in one plane and the major portion of movement could be restricted to the shoulder joint, thus limiting the number of muscles involved and increasing the accuracy and ease of photographing. Previous studies also indicated that accuracy could be improved statistically with a relatively short practice period (Hobart, 1972).

The 20 subjects were given a complete explanation of the study and were allowed one practice toss to get a general understanding of the type of throw and the force necessary to propel the ball to the target. Each subject made 104 tosses; data were gathered during Tosses 1–4, 26–29, 51–54, 76–79, and 101–104. To avoid fatigue, the subjects were given a 3- to 5-min rest period after Trial 54.

The target, 72 in sq, was secured at an angle of approximately a 35° to the floor, with the near edge placed 3 ft from the base of the throwing chair (Figure 1). A metal measuring tape was attached to the center of the target in a manner that allowed it to be rotated through 360°. Prior to each toss, the ball was dampened, a process which caused it to leave a visible wet mark at the point where it struck the target. The distance as measured to the inside edge of this mark from the center of the target comprised the score for each toss. Any trials that missed the target completely were scored as 60 in.

Beckman miniature surface electrodes were used to monitor the electrical activity from the anterior and posterior deltoids (Hobart, 1972). The electrode sites were prepared in the normal manner, and the electrodes were

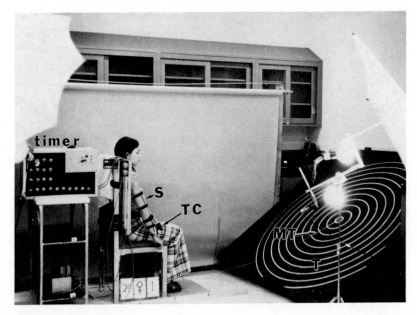

Figure 1. The testing site including target (*T*), splint (*S*), throwing cup (*TC*), and measuring tape (*MT*).

placed over the middle of the muscle belly. Resistance measurements of the skin were taken and values of under 5,000 ohms were maintained. The electrical output was processed and recorded by a two-channel Beckman Type R Dynograph equipped with 9852A integrating couplers. The amplifier sensitivity of the channel recording from the anterior deltoid was set at 1.0 mv per cm, and that of the posterior deltoid channel was 0.5 mv per cm. The higher sensitivity of the latter was necessary in order to extract a reasonable signal size from the muscle. The integrated signal was fed into a voltage-to-frequency converter and then to a digital counter, which gave a numerical value proportional to the area under the integrated curve. This value is similar to the electromyogram (EMG) units derived when a polar planimeter is used to analyze integrated EMG curves. No attempt was made to convert the values because each subject was compared only with himself.

The EMG and cinematography (Vorro and Hobart, 1974) were synchronized through the use of a specially constructed electronic timer (Figure 1), (Hobart et al., 1973). The event marker of the EMG was controlled by the timer so that a mark was placed on the EMG record every 0.10 sec. The paper speed of the EMG was 250 mm per sec; thus, each millimeter on the paper represented 0.004 sec. This procedure allowed the EMG record and the film to be correlated with the 0.004 sec (Figure 2).

RESULTS AND DISCUSSION

Acquisition of the skill was assessed by a gain in accuracy, that is, a reduction in the distance as measured from the center of the target for each toss. The data were submitted to an analysis of variance corrected for repeated measures with the 0.05 level set for statistical significance. It was apparent from the data presented in Figure 3 that repeated practice produced a significant improvement or reduction in throwing error. An analysis of the means by the

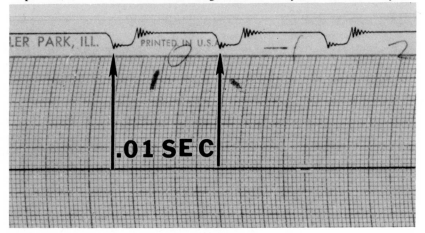

Figure 2. EMG record showing the timing marks.

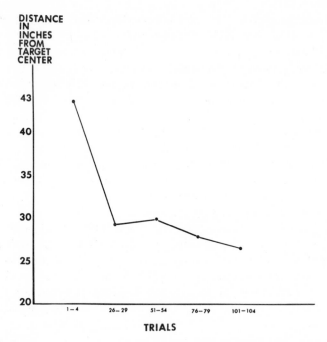

Figure 3. Summary of the mean reduction in throwing error during the practice period.

Student-Newman-Keuls test revealed that significant changes occurred between Trials 1–4 and the remaining test trials.

Two preliminary steps were taken before the EMG data were analyzed. First, the beginning time of the toss and the point of ball release were transposed from the synchronized film record to the EMG record. Second, the points where the pens of the recording unit rose 1 mm above the base line were marked on the EMG record. With these steps completed, the actual analysis of the data was undertaken.

In order to detect any changes that might have taken place in the relationship between the onset of muscle activity and the beginning of movement, the number of millimeters between them was counted. A direct conversion to time was made, because under the conditions of this experiment 1 mm was equal to 0.004 sec. The means of the test trials were analyzed statistically and a significant F ratio ($\alpha < 0.05$) indicated that, as practice progressed, there was a significant change in the premovement time of the

Table 1. Analysis of variance of premovement activity of the posterior deltoid

Source	df	MS	F	p
Trials	4	0.0021	3.352	< 0.05
Residual	95	0.0006		

Table 2. Means of the time of the onset of premovement activity
of anterior and posterior deltoid

Muscle	Trials				
	1–4	26–29	51–54	76–79	101–104
Anterior deltoid	0.100	0.104	0.108	0.106	0.108
Posterior deltoid	0.038	0.061	0.061	0.062	0.059

posterior deltoid (Table 1). The posteriori analysis by the Student-Newman-Keuls test revealed that Trials 1–4 differed significantly from the remaining test trials (Table 2). Thus, the greatest change occurred very early in the task. The analysis indicated that, as the accuracy of the task improved, activity of the posterior deltoid commenced sooner before the beginning of movement and closer to the time of beginning activity of the anterior deltoid (Figure 4).

The time between the beginning activity of the antagonistic muscles was measured in an effort to reveal any modifications in this relationship as the skill was acquired. The statistical analysis indicated that practice resulted in a significant change (Table 3). The analysis of the means (Table 4) by the Student-Newman-Keuls test revealed that Trials 1–4 differed significantly from the remaining trials.

Thus, after the initial reduction in the time between the beginning activity of the two muscles at Trials 26–29, there was very little change throughout the remaining practice periods (Figure 4).

Time from the onset of activity to peak activity was measured. These measurements were taken to establish changes that practice might have made in the ability of the subjects to mobilize their muscles to the maximal level necessary to accomplish the task. Although it was not statistically significant, there was a general trend for both muscles to reach peak activity earlier (Table 5 and Figure 4).

The total integrated activity of the anterior and posterior deltoid was measured, and no significant changes were found either during or after practice (Table 6 and Figure 4).

The time between the beginning activity of the anterior deltoid muscle and the premovement activity of the posterior deltoid exhibited a significant difference between Trials 1–4 and the remaining test trials. A significant

Table 3. Analysis of variance of the time
between the beginning activity of anterior
and posterior deltoid

Source	df	MS	F	p
Trials	4	0.00175	5.336	< 0.01
Residual	95	0.00033		

Table 4. Means of time between beginning activity of anterior and posterior deltoid

		Trials			
	1–4	26–29	51–54	76–79	101–104
Means	0.0685	0.0469	0.0488	0.0456	0.0509

reduction in throwing error occurred at the same time. The earlier beginning
of the posterior deltoid and the significant decrease in the time between the
beginning activity of the two muscles are an indication that, as the skill was
acquired, modifications occurred in muscle timing or coordination (Figure 4).
These findings support the earlier works of Hobart (1972), Kamon and
Gormley (1968), and Person (1958).

Neither muscle exhibited a significant change in total activity or the time
necessary to reach peak activity as the skill was acquired. Thus, it seems that
modification of these aspects was not necessary for a significant reduction in
throwing error (Brush, 1966). These results vary from those of earlier studies
that found changes in these measures (Finley and Wirta, 1967; Hobart, 1972;
and Payton and Kelley, 1972). A possible explanation for the contradiction

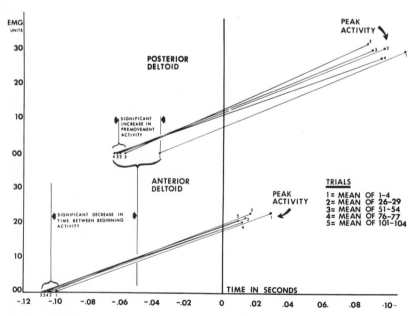

Figure 4. Summary of the relationships of the onset of muscular activity, peak activity,
and total electrical output of the muscles.

Table 5. Means of the time necessary to each peak activity

Muscle	Trials				
	1–4	26–29	51–54	76–79	101–104
Anterior deltoid	0.0284	0.0171	0.0152	0.0133	0.0106
Posterior deltoid	0.1069	0.0989	0.0911	0.0973	0.0897

might be that the muscles most involved in the task possibly were not the two muscles studied in these investigations.

Some previous investigations (Hobart, 1972; Payton and Kelley, 1972) were limited to pre- and postlearning analysis. Although these investigators found changes, they were not able to state at which point during practice the modifications occurred. The present study, in an attempt to overcome this limitation, gathered data at five separate points during the practice period. The results indicated that all significant changes that were found occurred within the first 25 practice throws. After these initial major modifications were made, the remaining adjustments were minor.

As indicated by Vorro and Hobart (1974), the major change found in the performance of the task was a decrease in the angle of release that probably resulted in the decrease in resulting movement time, because no changes in velocity were noted. All of these modifications also occurred during the first 25 practice throws.

Within the limitations imposed by the present investigation, the analysis of the data appear to support the following conclusions with respect to the learning of this task. The modifications in the timing or coordination of antagonistic muscles resulted in a decrease in the angle of ball release and thus in the improvement in the accuracy of the toss. These changes all were evident very early in the practice period.

Table 6. Means of total activity of anterior and posterior deltoid

Muscle	Trials				
	1–4	26–29	51–54	76–79	101–104
Anterior deltoid	23.525	23.340	22.375	20.795	21.145
Posterior deltoid	30.870	31.425	30.765	28.025	31.165

REFERENCES

Brush, F. C. 1966. Patterns of movement and associated electrical muscle activity in college women of high and low motor ability. Unpublished Ph.D. dissertation, The University of Maryland, College Park, Md.

Finley, F. R., and R. P. Wirta. 1967. Myocoder studies of multiple myopotential response. Arch. Phys. Med. 48: 598–601.

Finley, F. R., R. P. Wirta, and K. A. Cody. 1968. Muscle synergies in motor performance. Arch. Phys. Med. 49: 655–660.

Hobart, D. J. 1972. Electromyographic and cinematographic analysis of the modifications occurring during the acquisition of a novel throwing task. Unpublished Ph.D. dissertation, The University of Maryland, College Park, Md.

Hobart, D. J., J. Diggs, D. L. Kelley, and D. V. Provenza. 1973. An electronic timer for synchronized electromyography and cinematography. Amer. J. Phys. Med. 52: 243–249.

Kamon, E., and J. Gormley. 1968. Muscular activity pattern for skilled performance and during learning of a horizontal bar exercise. Ergonomics 11: 345–357.

Payton, O. D., and D. L. Kelley. 1972. An electromyographic study of motor learning. Phys. Ther. 52: 261–266.

Person, R. S. 1958. An electromyographic investigation on coordination of the activity of antagonist muscles in man during the development of a motor habit. Pavlov J. Higher Nerv. Activ. 8: 13–23.

Vorro, J. R., and D. J. Hobart. 1974. Electromyographical analysis of the intermittent modifications occurring during the acquisition of a novel throwing skill. In: R. C. Nelson and C. A. Morehouse (eds.), Biomechanics IV, University Park Press, Baltimore, and S. Karger, Basel.

Author's address: *Dr. Donald Hobart, Department of Anatomy, University of Maryland School of Dentistry, Baltimore, Maryland 21201 (USA)*

Aerodynamics of the human body

J. R. Shanebrook and R. D. Jaszczak
Union College, Schenectady

The purpose of this study was to present a model for the determination of aerodynamic forces on the human body during sport activities.

THE MODEL

The human body was modeled by a series of conjugated circular cylinders to simulate the trunk and appendages and a single sphere to simulate the head. Figure 1 illustrates the model (not drawn to scale) for the case of a human body in free-fall descent in which arms and legs are fully extended. Each cylinder is numbered for future reference. It is noted that each leg is modeled by two cylinders of different diameters where the upper cylinders (*3* and *4*) extend from the base of the trunk to the knee.

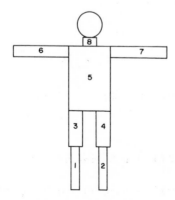

Figure 1. Model of human body (not drawn to scale) for determination of aerodynamic forces during free-fall descent through the earth's atmosphere. See text for detailed explanation.

For a given human body, the dimensions of the cylinders and sphere are determined by maintaining an equivalence between the body and its model of overall height and total area projected normal to the free-stream direction. Also, the length and area (projected normal to the flow) of each body part (except the head and neck regions) are preserved in the model, leaving only the diameter of the equivalent cylinder to be determined from the ratio of area to length. The sphere diameter is found by maintaining an equivalence between the circular projected area of the sphere and the area of the head region normal to the flow. Because a sphere has only one characteristic dimension (its diameter), there is a small discrepancy between the height of the head region and the height of its equivalent sphere. This is corrected by adjusting the height of Cylinder 8 (neck region) such that the overall height of the human body and its model correspond and the projected areas of the neck region and Cylinder 8 are equal.

With the dimensions of the model known, the aerodynamic forces on each part of the model can be determined by the following steps, assuming that the speed of motion is known, as well as the local atmospheric conditions of temperature and pressure.

1. The Reynolds number, R_d, for the flow past the cylinder or sphere of diameter, d, is found from the relation,

$$R_d = \frac{Vd}{\nu}$$

where V is the flow speed and ν is the kinematic viscosity of air.

2. The drag coefficient, C_D, is found from published experimental results expressing drag coefficient as a function of Reynolds number. All calculations reported here are based on the circular cylinder and sphere data presented in the works of Hoerner (1965), Jacobs (1929), Schlichting (1960), and Welsh (1953).

3. The drag force, D, then follows from the relation,

$$D = C_D \left(\tfrac{1}{2} \rho V^2 A \right)$$

where ρ is the air density and A is the area of the cylinder or sphere projected normal to the free-stream direction.

APPLICATION TO FREE—FALL DESCENT

Webster (1947) presented results for the terminal speed at sea level for human bodies of various weights in free-fall descent through the earth's atmosphere. As mentioned previously, this information is sufficient to calculate the aerodynamic forces on the various parts of the descending body, provided

Table 1. Results for American male in 2.5 percentile of population (V = 192 ft/sec)

Cylinder	Area (in^2)	d (in)	$R_d \times 10^{-6}$	C_D	D (lbs)
1, 2	54.0	2.96	0.296	0.9	2(14.5)
3, 4	47.7	4.08	0.407	0.6	2(8.52)
5	264.0	10.63	1.06	0.35	27.6
6, 7	81.4	3.07	0.306	0.9	2(22.1)
8	8.66	3.14	0.314	0.85	2.20
Sphere	40.4	6.95	0.695	0.11	1.33

that the dimensions of the corresponding model, shown in Figure 1, can be determined. Dreyfuss (1959) has presented detailed scale drawings and body weights of adult American males representing 2.5 (small), 50 (average), and 97.5 (large) percentiles of the population. These scale drawings were used to determine areas and lengths such that a complete dimensioned model for each of these three representative bodies was determined. Because the terminal speed for each body is known from the results of Webster (1947), the force on each part of the three models can be calculated. Tables 1–3 summarize the results found for each of these three bodies. It is noted that three entries in the last column of each table have a multiplicative factor of two, to account for the pair of cylinders considered, with the number in parentheses being the drag force on one of the cylinders.

By adding the drag forces in the last column of each of these tables, the total drag on each body can be determined. This result is of particular interest because it indicates the accuracy of the model. That is, at terminal speed, the total drag force on the body is equal to its weight. The total drag on the body listed in Table 1 is calculated to be 121.37 lbs, which compares favorably with the weight of 127.7 lbs. The weight of the body listed in Table 2 is 161.9 lbs. whereas the total drag force is found to be 133.81 lbs. Both of these comparisons are reasonable, because Hoerner (1965) indicates that

Table 2. Results for American male in 50 percentile of population (V = 199 ft/sec)

Cylinder	Area (in^2)	d (in)	$R_d \times 10^{-6}$	C_D	D (lbs)
1, 2	64.5	3.26	0.339	0.8	2(16.6)
3, 4	67.7	5.25	0.543	0.3	2(6.54)
5	312.0	11.9	1.23	0.35	35.2
6, 7	95.4	3.35	0.346	0.8	2(24.6)
8	11.0	4.4	0.456	0.4	1.42
Sphere	48.3	7.85	0.813	0.11	1.71

Table 3. Results for American male in 97.5 percentile of population (V = 208 ft/sec)

Cylinder	Area (in^2)	d (in)	$R_d \times 10^{-6}$	C_D	D (lbs)
1, 2	79.0	3.7	0.401	0.7	2(19.5)
3, 4	90.3	6.15	0.666	0.35	2(11.1)
5	360.0	13.6	1.47	0.4	50.6
6, 7	125.0	4.05	0.439	0.55	2(24.2)
8	11.7	4.69	0.508	0.3	1.23
Sphere	61.0	8.81	0.95	0.13	2.79

clothing can contribute as much as 10 percent to the total drag on the body. However, for the body listed in Table 3, which has a weight of 208.9 lbs, the total drag force is found to be only 164.22 lbs.

The authors feel that the most significant source of error in this aero-dynamic force analysis of the human body is in the determination of cylinder drag coefficients from published graphs of experimental data. Nearly two-thirds of the cylinders listed in Tables 1–3 have Reynolds numbers in the range 2×10^5 to 5×10^5. Unfortunately, in this range of Reynolds numbers, the drag coefficient decreases the most rapidly, and, in addition, there are relatively few data points available for interpolation purposes. As discussed by Schlichting (1960), this sudden decrease in drag coefficient with increasing Reynolds number is due to transition in the boundary layer from laminar to turbulent flow. That is, a turbulent boundary layer can penetrate further into a region of adverse pressure gradient before separating from the body because of the additional momentum transported to the layers near the wall by turbulent mixing. Thus, in this range of Reynolds numbers, the decrease in drag coefficient is due to a downstream shift in the position of separation, which in turn reduces the size of the wake region and, consequently, the pressure drag on the cylinder. It should be noted also that the Reynolds number at which transition begins depends on certain properties of the flow field as well as on the physical condition of the cylinder surface.

An interesting observation from Tables 1 and 2 is the relatively small drag force on the trunk (Cylinder 5) in comparison with the drag force exerted on the arms (Cylinders 6 and 7) of each model. This is due to the previously discussed sudden drop in drag coefficient experienced by circular cylinders during boundary layer transition. That is, the arms have a Reynolds number close to that corresponding to laminar flow conditions, whereas the trunk Reynolds number is in the fully turbulent flow regime. However, in Table 3, the Reynolds number for the arms is very close to the fully turbulent condition, resulting in a relatively low drag force. If the flow conditions on the arms of the body in Table 3 were such that the boundary layer was laminar (e.g., clothing might exert a stabilizing influence on the boundary layer), then the increased drag coefficient (e.g., 0.8 instead of 0.55) would bring the total drag force on the body into more reasonable agreement with the weight of the body (the total drag then would be about 186 lbs).

CONCLUSIONS

In view of the favorable comparisons with the body weights in Tables 1 and 2, it is concluded that the present model offers a promising means for determining the aerodynamical forces on the human body during sport activities, especially when the Reynolds numbers are in the fully laminar and/or fully turbulent flow regimes where dependable drag coefficient data are available. The model currently is being employed to analyze the air resistance of a sprint runner.

REFERENCES

Dreyfuss, H. 1959. The Measure of Man, Human Factors in Design. Whitney Publications, New York.
Hoerner, S. F. 1965. Fluid-Dynamic Drag. Published by the author.
Jacobs, E. N. 1929. Sphere Drag Tests in the Variable Density Wind Tunnel. NACA TN 312.
Schlichting, H. 1960. Boundary Layer Theory. McGraw-Hill, New York.
Webster, A. P. 1947. Free-Falls and Parachute Descents in the Standard Atmosphere. NACA TN 1315.
Welsh, C. J. 1953. The Drag of Finite-Length Cylinders Determined from Flight Tests at High Reynolds Numbers for a Mach Number Range From 0.5 to 1.3. NACA TN 2941.

Author's address: *Dr. J. R. Shanebrook, Department of Mechanical Engineering, Union College, Schenectady, New York 12308 (USA)*

Relationships among measurements of explosive strength and anaerobic power

A. Ayalon, O. Inbar, and O. Bar-Or
Wingate Institute for Physical Education and Sport, Wingate

Traditionally, the term "explosive strength" (or explosive power) has been used to define the type of activity that requires a relatively short, all-out muscular effort. This type of strength has been related to the mechanical concept of power, that is, to the time rate of work performance. For example, it was assumed for years that the vertical jump was a measurement of explosive strength. Adamson and Whitney (1971) and Barlow (1970) contend that the vertical jump is not a measure of human power. It also has been suggested that similar results could be found in activities such as throws or in projections of the body. In other words, the vertical jump is more a measurement of impulse (f × t) than of power.

Power output also has been used to measure anaerobic activities. The maximal aerobic capacity, measured by maximal oxygen consumption, has been the subject of many investigations, and various methods (direct and indirect) have been suggested for its determination. On the other hand, and in spite of the fact that anaerobic types of activities are essential to many sports, less attention has been paid to anaerobic tests. The methods commonly used to estimate anaerobic power or anaerobic capacity are: oxygen debt, oxygen deficit, concentration of lactic acid in the blood or in the muscles, and the Margaria step test.

The purpose of this study was to assess the degree of the relationships among: (a) two tests of explosive strength, both of which measured the power exerted by the left leg in the propulsive phase of pedaling a bicycle ergometer (one with a constant load for all the subjects and the other with a load relative to their body weights), (b) the maximal number of revolutions of the pedals that the subjects performed in 30 sec while sitting upright, with resistance relative to their body weights, (c) the Margaria step test for anaerobic power, and (d) the maximal number of revolutions of the pedals that the subjects performed with their arms during 30 sec.

METHODS

Fifteen untrained male subjects, 19–21 years of age, with a mean weight of 71.93 kg, performed the five tests in one session, with rest periods of at least 30 min between tests.

The tests used, in order of administration, were as follows.

Explosive Power (Absolute)

This test measured the power exerted by the left leg in the propulsive phase of pedaling on a Fleisch bicycle ergometer against a constant force of 2.90 kg for all the subjects. Subjects were instructed to perform one all-out effort over 150° of motion. Two microswitches operated a stop clock (sensitivity of 0.01 sec) that measured the time involved in moving the pedal through 120°. Three trials were performed by each subject, and the average time was used for the computations. Power units were calculated for both explosive tests. The values reported in Table 1 were computed by using the load set on the bicycle ergometer. However, it was considered that the real power output for these tests should have been higher because of two factors: (a) the breaking of the inertia at the beginning of the movement, and (b) the work done to lift the weights in this type of bicycle ergometer.

Explosive Power (Relative)

This test was similar to the previous one, with the difference that the load given to each subject was relative to his body weight. The load used was 40 per kg of body weight.

Margaria's Step Test

This test was a modification of the W Max 30 of Cumming (1973) in that loads relative to the body weights of the subjects were applied, rather than a constant load as administered by Cumming. These loads were 40g. per kg of performed. The time employed by each subject from the fourth to the sixth jump was recorded by a stop clock (sensitivity of 0.01 sec) operated by two mat switches that opened and closed the circuit at foot contact.

Computations of power output were carried out according to the body weight, the vertical distance covered, and the time involved in covering the four steps (two jumps at two steps per jump).

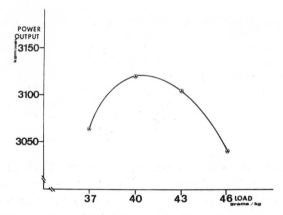

Figure 1. Total power output for the 30-sec leg pedaling test with different loads per kg of body weight.

Thirty-sec Leg Pedaling Test

This test was a modification of the W Max 30 of Cumming (1973) in that loads relative to the body weights of the subjects were applied, rather than a constant load as administered by Cumming. These loads were 40 g per kg of body weight. It was found in a pilot study (n = 5) that this load yielded the highest power output for this task and population, as summarized in Figure 1.

The test was performed on a Fleisch bicycle ergometer, and the subjects were instructed to pedal at maximal speed for 30 sec. The number of revolutions completed every 5 sec was recorded. The total power output for the 30 sec was computed, as was the power output for the fastest 5-sec period of each performance.

Thirty-sec Arm Pedaling Test

This test was similar to the 30-sec leg pedaling test, with the difference that the subjects performed the activity with their arms, in a sitting position. A load of 20 g per kg of body weight was used.

RESULTS AND DISCUSSION

Values of power output obtained for the different tests in this study are presented in Table 1.

An analysis of the intercorrelations was performed among all the tests, and the results are presented in Table 2. All of the coefficients were significant at the 0.01 level.

The correlation coefficients found between either explosive power test

Table 1. Data means, standard deviations, standard errors, and units

Test	X̄	SD	SE	units
Explosive power (relative)	2,820.77	507.60	131.58	kgm/min
Margaria's step	6,160.76	664.53	171.58	kgm/min
30 sec leg	3,607.30	240.72	60.51	kgm/min
30 sec arm	1,854.50	181.70	46.90	kgm/min
Weight	71.93	7.87	2.03	kg

and the 30-sec leg pedaling test seemed to indicate that in the latter, contrary to the results obtained in studies related to the vertical jump (Adamson and Whitney, 1971; Barlow, 1970), there is some relationship between mechanical power output, as measured by the explosive power test, and the results achieved in the task, as measured by the 30-sec leg pedaling test.

A fairly strong correlation was found between the arm and leg pedaling tests. This relationship seems to lead to the assumption that there is some kind of generality in the anaerobic capacity of man. In other words, the performance in the leg test can give an indication of the performance in the arm test and vice-versa as far as untrained young adults are concerned.

The significant values obtained in the analysis of intercorrelations seem to grant the assumption that all the tests used have some common elements. More simply, tests of anaerobic power or capacity imply the use of successive explosive power acts, whereas, in the explosive power tests, the subject has to perform an all-out muscular effort. In the Margaria step test (which measures anaerobic power), two climbing steps are involved between the eighth and 12th stairs, and in the 30-sec tests (which measure anaerobic capacity), the activity is based on the summation of power acts.

The high correlation between both explosive tests indicates that either of them could be used for the determination of explosive power. In both tests, it was observed that the acceleration of the movement was not constant, as can be seen by the parabolic shape of the curve in Figure 2.

Test-retest reliability for all the tests used are shown in Table 3.

Table 2. Intercorrelations among the variables

	Test						
1	Explosive power (absolute)						
2	Explosive power (relative)	0.91					
3	Margaria's step	0.78	0.75				
4	30 sec arm	0.74	0.75	0.82			
5	30 sec leg	0.73	0.75	0.77	0.81		
6	5 sec arm	0.78	0.76	0.84	0.94	0.80	
7	5 sec leg	0.67	0.68	0.79	0.72	0.79	0.73

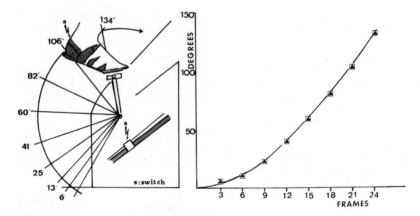

Figure 2. Time rate. Graph of the movement in the explosive power tests. The position of the pedal every three frames (0.053 sec) is shown by the radial lines in the schematic and plotted against frames in the graph.

Table 3. Test-retest reliabilities for the tests

Test	Reliability
Explosive power	0.92
Explosive power	0.93
Margaria's step	0.85
30 sec leg	0.91
30 sec arm	0.93

CONCLUSIONS

The explosive power tests and the 30-sec pedaling tests were found to be feasible and reliable for this population.

It was concluded that there are fairly strong interrelationships among the tests used in this study and that the tests of explosive power may be representative of the performance of a larger and more complex anaerobic activity.

A fairly strong relationship was demonstrated between the anaerobic capacity indicators of arm and leg performance.

REFERENCES

Adamson, G. T., and R. J. Whitney. 1971. Critical appraisal of jumping as a measure of human power. *In:* Biomechanics II, pp. 208–211. University Park Press, Baltimore, and S. Karger, Basel.

Barlow, D. A. 1970. Relation between power and selected variables in the vertical jump. *In:* John Cooper (ed.), Biomechanics. The Athletic Institute, Chicago.

Cumming, G. R. 1973. Correlation of athletic performance and aerobic power in 12 to 17 year old children with bone age, calf muscle, total body potassium, heart volume and two indices of anaerobic power. *In:* Proceedings of the 4th International Symposium on Pediatric Work Physiology, pp. 109–134. Wingate Institute, Netanya, Israel.

Margaria, R., P. Aghemo, and E. Rovelli. 1966. Measurement of muscular power (anaerobic) in man. J. Appl. Physiol. 21: 1662–1664.

Author's address: *Dr. Alberto Ayalon, Department of Research, Wingate Institute for Physical Education and Sport, Wingate Post (Israel)*

Effects of work surface angles on movement patterns of the arm in performing a simple manual task

M. Less and W. Eickelberg
Adelphi University, Garden City

A number of studies utilizing work stations have been recorded in the literature. Extensive anthropometrical studies involving ranges of body measurements were conducted on work stations by the armed forces during World War I, World War II, and the Korean War, and these data are readily available.

Several investigators have analyzed kinesiological variables and attempted to establish certain movement patterns of hand and arm motions as related to different conditions on work surface angles (Barnes and Mundel, 1939; Goldman, 1955; Reichard, 1967; Roberts, 1967; Tichauer, Mitchell, and Winters, 1962; Wall, 1967; Wyatt, 1965). These studies were undertaken with quite primitive work stations that did not allow for small increments in angular changes in the three-dimensional planes of movement. A new, automated geometric station (Krobath, Eickelberg, and Less, 1973) that allows for small angle changes and accurate performance recording was utilized in a number of recent studies that detected certain trends in improved performance in accuracy and energy cost variables as the surface angle increased from 0 to 18° (Ciofalo, 1972; Eickelberg, 1971; Less, Eickelberg, and Palgi, 1973).

Eickelberg (1971), using an earlobe plethysmograph and digital electropneumotachograph and following Brouha methodology for assessment of cardiac cost, reported reductions in cardiac cost as the work surface angle increased from 0 to 6, 12, and 18°, although only the differences between the 0 and 18° angles were approaching a level of significance.

Production efficiency, as measured by accuracy in a simple tracking task, also was improved steadily as the angle increased, although differences were not found to be significant. It was suspected that the tracking task was not sufficiently difficult to detect significant improvements as the work surface angle increased. Ciofalo (1972) repeated this experiment and reported similar results. Less et al. (1973) increased the difficulty level of the tracking task by shortening its traveling time cycle and observed significant improvement in production efficiency in females when surface angle increased from 0 to 6, 12, and 18° (F = 2.71; $p < 0.10$).

This study was undertaken in order to identify possible trends in movement patterns in an attempt to explain the reduction in energy cost and increased accuracy when work surface angles are increased from 0 to 18°.

APPARATUS

An automated geometric work station was utilized in this study (Figure 1). The work place consisted of a flat work surface, adjustable for height, angle, and distance from subjects. The variable tracking device required subjects to move two small cylinders, 7/16 in in diameter and 3/16 in high, from two proximal light targets to two distal light targets, and return them in a 3-sec cycle. Both sound and light stimuli were used as directives; audition was included in this experiment because auditory reaction time is faster than visual reaction time. The light targets were sensing disks with a central target area. The sensitivity of the four targets was adjustable, thus allowing the

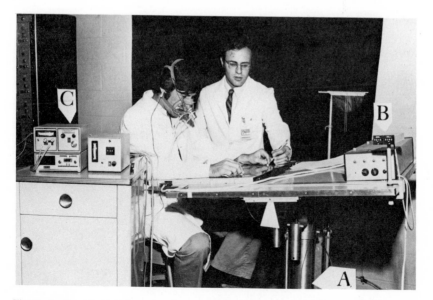

Figure 1. An automated geometric work station with electropneumotachograph and plethysmorgraph recording.

diameter of acceptability to be selected. The near targets were separated by a distance of 2 in, whereas the far were 15.5 in from their respective near targets on a line of an angle of 30° from the central longitudinal axis. The far targets were separated from each other by a horizontal distance of 13.5 in. This arrangement allowed the front targets to be observed simultaneously by both eyes but required separate viewing of the far targets, thus approximating a standard industrial task.

Movement patterns were assessed by a chronocyclographic technique. A simple, variable, 6-V flashing light element was mounted laterally on the wrist joint at the point of articulation of the ulna with the carpals and was adjusted to flash 10 times per sec. This enabled the investigators to analyze movement patterns with regard to horizontal and vertical displacement. The ambient light of the room was reduced, and the movement was photographed with a twin-lens reflex camera (Yashica) with an open shutter and a diaphragm opening of f/5.60. Verichrome Pan film (Eastman Kodak, Rochester, N.Y.) with an ASA rating of 125 was used and developed in D-19 for 6 min, a process which secured high levels of contrast (Figures 2 and 3).

PROCEDURES

Twelve male college students, ranging in age from 19 to 21 years, in weight from 152 to 187 lbs, and in height from 5 ft, 6 in to 6 ft, 2 in, were selected

Figure 2. Light element flashing at 10 times per sec indicates straight line motion of ulna-carpal joint at work surface angle of 18°.

Figure 3. Light element flashing at 10 times per sec indicates arc motion of ulna-carpal joint at work surface angle of 0°.

at random at Adelphi University. All instructions were standardized and given by the same instructor. When the subject arrived at the research laboratory, his seat was adjusted to a comfortable height, and then a 5-min trial was given to familiarize him with the task.

The task required moving two target pieces, symmetrically and simultaneously, from the starting point of the near targets to the lighted far targets, placing them within 0.33 in of the center of the target diameter. An audible tone (800 cycles per sec) indicated the successful completion of this step. If the step was not completed within 14 sec, a low-frequency tone (400 cycles per sec) indicated an unsuccessful completion. Providing subjects with the higher tone as an indication of successful movement was used because information feedback generally is considered to aid productivity. The numbers of successes and failures were recorded on a two-digital readout device in order to give the subjects the impression that production efficiency actually was measured. This was done in order to approximate the testing conditions of previous studies.

After the subject had performed the required task a number of times and a movement pattern had been established, the camera shutter was opened to record movement from proximal to distal targets. This procedure was repeated with the surface at the two angles, 0 and 18°, randomly.

RESULTS

The two variables that were under investigation were the degree of curvature and curve length of wrist movement in 0 and $18°$ work surface angles. Horizontal and vertical values for each light flash were secured and analyzed by a computer that was programmed to approximate the length of the curve by the polygonal line through the points $x_1, y_1 \ldots x_n, y_n$. The curve length was multiplied by 2.6, which was a factor indicating the ratio between actual length and photographically projected length. Mean curve length for $0°$ was 20.105 in with a standard deviation of 0.738 in, whereas mean curve length for $18°$ was 17.105 in with a standard deviation of 0.9022. An F-test for equality of variance ($F = 1.9; p < 0.05$) indicated the validity for utilizing a t-test for small sample means with equal variance. The t test indicated that mean curve length at $0°$ was significantly higher than mean curve length at $18°$ ($t = 8.341; p < 0.001$).

The degree of curvature of wrist movement was assessed by a computer program that was designed to calculate K (degree of curvature), as given by James and James (Mathematical Dictionary, 1964). Average curvature of a curve in a plane is defined as the ratio of the change in inclination of the tangent, over a given arc, to the length of the arc, i.e., the derivative of $\tan^{-1}$ $\frac{dy}{dx}$ with respect to the arc. In rectangular Cartesian coordinates, the curvature K is given by:

$$K = \frac{\dfrac{d^2y}{dx^2}}{\left[1 + \dfrac{dy}{dx}^2\right]^{3/2}}$$

For each 3 points (x_{i-1}, y_{i-1}), (x_i, y_i), (x_{i+1}, y_{i+1}), a finite difference of curvature was obtained as follows:

$$\frac{d^2y}{dx^2}_i = y_{i-1}\left(\frac{2}{(x_{i-1}-x_i)(x_{i-1}-x_{i+1})}\right) + y_i\left(\frac{2}{(x_i-x_{i-1})(x_i-x_{i+1})}\right)$$

$$+ y_{i+1}\left(\frac{2}{(x_{i+1}-x_{i-1})(x_{i+1}-x_i)}\right)$$

$$\frac{dy}{dx}_i = y_{i_1}\left(\frac{x_i-x_{i+1}}{(x_{i-1}-x_i)(x_{i-1}-x_{i+1})}\right) + y_i\left(\frac{(x_i-x_{i-1})+(x_i-x_{i+1})}{(x_i-x_{i-1})(x_i-x_{i+1})}\right)$$

$$+ y_{i+1}\left(\frac{x_i-x_{i-1}}{(x_{i+1}-x_{i-1})(x_{i+1}-x_i)}\right).$$

Then:

$$k_i = \frac{\left(\dfrac{d^2y}{dx^2}\right)_i}{\left[1 + \dfrac{dy}{dx}_i^{\,2}\right]^{3/2}}$$

The final approximation was the average of $k_2 \ldots k_{n-1}$. Mean curvature for $0°$ work surface angle was -0.19108 with a standard deviation of 0.04474, and mean curvature for the $18°$ work surface angle was -0.03802 with a standard deviation of 0.04299. The t test indicated that this mean degree of curvature at $0°$ was significantly higher than that at $18°$ ($t = 8.181$; $p < 0.001$).

DISCUSSION

Results indicated that when subjects moved their hands from proximal targets to distal target at an $18°$ work surface angle, a distance of 16 in, their hands moved in basically straight lines (Figure 2), as seen from the degree of curvature and the mean movement length, which was approximately 17 in. When the surface angle was flat ($0°$), the hands moved in a curve-like stroke (Figure 3), covering in the process an appreciably longer distance (approximately 20 in). This additional distance could have caused negative effects on production efficiency, which was measured in terms of accuracy in hitting traveling lights from proximal to distal targets. Because the traveling time was fixed to 3-sec cycles, the curve-like movement, which was 3 in longer than the straight movement, had to be executed faster; thus, the target was hit with faster velocity. If the movement velocity was equal to that at the $18°$ angle, then the subjects had less time to aim at the target. The greater distance also could be one of the contributing factors in greater energy cost when work surface angle is at $0°$ as opposed to $18°$.

The reasons for the differences in the movement pattern at 0 and $18°$ have not been established as yet. Electromyographic analysis will be needed before definite kinesiological and biomechanical factors that might contribute to these patterns are identified.

REFERENCES

Barnes, R. M., and M. E. Mundel. 1939. A study of simultaneous and symmetrical hand motions. Univ. Ia. Stud. Engng. Bull. 17.

Ciofalo, C. 1972. A study of work surface angles as related to production efficiency and physiological cost. *In:* First Notice. Insurance Company of North America, Philadelphia.

Eickelberg, W. 1971. An investigation of the effect of work surface angles on production efficiency. Paper presented at the Insurance Co. of North America MEND Research Convention, Philadelphia.

Goldman, J. 1955. Development of and testing of an electronic method for determining acceleration, constant velocity and deceleration of body motions. Unpublished Ph.D. dissertation, Washington Univ., St. Louis, Mo.

Krobath, H., W. Eickelberg, and M. Less. 1973. A description of an automated geometric work station. Microtecnic (Switzerland).

Less, M., W. Eickelberg, and S. Palgi. 1973. Effects of work surface angles on production efficiency of females on a simple manual task. Percept. Motor Skills 36.

Reichard, F. G. 1967. A kinesiological evaluation of parallel versus symmetrical patterns in simultaneous hand and arm motion. Unpublished master's thesis, Texas Technological College, Lubbock, Tex.

Roberts, R. 1967. A study of factors affecting the move and position elements of a small component assembly task. Unpublished master's thesis, Texas Technological College, Lubbock, Tex.

Tichauer, E. R. 1965. The biomechanics of the arm-back aggregate under industrial working conditions. American Society of Mechanical Engineers, Paper No. 65-WA/HUF-1.

Tichauer, E. R., T. B. Mitchell, and N. H. Winters. 1962. A comparison between the elements "move" and "transport" in MTM and work factor. Microtecnic (Switzerland) 16: 1–6.

Tufts College Institute of Applied Experimental Psychology. 1960. Handbook of human engineering data. 2nd Ed., rev. U. S. Naval Training Device Center, Port Washington, N. Y.

Wall, W. R. 1967. Kinesiological analysis of a simple assembly task. Unpublished master's thesis, Texas Technological College, Lubbock, Tex.

Wyatt, H. R. 1965. Optimal three-dimensional work place for the seated worker. Unpublished master's thesis, Texas Technological College, Lubbock, Tex.

Author's address: *Dr. Menahem Less, Human Resources Center, Albertson, Long Island, New York 11507 (USA)*

Seminar
activities

1. The Penn State Nittany Lion.

2. J. Orvis Keller Building, center for administrative and professional activities.

3. Kern Graduate Building. Convenient location for meals and site of International Coffee Hour.

4. Waring Hall. Seminar participants stayed here.

5. Biomechanics Laboratory. The research and graduate programs in biomechanics function within this majestic structure.

Special events during the seminar

The scientific program was complemented by a number of special events that included both professional and social activities.

PROFESSIONAL ACTIVITIES

Nearly all participants took advantage of a series of small group tours that were integrated into the official program. Daily tours were conducted each noon through the Biomechanics and Human Performance Laboratories. Some individuals also visited the Adage graphic facility at the University Computation Center.

On Tuesday evening, an Open House at the Biomechanics Laboratory enabled individuals to share ideas and to engage in more detailed discussions with members of the Biomechanics Laboratory staff concerning specific research projects and equipment.

Approximately 60 persons visited the Milton S. Hershey Medical Center of the Pennsylvania State University on Wednesday afternoon. This facility, located in Hershey, Pa., is one of the newest medical centers in the United States, yet it already is renowned for its outstanding research and teaching facilities and programs.

SOCIAL AND RECREATIONAL ACTIVITIES

A program of social activities occupied many leisure hours, beginning with the International Coffee Hour held in the Penn State International Center on Sunday afternoon, August 26th. This was followed by an informal reception on Sunday evening. On Monday, all Seminar participants and their families attended a picnic at Stone Valley, Penn State's outdoor recreation area.

Participants were able to engage in their favorite sports activities during free times because the outstanding sports facilities of Penn State were available to them. Many escaped the warm weather by visiting the outdoor swimming pool; others chose to play badminton, basketball, tennis, squash, or volleyball. More than 40 individuals played in a Seminar Golf Tournament on Wednesday afternoon.

While the participants were attending scientific sessions, their families were kept busy with a tour of the University campus, a visit to two local churches of modern architectural design, followed by coffee at the home of the Seminar Chairman, and a visit to the Amish Market Day at Belleville, Pa.

A reception in the Assembly Room at the Nittany Lion Inn, where the group was welcomed by Dr. John Oswald, Penn State University President, preceded the final social event of the week. This was a banquet in the Main Dining Room of the Inn, after which several awards were presented. Drs. Hebbelinck, Jokl, and Wartenweiler were given Nittany Lion statuettes by the Organizing Committee in recognition of their outstanding contributions to biomechanics at the international level. Other special awards were presented to various Seminar participants, including several prizes to participants in the Golf tournament. It was announced that the University of Jyväskylä, Finland, will host the Fifth International Congress on Biomechanics.

1. Examining the research equipment in the exhibit area.

2. Coffee breaks provided for personal interaction—an important aspect of the Seminar.

3. Browsing in the Book Exhibit, a popular activity.

4. Biomechanics Laboratory demonstration: Cinematography.

5. Biomechanics Laboratory demonstration: Evaluation of football helmets.

6. Biomechanics Laboratory demonstration: Walking.

7. Hershey Medical Center, site of professional tour.

8. Waiting for transportation to Stone Valley Recreation Area.

9. Fellowship at the picnic.

10. The outdoor swimming pool—popular refreshing activity.

11. Dr. John Oswald, University President (*right*) visits with Dr. and Mrs. Jurg Warten-weiler (Switzerland) at Banquet Reception.

12. Seminar Chairman Dr. Richard C. Nelson (*right*) (USA) presents award to Dr. Herbert Hatze (Union of South Africa) as the first registrant.

13. Dr. Doris I. Miller (Canada) receives golf prize (maximum number of strokes) from Vice Chairman, Dr. Chauncey A. Morehouse (USA).

Founding of International Society of Biomechanics

A highlight of the Fourth Seminar was the founding of the International Society of Biomechanics (ISB). The idea for such an organization was discussed at the Third Seminar in Rome in 1971, and at that time a committee under the direction of Dr. Jurg Wartenweiler of Zurich was formed to prepare the constitution and to make plans for the founding meeting. On Wednesday evening during the Fourth Seminar, the constitution was reviewed and discussed. The following day, the constitution was ratified, and officers and members of the Executive Council were elected. These included: Dr. Jurg Wartenweiler (Switzerland), President; Dr. Richard C. Nelson, (USA), Vice President; Dr. Jaap Vredenbregt (The Netherlands), General Secretary; Dr. Hans Debrunner (Switzerland), Treasurer. The Council members are: Dr. John V. Basmajian (USA), Dr. S. Bouisset (France); Dr. Wolfgang Gutewort (GDR); Dr. Marcel Hebbelinck (Belgium); Dr. Ernst Jokl (USA); Dr. Hideji Matsui (Japan), and Dr. Gunther Rau (BRD).

While the ISB grew out of the interest of those in the field of sport and physical education, its program is broad in scope to encourage the participation and interaction of researchers representing a variety of fields concerned with the biomechanics of human movement. These include anatomy, engineering, biology, medicine, ergonomics physiology, psychology, electromyographical kinesiology, and others. The purpose of the new society is to promote, on an international level, the study of biomechanics of human movement. It will encourage international communication through a regular newsletter and stimulate personal contacts through sponsorship of the biennial international seminar and support of meetings in related areas. To obtain information about the ISB including procedures for becoming a member one should write to: Prof. Dr. J. Wartenweiler, Laboratory of Biomechanics, Swiss Federal Institute of Technology, Plattenstrasse 26 8032 Zürich, Switzerland.

The first act of the ISB was to award the Fifth International Congress on Biomechanics to Jyväskylä, Finland. Dr. Paavo V. Komi will serve as Chairman. The Congress will be held June 29 to July 3, 1975. Information concerning this event can be obtained by writing Dr. Paavo Komi, Department of Kinesiology, University of Jyväskylä, Jyväskylä, Finland. Because of the ideal location and the expanding interest in biomechanics, a highly successful Fifth Congress is anticipated.

1. Dr. Ernst Jokl (USA), President of the Research Committee of the Council on Sport and Physical Educatio (UNESCO).

2. Dr. Jurg Wartenweiler (Switzerland) proclaims the International Society on Bio-mechanics to be officially founded.

3. Richard C. Nelson (*left*) (USA), Seminar Chairman and ISB Vice President, introduces Dr. Marcel Hebbelinck (Belgium).

4. Dr. Paavo V. Komi (Finland) invites participants to attend the Fifth International Congress on Biomechanics.

Commercial exhibitors

The Organizing Committee wishes to thank the following companies for their active participation during the Seminar. We also acknowledge their financial contributions, which made possible various social activities.

CO-SPONSORS

Instrumentation Marketing Corporation
820 Mariposa Street
Burbank, California 91506
Representatives: Vince Koenig and Glen R. Bridges

Kristal Instrument Corporation
2475 Grand Island Boulevard
Grand Island, New York 14072
Representatives: Ronald F. Lochocki and G. H. Gautschi

Lumex Inc., Cybex Division
100 Spence Street
Bayshore, New York 11706
Representative: David Hillery

CONTRIBUTOR

Redlake Corporation
2991 Corvin Drive
Santa Clara, California 95051
Representative: Ronald A. Gies

EXHIBITORS

Lafayette Instrument Company
P. O. Box 1279
Lafayette, Indiana 47902
Representative: Leon Littleton

Numonics Corporation
    Route 202 and Hancock Street
    North Wales, Pennsylvania 19454
Representative: Rolland H. Henderson

Visual Instrumentation Corporation
    239 West Olive Avenue
    Burbank, California 91502
Representative: L. J. Girola

Appreciation also is expressed to the 30 publishing companies for their valuable contribution of more than 85 publications. These formed a comprehensive exhibit of the most recent literature on biomechanics and related subjects, a popular feature of the Seminar.

Acknowledgments

The Organizing Committee wishes to acknowledge the encouragement and support provided by The Advisory Committee, The College of Health, Physical Education, and Recreation, and The Pennsylvania State University in staging the Fourth International Seminar on Biomechanics. Special recognition is due Mr. Wally Lester, Conference Coordinator in Continuing Education, who handled most of the administrative details, Mrs. Josephine Cleary, Secretary in the Biomechanics Laboratory, for her major contributions in the preparation of the Seminar materials, including the manuscripts for this publication, and in handling the majority of the Seminar correspondence. Thanks also are extended to John Palmgren and Kenneth Petak, Research Engineers, to Mrs. Linda Vongehr, part-time secretary, and to the several graduate assistants who assisted in the laboratory tours, served as assistants to the session chairmen, and assumed the responsibility for many of the minor details associated with the Seminar.

A successful Seminar would not have been possible without the willing cooperation and efforts of many individuals. The Organizing Committee wishes to thank publicly all of these persons for their contributions to a very rewarding professional endeavor.

Author Index